Environmental, Health, and Safety Auditing Handbook

McGraw-Hill Environmental Engineering Books

American Water Works Association
WATER QUALITY AND TREATMENT

Baker
BIOREMEDIATION

Brunner
HAZARDOUS WASTE INCINERATION

Chopey
ENVIRONMENTAL ENGINEERING FOR THE CHEMICAL PROCESS INDUSTRIES

Corbitt
STANDARD HANDBOOK OF ENVIRONMENTAL ENGINEERING

Freeman
HAZARDOUS WASTE MINIMIZATION

Freeman
STANDARD HANDBOOK OF HAZARDOUS WASTE TREATMENT AND DISPOSAL

Jain
ENVIRONMENTAL IMPACT ASSESSMENT

Harris and Harvey
HAZARDOUS CHEMICALS AND THE RIGHT TO KNOW: AN UPDATED GUIDE TO COMPLIANCE WITH SARA TITLE III

Levin and Gealt
BIOTREATMENT OF INDUSTRIAL AND HAZARDOUS WASTE

Kolluru
ENVIRONMENTAL STRATEGIES HANDBOOK

McKenna and Cunneo
PESTICIDE REGULATION HANDBOOK

Majumdar
REGULATORY REQUIREMENTS OF HAZARDOUS MATERIALS

Seldner
ENVIRONMENTAL DECISION MAKING FOR ENGINEERING AND BUSINESS MANAGERS

Waldo and Hines
CHEMICAL HAZARD COMMUNICATION GUIDEBOOK

Willig
ENVIRONMENTAL TQM

Environmental, Health, and Safety Auditing Handbook

Lee Harrison   Editor in Chief

Second Edition

McGraw-Hill, Inc.

New York   San Francisco   Washington, D.C.   Auckland   Bogotá
Caracas   Lisbon   London   Madrid   Mexico City   Milan
Montreal   New Delhi   San Juan   Singapore
Sydney   Tokyo   Toronto

Library of Congress Cataloging-in-Publication Data

Harrison, Lee
  Environmental, health, and safety auditing handbook / Lee Harrison,
   editor.—2nd ed.
       p.   cm.
    Rev. ed. of: The McGraw-Hill environmental auditing handbook, New
  York : McGraw-Hill,   c1984.
    Includes bibliographical references.
    ISBN 0-07-026904-1
    1. Industrial management—Environmental aspects—United States—
  Handbooks, manuals, etc.   2. Environmental auditing—United States—
  Handbooks, manuals, etc.   1. Harrison, L. Lee.   II. McGraw-Hill
  environmental auditing handbook.
  HD30.255.E58   1994
   363.7′06—dc20   94-39110
                                                                                  CIP

2 3 4 5 6 7 8 9 0   DOC/DOC   9 0 9 8 7 6 5

ISBN 0-07-026904-1

*The sponsoring editor for this book was Gail F. Nalven, the editing supervisor was
Mitsy Kovacs, and the production supervisor was Donald Schmidt. This book was
set in Palatino by McGraw-Hill's Professional Book Group composition unit.*

Printed and bound by R. R. Donnelley & Sons Company.

 This book is printed on recycled, acid-free paper containing a
minimum of 50% recycled de-inked fiber.

Contents

Part 1. Introduction

Part 2. The Benefits of an EHS Auditing Program

Part 4.   Establishing Your Own EHS Audit Program

viii   Contents

Part 5. Safety Auditing

12. Safety Management Systems Audits in the Chemical Process Industry   307
Ray E. Witter

Part 6. Industry-Specific Approaches to EHS Auditing

13. Manufacturing: Allied-Signal's Health, Safety, and Environmental Audit Progrm   377
Ralph L. Rhodes

14. Manufacturing: Eastman Kodak's Health, Safety, and Environmental Assessment Program   393
Alfred E. Fields

Contributors

Stein-Ivar Aarsaether   Norsk Hydro, Oslo, Norway (CHAP. 24)

Judd Bernstein   Vice President, Chemical Bank, New York, New York (CHAP. 21)

James W. Bobo   Tennessee Valley Authority, Knoxville, Tennessee (CHAP. 17)

Scott M. Brown   P.E., General Mills, Minneapolis, Minnesota (CHAP. 22)

Peter Chatel   Senior Environmental Auditor, the Coca-Cola Company, Atlanta, Georgia (CHAP. 8)

Carol Clayton   Wilmer, Cutler & Pickering, Washington, D.C. (CHAP. 11)

Paul D. Coulter   Director, Compliance Audit Program, Union Carbide Corporation, Fairfield, Connecticut (CHAP. 6)

Maryanne DiBerto   Arthur D. Little, Inc., Cambridge, Massachusetts (CHAP. 10)

Ronald DiCola   Environmental and Safety Compliance Assessment Technical Manager, AT&T, Basking Ridge, New Jersey (CHAP. 20)

Elizabeth M. Donley   Donley Technology, Garrisonville, Virginia (CHAP. 9)

Allen Dressler   Quality Assurance Manager, Environmental Engineering and Pollution Control Div., 3M, St. Paul, Minnesota (CHAP. 15)

Harry Fatkin   Director of Health, Safety, and Environmental Affairs, Polaroid, Cambridge, Massachusetts (CHAP. 16)

Alfred E. Fields, Ph.D.   Director, Corporate Health, Safety, and Environmental Assessment Program, Eastman Kodak Company, Rochester, New York (CHAP. 14)

David C. Fuller   Tennessee Valley Authority, Knoxville, Tennessee (CHAP. 17)

J. Ladd Greeno   Vice President, Arthur D. Little, Inc., Cambridge, Massachusetts (CHAP. 3)

Judith C. Harris, Ph.D.   Arthur D. Little, Inc., Cambridge, Massachusetts (CHAP. 10)

Gilbert S. Hedstrom   Manager, Environmental, Health, and Safety Auditing, Arthur D. Little, Inc., Cambridge, Massachusetts (CHAPS. 2 AND 5)

Robert D. Kennedy   Chairman and Chief Executive Officer, Union Carbide Corp., Danbury, Connecticut (CHAP. 1)

Christopher J. Keyworth   Senior Project Manager, ENSR, Acton, Massachusetts (CHAP. 23)

James E. Lukaszewski   Chairman, The Lukaszewski Group, White Plains, New York (CHAP. 4)

Madonna E. Martin   Tennessee Valley Authority, Knoxville, Tennessee (CHAP. 17)

John Nagy   Director, Environmental Audit Department, WMX Technologies, Oakbrook, Illinois (CHAP. 19)

Jane E. Obbagy   Arthur D. Little, Inc., Cambridge, Massachusetts (CHAP. 3)

Ralph L. Rhodes   Director, Corporate Health, Safety, and Environmental Audit, Allied-Signal, Morristown, New Jersey (CHAP. 13)

Daniel Schmid   Auditing Supervisor, Environmental Engineering and Pollution Control Division, 3M , St. Paul, Minnesota (CHAP. 15)

John R. Thurman   Manager of Environmental Audits, Tennessee Valley Authority, Knoxville, Tennessee (CHAP. 17)

E. O. Villeneuve   Vice President, Environmental Projects and Audits, Noranda, Toronto, Ontario, Canada (CHAP. 18)

Jeffrey D. Watkiss   Bracewell & Patterson, Washington, D.C. (CHAP. 11)

P. Kathleen Wells   Wilmer, Cutler & Pickering, Washington, D.C. (CHAP. 11)

Axel Wenblad   Corporate Manager, Environmental Auditing, AB Volvo, Gothenburg, Sweden (CHAP. 25)

David Wheeler   General Manager, Environment, Health & Safety, The Body Shop International, Watersmead, Littlehampton, West Sussex, England (CHAP. 7)

Ray E. Witter   for the Center for Chemical Process Safety, American Institute of Chemical Engineers, New York, New York (CHAP. 12)

Preface

Over a decade ago, when the first edition of this handbook was published, environmental, health, and safety (EHS) auditing was in its infancy. Only a handful of the largest and most prestigious industrial corporations had EHS auditing programs, and the financial community had virtually no interest in the subject.

How times have changed! EHS auditing is now a standard function at industrial companies of any size. Financial institutions, through hard experience, have become particularly aware of the staggering magnitude of environmental liabilities, and most see EHS auditing as a tool that must be employed to limit their exposure.

This volume, the second edition of the *Environmental, Health, and Safety Auditing Handbook*, has been expanded and updated to reflect these changes—as well as changes in auditing processes and techniques in the United States and Europe—to provide a valuable reference work not only for the practicing environmental auditor and the corporate environmental manager but also for everyone with responsibility for compliance with federal, state, and local laws and corporate environmental policy objectives—from line managers in industrial facilities, to the public affairs director, to corporate counsel, and ultimately to the chairman of the board.

Indeed, it is the board chairman who must communicate to everyone in the corporation—the board of directors, the stockholders, and other corporate stakeholders—that environmental excellence is a top priority of the corporation and that EHS auditing is critical to achieving corporate environmental goals and objectives. In the opening chapter, Robert Kennedy, the chairman of Union Carbide Corp., explains how this was accomplished in his company. He tells how the formation of a new corporationwide environmental organization "sent a clear signal to our whole company [that] top management was clearly committed to environmental action and leadership. And because top management's concern was clear from the outset, environmental action and leadership has over the years become deeply embedded in our company culture."

Throughout the 1980s and into the 1990s, as other corporations followed Union Carbide's lead, became aware of the benefits of EHS auditing, and adapted auditing processes and techniques to their own individual corporate environments, the auditing process itself underwent an evolution. In the early 1980s the companies that audited their industrial operations were largely those that wanted to solve immediate environmental problems—where there was substantial risk to workers or the surrounding community—or to measure compliance with federal and state laws and regulations. Now, such auditing is considered standard practice in most companies in the oil, chemical, mining, and electric power generation industries of North America and Western Europe.

A second and more recent trend in EHS auditing, however, focuses on internal corporate environmental management systems—the ability of such systems to manage and control a diverse range of issues of environmental consequence.

A third approach to EHS auditing, and one that is only now emerging, is the practice of auditing for sustainability. As described by David Wheeler, general manager, environment, health, and safety of The Body Shop, "It is a holistic approach predicated on a clear world view and an understanding of the need for a paradigm shift in business culture." Corporations that audit for sustainability, says Wheeler, "are especially keen to link environmental performance to wider issues of global ecology." The overriding factor for The Body Shop, he says, "is the perception of a moral obligation to drive towards sustainability in business. It is impossible to measure progress towards this ideal without a systematic process of data gathering and public reporting. Hence, auditing activities are considered absolutely essential to the company's long-term mission to become a truly sustainable operation—effectively to replace as many of the planet's resources as are utilized."

The Body Shop, Wheeler says, "wants to play a full part in handing on a safer and more equitable world to future generations. It is the strong belief of The Body Shop that the moral burden of achieving sustainability in business will become the principal driving force behind environmental auditing and environmental management systems in the 1990s."

The Declaration of the Business Council for Sustainable Development sounds a similar theme in its opening paragraph "Business will play a vital role in the future health of this planet. As business leaders, we are committed to sustainable development, to meeting the needs of the present without compromising the welfare of future generations."

In his book, *Changing Course*, Stephan Schmidheiny, a Swiss industrialist and chairman of the Business Council for Sustainable Development, states, "There is an inescapable logic in the concept of sustainable development." But he also acknowledges, "Business is a large vessel; it will require great common effort and planning to overcome the inertia of the present destructive course, and to create a new momentum toward sustainable development."

It is hoped that this second edition of the *Environmental, Health, and Safety Auditing Handbook* will contribute to this effort in some small way.

Lee Harrison

PART 1
Introduction

1

Have Laptop, Will Travel:

A Company Auditing Program

Robert D. Kennedy

Chairman and Chief Executive Officer,
Union Carbide Corp.

Today we know that our world is a fragile planet, one that we can damage irreparably if we don't take proper care of it. We also know that such care is an immensely difficult task, given expanded population pressures and worldwide demand for higher living standards that can only be met by increased industrialization.

In this situation, it is only to be expected that international organizations, governments, and the public generally are constantly pressing industry to improve its health, safety, and environmental record.

For industry, the challenge is not only to meet governmental standards, but, where possible, to go beyond them—because today we know that expenditures for health, safety, and the environment are more than costs of doing business to be grudgingly paid. Increasingly, "green" business is becoming good business.

In this greening of business, health, safety, and environmental auditing has become a powerful tool, with its own theory and recognized techniques. At Union Carbide, we are proud of our contributions to the body of knowledge laid out in this book. Some reflections on our experience, and on some of the lessons we have learned, may be appropriate here.

A Culture of Concern

Up to the Bhopal disaster in 1984, we had a good health, safety, and environmental record during 30 years of worldwide operations, achieved in large part through the technical skills and advice furnished to the company's 20-odd businesses by a large corporate department.

But Bhopal forced us into making an intensive review of our policies worldwide. Our first conclusion—that shortcomings were mainly in our overseas plants in countries with no strong regulatory structure—was overoptimistic, as a couple of domestic incidents soon proved.

So we went back to the drawing board, and, with the help of Arthur D. Little, Inc., our management consultants, we developed a new health, safety, and environment (HSE) program. The vice president for HSE reported directly to Warren Anderson, my predecessor as chief executive officer (CEO), and to the Health, Safety, and Environmental Affairs Committee of the Board of Directors, chaired by Russell Train, chairman of the World Wildlife Fund and a former head of the U.S. Environmental Protection Agency.

The formation of this new organization sent a clear signal to our whole company: top management was clearly committed to environmental action and leadership. And because top management's concern was clear from the outset, it has over the years become deeply embedded in our company culture. All our people know they are part of that culture, and they know the importance of stringent health, safety, and environmental protection in every aspect of our business.

As CEO, I have tried to strengthen that culture of concern. Our employees know that I spend a great deal of time on health, safety, and environmental affairs. They know that I insist on being informed immediately about any serious incident. And top management as a whole continues to make it plain that we expect concern for health, safety, and the environment to be part of everyday decision making.

Such companywide concern is, I believe, crucial to the success of any auditing program.

Clear, Concise Policies

A second essential is clarity and ease of communication. This may sound obvious, but we know from experience that it is not. In 1985 we recognized that part of our problem was a whole shelffull of manuals providing complex and detailed guidelines for company operations, written at various times and in different styles. They often left line managers in doubt about what was expected of them.

Our solution was to junk the manuals and to substitute a simple set of easily understood requirements. Our manual on health, safety, and the environment, which is updated regularly, tells plant managers clearly what they are expected to achieve in terms of personnel safety of programs, emissions reduction, pollution prevention, operational safety risks, operating procedures, and other factors. Figuring out how to achieve these required goals is left to each individual Union Carbide facility.

As a result, plant managers around the world enjoy additional flexibility in running their operations, while knowing clearly what the rules are, and therefore what a visiting audit team will expect to find.

Our audit teams also find that clearly stated standards and procedures make their job much simpler. They know that they can concentrate on reviewing policy implementation instead of having to decide on an ad hoc basis what the policy is. And this clear delineation between policy setting and implementation helps to ensure fair and equal treatment for every unit in our organization when the auditors call.

I might add that some competing multinational companies have recognized the success of our methods by asking for copies of our manual. We have been glad to supply them.

Establishing Trust

Top management concern plus clear policies have indeed provided a framework for an effective auditing program. But I do not want to give the impression that our auditing program has been all clear sailing. For some companies in the early stages of an audit program, a review of our early experience may be in order.

In the wake of Bhopal, our board of directors wanted above all else to be able to assure our stockholders that everything possible to avoid any future disaster was being done and that an auditing group, independent of line management, was in place.

Inevitably, that concern cast our auditing group in the unenviable role of police, intent on finding out what was being done wrong, rather than consultants providing advice on how to improve performance. Their arrival was too often viewed with the same lack of enthusiasm that greets a group of financial auditors descending on a company in trouble at the end of the year, but with the additional complication that our auditors might arrive any time.

There was, of course, ample justification for the police approach. Union Carbide was widely viewed as a company that might not survive another major disaster. As a result, some plant managers were reluctant to receive our auditors. And line management's traditional fear that visitors from company headquarters did not understand problems at the local level made matters worse.

In this situation, our auditors faced a difficult task convincing plant management that they would be judicious and fair in applying audit guidelines. They eased apprehensions by including a review procedure for resolving disputes. Should plant management disagree with any audit findings, the plant could seek an independent review by the corporate Health, Safety, and Environment (HS&E) group.

The procedure has worked well; nowadays, most disputes are resoved before the audit team leaves the plant. The review procedure, and the professionalism of the auditors, has convinced plant management that, while deficiencies would be duly noted, neither would auditors hesitate to confirm a plant's compliance with HS&E requirements.

The plants also acknowledge and applaud the audit program's role in raising awareness of HS&E standards at Union Carbide's facilities around the world.

Trust is indeed an important ingredient in a successful auditing policy, and today I think we have it. Surveys show that plant managers view our audit program as a valuable tool in gauging their own performance and improvement. And the growth of that trust has played a role in increasing the number of plants that fully meet all requirements and the near-elimination of failing grades.

Our Program Today

Let me describe our present auditing program briefly. Today we have six full-time auditors of varying ages and fields of expertise. Each year, they audit some 30 of Union Carbide's more than 150 wholly owned and partially owned facilities in North and South America, Europe, the Middle East, South and Southeast Asia, and Australia, aided where necessary overseas by local experts and interpreters.

Since the program began, staff members have performed more than 1000 audits in more than 30 countries. Armed with computerized protocols and laptop computers and printers, they are on the road for most of the year. Two auditors may complete an audit of a smaller facility in a single day; five auditors may take up to 2 weeks to audit a major facility.

The timing of a particular visit is determined by a mathematical model that balances various factors including the facility's hazards, performance in an earlier audit, and the number of its employees and contractors.

Auditors visiting a site will examine functional areas from each of three perspectives: health and safety, environment, and product responsibility. They note exceptions to Carbide standards and determine whether or not the exceptions have the potential to cause harm, or otherwise warrant immediate management attention.

Since 1988, our auditors have been greatly helped by the Episodic Risk Management System (ERMS), which utilizes state-of-the-art risk management technology to identify and rank potential risks to the public. The system enables us to establish priorities for reducing risk, allocating resources where they are most needed, and ensuring that risk decisions worldwide are made on a consistent basis.

When an audit is completed, the plant receives one of four possible grades, ranging from a high of M (meets governmental and internal compliance standards) to a low of RSI (requires substantial improvement). Intervening grades, from highest to lowest, are SM (substantially meets standards) and GM (generally meets standards).

To achieve an M rating, a facility must meet a truly high standard: at least 75 percent of the functional areas audited must be in the M category and none in the GM or RSI category, and there must be no information that suggests that an M rating may not be appropriate. Other overall ratings are awarded in a similar way.

Auditors give their reports to line management before leaving the facility at the end of the audit. Overall audit results are shared with me and our board of directors committee. When improvement at a facility is indicated, the business group involved tells us its specific correction plans and reports back on results. We hold line managers responsible for health, safety, and environmental protection in their performance goals, in their yearly job appraisals, and in calculating the variable component of their annual compensation.

This total audit process has rewarded us in many ways, not least in some discoveries we never expected. In one country, an audit team found that standards at a company plant declined when it was announced that the plant would soon be closed, causing regular employees to leave to get other jobs and forcing the plant to employ workers who were less well trained. As a result of this experience, it is now standard practice for us to recruit qualified personnel from elsewhere in Union Carbide to fill critical vacancies at a plant in the final few months before closure.

Today, we think that our health, safety, and environment program—and particularly our audit program—are among the world's best. Of course, we know that a single bad slip can cancel the work of the past 9 years of progress.

But I know we are doing something right. Some 50 other companies have so far sent representatives to see how we do our job, including not only major chemical companies but also an aircraft company and surface transportation companies.

Also, Arthur D. Little (ADL) audits our auditors' performance by sending out its own auditors to visit facilities along with ours. They report independently to our board of directors and thus serve as a check on our own staff's performance.

So far, I am pleased to add, ADL has given us high ratings for our overall HSE procedures—in particular for the scope of our audit program, for its ability to monitor potential problems and strengthen compliance with new regulations and internal standards as necessary, and for the detail and clarity of its reporting.

Looking to the Future

So far, then, so good. But clearly this is no time for us, or for any company, to relax efforts. Indeed, the need for close auditing of companies' health, safety, and environmental performance can be expected to become greater than ever.

At Union Carbide, we recognize this need. We continue to develop new quality programs to help us do a better and more efficient job of auditing company installations. We are using new techniques to improve our final audit format and content.

We are aware, too, that the whole concept of company self-auditing, of which we were among the pioneers, has come of age. The Environmental Auditing Roundtable, the original organization of professionals in the field, will soon publish a set of standards for auditor proficiency, diligence, and ethics. The U.S. Environmental Protection Agency has a stated policy of encouraging company environmental auditing.

Such actions will make it increasingly clear to all companies whose products and processes may harm the environment that sophisticated professional audit programs are a must. So, too, will the unrelenting pressure that the public puts on us to reach higher and higher standards of excellence in these fields.

Today, of course, the chemical industry has undertaken a major initiative to improve health, safety, and environmental performance and, thereby, improve the public's perception of our industry—the Chemical Manufacturers Association's Responsible Care® program.

In particular, Responsible Care's Product Stewardship Code, promulgated in 1992, has extended member companies' responsibilities to areas not previously covered, such as product design and dealings with customers, suppliers, and third-party manufacturers. The code effectively covers all stages of a product's life cycle from design to disposal. As such, the code dramatically extends the range of issues to be addressed by company HSE auditors.

Overseas, we expect a growing role for our audit teams in that our plants are adopting corporate standards based on Responsible Care requirements, even though these may be substantially higher than those in some countries in which they operate. We will continue to exert our influence as shareholders in our overseas affiliate companies to require them to meet these standards both because it is right and because in the long run it makes good business sense.

While this policy may put us at a competitive disadvantage in some countries in the short run, we know that worldwide concern for environmental quality will continue to grow. That concern will be reflected in stricter government controls. When that happens, Union Carbide and other companies that have adopted standards in excess of national requirements will have the edge, since they will not have to pay the cleanup bills that their less far-sighted rivals will have to meet.

In short, the future of companies seriously concerned about health, safety, and environmental excellence looks bright—if we watch our own products and processes every step of the way. A generation ago, our school books taught us that the price of liberty is eternal vigilance. Today, we know that the price of worldwide health, safety, and environmental excellence can be no less.

2

Environmental, Health, and Safety Auditing Comes of Age

Gilbert S. Hedstrom
Arthur D. Little, Inc.

Environmental, health, and safety (EHS) auditing has grown rapidly in the past two decades from an internal evaluation tool used independently by a few companies to a practice nourished by its own professional associations and literature and actively recognized in government regulations and guidelines. More than a thousand companies in North America, Europe, Latin America, and Asia now have formal environmental, health, and safety audit programs in place.

Two factors fed the early growth of EHS auditing and remain central to its purpose today:

- The recognition within companies that they can benefit from reviewing specific operations for EHS shortcomings and departures from standards

- The growth of a complex and demanding regulatory system for EHS matters

Both of these factors helped to motivate industry's first comprehensive independent EHS audit, conducted in 1977 by Arthur D. Little for AlliedSignal, then called Allied Chemical Co. Allied decided to undertake the audit as part of its response to a serious environmental incident at a contract packager's plant. The audit surveyed sites, records, operations, and personnel at 35 Allied plants worldwide to determine the level of the company's compliance with government regulations and internal EHS policies and procedures. The audit showed that Allied's EHS programs and procedures met its corporate standards. To provide for continued assurance and improvement, Allied then established a corporatewide EHS auditing program—among the first of its kind—in 1978.

By the end of the 1970s, a number of companies had developed audit programs, using EHS audits to measure their compliance with government regulations and internal standards. Leading companies today have expanded their expectations for EHS audits beyond compliance to include auditing against "good industry practices." But measuring compliance remains a core theme of the discipline and a key goal of most EHS audits.

The late 1980s saw the appearance of a third factor in the growth of EHS audits: rising public expectations that companies fully communicate their environmental performance. In recent years, a number of chemical, petroleum, and manufacturing companies have sought to respond to this pressure with public environmental reports. Norsk Hydro U.K.'s environmental report in 1990 was regarded as a groundbreaking effort in this area because of its candid coverage of shortcomings and its verification by an independent auditor. Other companies that have published relatively detailed environmental reports include Union Carbide, WMX Technologies, Olin, Monsanto, Noranda, and Rhône-Poulenc. There is much discussion today in the environmental auditing community and in industry about appropriate goals and strategies for reporting environmental performance—and about the risks involved. Nonetheless, the reports now available show a recognition that the information a company gathers about its environmental performance can play an important role not only within the company, but with all of a company's stakeholders. For many companies, the environmental audit stands out among performance measurement tools as an important way to measure compliance.

The U.S. and European regulatory communities have also increased pressure on companies to provide measurements of environmental performance. In 1979, the U.S. Environmental Protection Agency considered, then decided against, mandatory independent environmental audits. Since then, U.S. regulations have begun to ask companies to make certain records public. For example, the Superfund Amendments and Reauthorization Act (SARA), Title III, Sec. 313 calls on companies to provide information about quantities of released toxic substances.

In Europe, the regulatory situation for company environmental disclosure varies considerably from country to country. But with the advent of the European Union's (EU's) Eco-Management and Audit Regulation, published in its final version in June 1993 and scheduled to take effect in 1995, these variations will begin to lessen. The regulation sets out environmental management standards that, although voluntary, are likely to become an influential international norm. The regulation requires participating companies to conduct environmental audits and to provide an environmental statement to their country governments, which may "disseminate it as appropriate to the public." The statement covers a broad range of environmental concerns, including pollutant emissions, waste generation, noise, and consumption of raw materials, energy, and water.

Now, as environmental management makes the transition in leading companies from a stand-alone function seeking to graft its priorities onto other business processes to an ingrained function within and across business processes, environmental auditing is being driven by a fourth factor: management's increasing confidence in the discipline's value as a measurement tool for assessing and helping to change and improve EHS performance.

In a sense, environmental auditing has come full circle. It was industry that first saw the need to review environmental aspects of operations, and it is industry now that expects environmental auditing to play a key role in the next phase of environmental management. As in the past, too, the expectations of government and the public will continue to shape aspects of EHS auditing. These expectations will tend to sharpen the demand for universal principles and standards in the discipline.

A Growing Profession

As industry needs and external pressures have given shape to environmental, health, and safety auditing in its first two decades, the discipline has developed a range of characteristics comparable to those in other professions. These include:

A Body of Knowledge.   Since the early 1970s, experience, expertise, and knowledge have been developed about methods, techniques, and problems in environmental auditing. Companies have certainly gathered a large and valuable store of information in the records and reports of their environmental audit programs. Much has also been written in several handbooks, in hundreds of forums and seminars, and in a growing file of articles in environmental and business periodicals (see Readings in Environmental Auditing, at end of this chapter).

A Consistent Rationale.   As environmental auditors developed a core group of common practices, some consensus on principles and standards for EHS auditing has recently emerged. Building on earlier expressions of environmental auditing principles such as Arthur D. Little's *Principles for Conducting Environmental, Health, and Safety Audits,* the Environmental Auditing Roundtable—the largest North American association of environmental auditors—published "Standards for Performance of Environmental, Health, and Safety Audits" in 1993 (see Part 3, Chap. 5, "Environmental, Health, and Safety Auditing Standards").

Competing Theories.   Since the late 1970s, discussions among environmental auditing practitioners have focused on matters such as the distinction between compliance audits and other types of audits, the extent to which legal protection of audit findings might be appropriate, and efforts to define audit standards and determine approaches to auditor certification. In recent years, perspectives on each of these issues have evolved considerably.

Basic Research.   Starting in the early 1980s, many companies, as well as the U.S. EPA, conducted research, through informal networking and formal benchmarking (including peer and independent reviews), about the components of a top-notch environmental audit program. Examining how audit programs are organized, staffed, designed, and implemented, companies found that no two programs were alike. Companies differed in their approach to staffing audits and determining their scope, in the timing and frequency of facility audits, in the style and format of audit reports, and in policies about who receives the report. Nonetheless, auditing professionals were able to identify common characteristics of leading audit programs (see Fig. 2-1). Moreover, their research has made an important contribution to the more recent effort to provide the environmental auditing community with consistent principles and standards.

Societies and Publications.   The 1980s saw the emergence of formal environmental auditing associations in the United States, Canada, Great Britain, and Sweden, as well as many informal affiliations and a wealth of conferences, seminars, studies, and surveys. The oldest association is the Environmental Auditing Roundtable, which first met in 1982 at Arthur D. Little. Two other groups are based in the United States, the Institute for Environmental Auditors and the Environmental Auditing Forum. In Canada, the Environmental Auditing Association grew in the early 1980s from a group sponsored by Arthur D. Little. Environmental auditors in Great Britain are represented within the Association of Environmental Consultancies, and the Environmental Auditors' Registration Association has established a scheme for registering environmental auditors. In Sweden, an auditing association that now numbers over 100 members developed in the early 1990s after the Federation of Swedish Industries

---

Objectives: Explicitly defined objectives can prevent varying interpretations by audit team members and misunderstandings among line managers.

Scope: Clearly defined boundaries for the program's overall scope make it possible to meet audit objectives with available resources.

Coverage: Decisions about coverage priorities enable organizations with many facilities to emphasize major facilities without overlooking others.

Approach: Matching audit approach to audit objectives enables the company to better focus specific audit activities on areas such as verifying compliance or confirming management systems.

Resources: Audit program effectiveness is a direct result of the expertise, proficiency, and training of the personnel who conduct the audit.

Organization: Senior management support is critical if the audit is to achieve its objectives, obtain accurate information about environmental performance, and contribute to company efforts to increase performance levels.

Figure 2-1. Characteristics of Effective Audit Programs. (*Source: Arthur D. Little*)

sponsored five training courses. In Germany, environmental managers from major companies first met in January 1993 to discuss environmental auditing issues, and especially the environmental statement provisions of the E.U.'s Eco-Management and Audit Regulation, which was still in draft form then. In Australia and several other countries, environmental auditors are currently considering forming associations. We also see an increasing volume of articles on environmental auditing in environmental, business, and general trade publications, and expect that the literature of auditing will continue to grow.

Practitioners.   The environmental auditing community now encompasses a broad range of practitioners, ranging from individuals who have been involved in auditing—within companies or as consultants—since its beginnings in the 1970s to younger part-time auditors for whom the work is one facet of their EHS responsibilities. This range of practitioners has come to share an increasingly consistent environmental auditing methodology, one which in turn has been shaped by the increasingly serious consequences when a consistent methodology is not followed.

The rapid overall development of environmental auditing as a profession also indicates that companies have quickly perceived its usefulness as a safeguard of business interests. Indeed, organizations with long-standing audit programs are beginning to regard them as essential functions, like internal financial auditing or legal affairs, not as functions that must repeatedly demonstrate their value in order to receive support.

From Here to the Year 2000

In the next few years, environmental auditing will primarily be influenced by the integration of environmental, health, and safety concerns throughout business processes. Increasingly, auditing will be shaped by changes in environmental management and operations. We expect to see these changes:

Environmental Auditing Will Take on Increasing Importance as a Tool for Improving Business Performance

Just as environmental, health, and safety management is becoming more ingrained within overall business management, so environmental auditing will continue to develop broader significance as a management tool. Much of the growth in environmental auditing in the next several years will be in small and midsized companies as they realize that they cannot afford to operate without these programs. Larger companies, for the most part, already recognize this.

As business managers recognize environmental auditing's management implications, environmental audit reports are being distributed more widely up the management chain. A 1993 survey by Arthur D. Little shows a rise from 43 to 54 percent in reporting of audit results to senior management in the last 3 to

5 years, a net increase of 25 percent. The survey also shows broader distribution at every level—a clear indication that the information in environmental audits is of greater importance now across the business process.

As companies move beyond compliance problems, environmental reports are gradually playing a larger role in business strategy. Increasingly, the environmental audit is taking on an expanded role as an important vehicle for providing assurance and informing corporate officers about the company's environmental, health, and safety progress—or lack thereof. While audit reports are not the only source of information about a company's environmental, health, and safety performance, they are among the most visible, and, as such, provide environmental, health, and safety professionals an opportunity for sharing valuable information with top management.

A recent survey of the contents of audit reports (Fig. 2-2) shows how they are including more information to help senior management with its responsibilities for environmental issues.

The 50 percent increases in executive summaries and in audit opinions are especially indicative of the increasing importance of audits to higher levels of management.

Focus Moves beyond Compliance Reviews to Evaluate Management Effectiveness

Most audit programs start with the explicit objectives of verifying compliance; some also seek to confirm the effectiveness of EHS management systems. At companies with a wide range of problems, auditors focus on the problem-finding mode. As performance improves, their focus shifts from identifying problems to verifying compliance and confirming the absence of problems (Fig. 2-3). This shift in emphasis tends to occur in companies with long-standing audit programs.

	Now (%)	3–5 years ago (%)
List of exceptions	88	75
Recommendations	82	66
Executive summary	73	48
Good practices	55	46
Audit opinions	41	27
Action plans	36	20

Figure 2-2. (*Source: Arthur D. Little*)

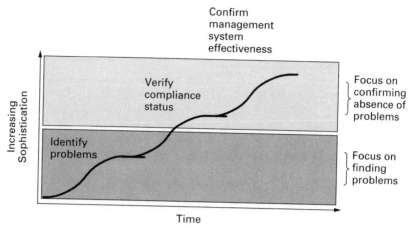

Figure 2-3. The shift in audit focus.

In the next few years, auditors will see changes in the nature of the typical audit program that reflect the natural evolution toward confirming the effectiveness of EHS management systems. The challenge for auditors, in some cases, will be to ensure that companies understand and take advantage of the potential for change and evolution in their audit programs.

The Scope, Depth, and Rigor of Audits Will Continue to Increase

Despite downsizing in many functional areas—even in EHS staffs—companies are maintaining their audit programs and plan to expand them. In an Arthur D. Little survey of Fortune 100 companies in 1992, 76 percent had increased audit scope in the last 5 years and 86 percent anticipated future increases. Companies that have only conducted audits of domestic facilities are broadening their program scope to include overseas locations. Companies whose programs focus only on environmental topics are expanding them to include health, safety, and product safety issues.

Perhaps the most significant change in the years ahead will be the continued increase in the depth and rigor of reviews. More companies will shift the orientation of their programs from assessment to compliance verification and from compliance verification to confirming the effectiveness of EHS management systems. Some companies have already begun to apply environmental auditing across the full range of business processes, from design and engineering through manufacturing and operations. In the future, if the concept of full-cost accounting gains ground, audits may look outward to suppliers, customers, and product waste and recycling streams as well.

Precedents for Auditor Liability in Environmental Noncompliance Cases Will Refocus EHS Auditors' Approach

Precedents have already been established for the liability of senior management in environmental cases, and pressure exists now to ensure the accuracy of environmental audit findings. The chief executive officer (CEO), the board, and senior management receive reports based on the audit; they make judgments about a company's environmental performance on that basis, and they can be held liable for the judgments they make.

The environmental auditing community will bear these precedents in mind as it continues to refine auditing standards—which can also function as criteria for areas in which auditors may be held accountable. These standards must focus not only on how the audit is to be conducted, but on what is audited, including company environmental performance, environmental technology, and environmental management systems.

Public and Private Sector Encouragement Will Make Audits a Threshold Requirement

Government agencies in the United States have been interested in environmental audits since the late 1970s. After it considered requiring audits in the early 1980s, the U.S. Environmental Protection Agency (EPA) decided on a policy of endorsing the audit concept, assisting interested companies in starting audits, and supporting research into audit fundamentals. In Europe, the Netherlands has instituted a voluntary audit program for industry that may eventually become mandatory if the industry response falls short of expectations. The EU's approval of the Eco-Management and Audit Regulation, which establishes criteria for voluntary audits, will create, in effect, an official audit standard. Companies that do not participate may be seen as environmental laggards. Eventually, the Eco-Management and Audit Regulation reporting requirements may become a model for guidelines, and even regulations, in other regions. If so, companies that participate early will gain a business edge over competition in Europe and elsewhere in the world.

Industry-sponsored encouragement is also making the environmental audit a threshold requirement for companies. For example, ICC's Business Charter for Sustainable Development, the chemical industry's Responsible Care® program, and the British Standard on environmental management systems all include environmental audit components. Eyeing developing industry standards on the one hand and government standards on the other, companies in the next few years will need to give careful thought to structuring audit programs that can be adapted efficiently as these standards become established (Fig. 2-4).

ICC's Business Charter for Sustainable Development

1. Recognize environment as a top corporate priority
2. **Integrate it into business management**
3. **Continuously improve performance**
4. Educate and motivate employees
5. **Reduce environmental impacts**
6. Improve products and services
7. **Advise customers**
8. **Operate facilities mindfully**
9. **Research impacts**
10. Use precautionary approaches
11. Promote principles with suppliers
12. **Develop emergency preparedness**
13. Transfer environmental technology
14. Contribute to the common effort
15. Foster openness and dialogue
16. Measure and report compliance

British Standard on Environmental Management Systems

1. **Establish environmental management system**
2. Define, document, and communicate policy
3. Define responsibility
4. Register regulatory requirements and manage response to communications
5. Specify environmental objectives
6. Set up a program for achieving objectives
7. **Maintain a manual and documentation**
8. **Implement operational controls**
9. Maintain environmental management records
10. Conduct environmental management audits
11. Review the management system

Figure 2-4. Areas of coverage in selected sets of environmental principles (Provisions in bold specify principles related to EHS auditing).

CERES Principles

1. Protection of the biosphere
2. Sustainable use of natural resources
3. Reduction and disposal of waste
4. Energy conservation
5. Risk reduction
6. Safe products and services
7. Environmental restoration
8. Informing the public
9. Management commitment
10. Audits and reports
11. Environmental ethic

Keidanren (Japan) Global Environment Center

1. **Manage business activities to realize an environmentally protective society**
2. Establish an environmental management system
3. Manage all activities with concern for the environment
4. Develop innovative technologies
5. Transfer those technologies
6. Manage emergencies
7. Publicize environmental protection information
8. Preserve community environment...promote dialogue
9. **Observe "Guidelines for Japanese Enterprises Operating Abroad"**
10. Contribute to public policies
11. Help to solve global problems

Figure 2-4. (*Continued*)

Auditor Certification or Registration Will Become Real

The issue of auditor certification has been discussed since the mid-1980s. There is a logical evolution from developing common audit practices to defining basic principles and standards to determining standards and training for auditor certification. Now that audit practices and principles are well-established and standards are rapidly being defined, the subject of auditor certification is certain to receive increased attention.

Europe may perhaps see auditor certification sooner than the United States. Annex III of the Eco-Management and Audit Regulation establishes "Requirements for the accreditation of environmental verifiers" and places the accreditation process in the hands of an official body that will be specified at a later date.

In the United States, the Clean Water Act Amendments stipulate that after audits required under certain conditions in the amended regulation, the EPA will assess the auditors' qualifications. Several states are examining the possibility of including auditor certification in environmental regulations. Many environmental audit practitioners believe that it would be premature to establish auditor certification programs before environmental audit standards are fully in place.

EHS Auditing Practices, Principles, and Standards Become Globalized

Internationally accepted auditing standards are rapidly coming into focus. The International Standards Organization is currently drafting an environmental management standard that includes provisions for auditing. It is expected to incorporate elements of recently completed national environmental management standards in Great Britain and Canada and of the EU's Eco-Management and Audit Regulation. The United States hopes to contribute to the ISO effort as well with draft standards prepared by the American Society for Testing and Materials and other organizations.

The movement toward internationally compatible environmental auditing standards has also been given impetus by environmental management charters such as the Business Charter for Sustainable Development, the Council for Environmentally Responsible Economies (CERES) Principles, and Japan's Keidanren Global Environmental Charter, as well as industry charters such as the Responsible Care® program.

At the same time, leading transnational corporations have announced their intention to apply one environmental standard wherever they operate in the world. The existence of widely accepted standards for environmental management, and for auditing, will make the documentation and communication of these global environmental management efforts simpler and more effective.

More Companies Use Full-Time Auditors

Companies with high-quality environmental audit programs staff them either with full-time professional auditors or with some full-time staff supported by part-time auditors. As the profession matures, and as stakeholder expectations for audit integrity and independent validation increase, more companies will use full-time auditors. But even full-time auditors will not necessarily make their entire careers in environmental auditing. Moreover, the part-time auditor who moves between auditing and other responsibilities will remain a key piece of the audit staff equation.

Line management's increasing environmental responsibilities will also increase the need for full-time auditors who can provide a clear accounting both of problems and of good performance and communicate it throughout the company.

Environmental Auditing Is Recognized as a Training Ground

Environmental auditing practitioners have long recognized that environmental auditing is an excellent training ground for EHS staff. Management, too, is beginning to see that the environmental audit discipline, like financial auditing, provides an excellent basis for learning about an organization's operations, processes, and personnel. Experience in environmental auditing helps management develop a strong, practical awareness of EHS issues, challenges, and best practices.

We expect that, increasingly, promising, well-trained candidates will spend 3 to 5 years in environmental audit programs as a way of preparing for other management positions. This training approach will also help companies integrate environmental management with traditional management. The approach also makes sense because many aspects of environmental auditing do not necessarily call for senior-level management skills.

One leading chemical company, recognizing the training value of EHS auditing, places top-notch line managers in the EHS auditing group. The expectation is that they will later return to plant manager positions with an excellent understanding of the nuts and bolts of EHS issues.

Environmental Auditing Increases in Importance as a Communication Tool

As more companies adopt a policy of openly and routinely communicating their environmental goals and performance in relation to those goals, environmental auditing will receive increased exposure as a source for what companies communicate. In this arena, the principles of environmental auditing must demonstrate their effectiveness, as companies will be staking their reputations on the information they make public. Yet companies will also be challenged at times. These challenges will add to the momentum for widely accepted environmental auditing standards that can help give all of a company's stakeholders assurance about a company's environmental information and its environmental performance.

Readings in Environmental Auditing

J. Ladd Greeno, Gilbert S. Hedstrom, and Maryanne DiBerto, *Environmental Auditing: Fundamentals and Techniques* (2d ed.), Arthur D. Little, Acorn Park, Cambridge, Mass.

Center for Environmental Assurance, *Principles for Conducting Environmental, Health, and Safety Audits,* Arthur D. Little, Acorn Park, Cambridge, Mass.

J. Ladd Greeno, Gilbert S. Hedstrom, and Maryanne DiBerto, *The Environmental, Health, and Safety Auditor's Handbook,* Arthur D. Little, Acorn Park, Cambridge, Mass.

Maryanne DiBerto, J. Ladd Greeno, Gilbert S. Hedstrom, Ralph L. Rhodes, Ann C. Smith, and William A. Yodis, *ICC Guide to Effective Environmental Auditing,* ICC Publishing, New York.

PART 2

The Benefits of an EHS Auditing Program

3

Environmental, Health, and Safety Auditing as an Assurance Tool for Corporate Officers and Directors*

J. Ladd Greeno and Jane E. Obbagy

Arthur D. Little, Inc.

As environmental issues become a more visible business consideration, not only are corporate officers becoming more familiar with environmental issues, many are starting to take a more active role in ensuring the success of their companies' environmental, health, and safety (EHS) programs. The reasons for this vary, but frequently stem from:

- Intense public scrutiny and media attention directed toward corporate management of EHS matters

---

*Some parts of this chapter are abstracted from an article by J. Ladd Greeno entitled "The Director as Environmental Steward," that appeared in *Directors & Boards,* vol. 18, no. 1, Fall 1993. Abstracted with permission.

- The threat of significant legal and financial liability and resulting loss of corporate reputation from a major EHS crisis

- A realization that environmental stewardship can contribute to competitive success and is increasingly a business strategy and planning issue

In addition, both the government and stockholders are beginning to press corporate officers to demonstrate that their companies are managing their environmental, health, and safety obligations effectively. In particular, companies are being pushed to provide for more disclosure of their EHS performance.

Few officers and directors today need to be convinced that boards should be considering environmental issues at least periodically. They are taking environmental concerns seriously not only because of compliance and liability concerns, but also because companies that understand strategic environmental issues stand to gain significant competitive advantage in the future. According to a survey of *Fortune* 500 companies by the Investor Responsibility Research Center, almost half of the 200 respondents have board-level committees responsible for environmental affairs. Just as significant is a new readiness, especially in the chemical industries, for line managers with environmental, health, and safety responsibilities to be seen as strong candidates for senior management positions. Still, for many directors, uncertainties remain about the right way to oversee corporate environmental performance.

Not surprisingly, therefore, senior executives are asking their environmental, health, and safety staffs to provide them with assurance that:

- No substantive compliance violations exist

- Current operations do not pose any potentially serious threats to human health and safety or the environment

- Systems are in place and functioning to manage EHS compliance obligations and risk appropriately both now and in the future

As a result, the environmental audit—particularly through the written audit report—is taking on an expanded role as an important vehicle for providing that assurance and informing corporate management and the board about the company's EHS progress or lack thereof. Audit reports are not the only source of information regarding the company's EHS performance. However, they are among the most visible and, as such, provide EHS professionals with an opportunity for sharing valuable information with top management.

Board Responsibilities

The environmental responsibilities of board members grow out of traditional board functions. In its *statutory and fiduciary* role, for example, the board oversees the corporation's responsible compliance with the spirit and letter of the law on environmental, health, and safety matters. In its *evaluative* role, the board judges corporate and management performance on behalf of the stockholders and other stakeholders, and with respect to all of the corporation's environmental, health, and safety exposures—and, increasingly, competitive opportunities created by environmental issues. In its *advice and counsel* role, the board guides management on environmental affairs through board committees or in discussions by the board as a whole. Last, in its *crisis resolution* role, the board takes action when a crisis threatens the corporation's well-being. In environmental matters, this role is only likely when senior management has either neglected to make the company's environmental performance a key business concern or has overlooked the need to have in place a responsive crisis management program that can be set into motion should a crisis of any kind occur.

In practical terms, directors fill the environmental stewardship role when they understand:

- The company's stance on key environmental management issues
- Environmental risks and liabilities inherent in the company's activities
- The company's environmental performance goals and measurements
- What questions to ask senior management when the environmental position is not clear or actions and goals seem mismatched
- The importance of senior management's plans and actions for effective environmental management and for finding competitive advantage in strategic environmental issues

The board's full understanding of each of these areas depends on strong reporting and discussion links between the board and senior management, so as to provide genuine assurance to board members that the company's environmental stance is appropriate and effective.

The Corporate Environmental Profile

Every company presents a different environmental profile, shaped by the apparent risks associated with its processes and products; the age, history, and condition of its facilities; the responsiveness of its management and operations culture to change; the regulatory framework within which the company operates; and many other factors. Our experience with environmental management issues at many companies has shown, however, that businesses share certain patterns and themes as they address environmental concerns.

A three-stage framework, summarized here, helps to characterize a company's environmental practice, management, and performance:

- *Problem solving.* Covers companies focusing on immediate and near-term environmental problems and issues. Management's primary reason for taking action on environmental issues is to avoid burdensome costs. No top management statement on corporate environmental expectations exists, and the board limits its involvement to considering major known or anticipated problems.

- *Managing for compliance.*   Defines about 80 percent of major North American and European companies today. These companies are building systems and skills for achieving goals and coordinating compliance efforts to avoid wasting resources. They tend to avoid large changes in the way things are done and to focus more on compliance with laws, regulations, or industry requirements than on future liabilities or opportunities. Senior management makes occasional statements to the company on environmental expectations, and the board hears occasional reports on environmental performance.

- *Managing for assurance.*   Describes companies that actively manage environmental, health, and safety risks *and* opportunities. Their goal is to protect company resources and the environment from harm, to manage risk, and, where appropriate, to achieve business advantage. These companies have clearly defined environmental goals and policies. Management systems help reinforce line managers' responsibility for environmental performance. Monitoring programs such as environmental audits independently assess the effectiveness of these management systems. Senior management, supported by the board, is the driving force behind environmental progress.

How directors understand a company's environmental stance depends on the quality and type of information they receive from senior management. It is generally well understood in corporate governance circles that directors, as stewards rather than managers, generally should not be involved in the details of line management. But when information from senior management does not give directors a good definition of the overall corporate environmental position or assurance of the company's environmental well-being, it becomes the board's responsibility to probe more deeply, to ask questions, and to make environmental issues a dynamic part of the boardroom agenda.

Environmental Information to Boards

Typically, boards of directors learn about company environmental matters through executive summaries, full internal reports forwarded to the board, presentations by company environmental professionals to a committee or the whole board, or presentations by senior management. These reports cover the company's position in areas such as:

- *Environmental compliance and enforcement.*   This information concentrates on past events and provides a retrospective view of the company's performance regarding regulations that have been actively enforced. An additional focus area is how well the company is monitoring and measuring its own performance against regulations.

- *Environmental policies, programs, and practices.* These reports focus on present activities, and many include coverage of pollution prevention efforts and other measures that go beyond compliance. Directors should be aware that a company's environmental programs at the facility level often lag behind the objectives set at the corporate level and understand what the gap may mean for liability issues.

- *Environmental audits and performance reviews.* The environmental audit is becoming a common vehicle for informing senior management and the board about a company's environmental progress. Environmental auditing is a growing discipline; directors should recognize the specific philosophy and objectives that shape a company's audit approach and determine what audit results really convey about the company's environmental performance.

- *Environmental risks, liabilities, expenditures, and liability reserves.* Directors' fiduciary responsibilities make this a high priority. Changes in the regulatory structure, such as the evolving natural resource damage assessment regulations, can have an enormous impact on potential liability. In the current climate, a single incident can lead to astronomical expenses. More commonly, companies spend large amounts on compliance, pollution, and strategic position regarding competitors and key environmental issues and trends.

Because of its specific focus, each type of report has its own strengths and weaknesses. The board can likely gain the fullest understanding of the company's progress and performance by receiving an appropriate combination of reports, but particularly including the audit report or its summary.

The environmental audit has emerged as one of the most important vehicles for providing assurance and information to senior management and the board about a company's environmental, health, and safety progress—or lack thereof. As the environmental auditing profession grows, methodology and best practices will become more codified. But today, audits at individual companies are designed to meet a variety of objectives, and companies make the decisions about the audit focus.

In reviewing board-level reports on environmental audit programs, directors should look for the specific objectives that define the audit and identify how the company is using its results—that is, to assist facility operation and management, to accelerate the development of companywide environmental management systems, to protect the corporation from liabilities, and so forth. It is no longer enough to assume that the existence of an audit proves the company's good environmental intentions. Directors may need to assess the appropriateness of the company's audit philosophy and the effectiveness of its response to audit findings. Some boards are assisted in this process by having the audit program director and/or an outside consultant report to the board on a regular, periodic basis about the philosophy, goals, progress toward goals, and findings of the audit program—and its part in the company's environmental assurance program overall.

As recipients of a range of company environmental communications, directors are in a good position to ask (and should ask) a range of questions of management, that include:

- *What are the company's financial liabilities from environmental matters? Are we protected?* Companies should review their environmental liability situation and ensure that adequate reserves have been set aside.

- *Does our company communicate environmental performance effectively and appropriately to stakeholders—customers, owners, employees, local community, and the wider public?* Stakeholders are becoming more sophisticated about environmental issues. Robert Kennedy, Chairman of Union Carbide, acknowledged this when he positioned his company by inviting stakeholders to examine the environmental performance record; "Don't trust us, track us." Directors should seek assurance that there are solid, defensible links between what the company says it does and what it does. The audit is a valuable tool in helping to provide this assurance.

- *What is the company doing to reduce its overall environmental impact? Will these initiatives meet stakeholders' needs?* As more companies appreciate the impact of strategic environmental issues on business processes, they are moving beyond managing for compliance. Moreover, transnational companies, anticipating the emergence of international standards, are working toward common levels of performance that are higher than the existing standards in many countries where they do business.

Audit Program Objectives and Benefits

The objectives and benefits to industry of environmental auditing are broad and varied. We define an objective of an audit program as an end toward which effort is directed over a period of time, and a benefit as an aspect or output of auditing that contributes significantly and positively to the achievement of company objectives. In short, companies work toward achieving audit program objectives so that they can gain the resulting value or benefit.

Objectives

There are a variety of audit program objectives, not necessarily mutually exclusive. The relative importance given to them, however, will influence the shape of the auditing program. For example, where the focus is to provide assurance, the program is typically supported by top management—the report may go to the board of directors or the chief executive officer (CEO) and there often is a scheduled reporting basis. By contrast, when the emphasis is on helping the facility manage, program support is typically at a lower level within the organization—the report may go to a vice president, yet there is often no scheduled reporting basis.

Virtually all environmental audit programs are to a significant degree compliance-oriented, established to provide a systematic, objective check on the extent to which a facility complies with the terms of its various environmental operating permits. Within the overall objective of identifying and documenting compliance status, a program may take a variety of forms, such as identifying and documenting compliance discrepancies and recommending steps to facilitate corrective action; helping facility management understand and interpret regulatory requirements, company policies, and guidelines; and identifying differences among or shortcomings at individual facilities, or patterns of deficiencies that may emerge over time. Some programs are explicitly established to provide assurance to senior management and its board that the facilities are in compliance, while others specifically focus on helping the facility manager understand and conform to regulatory requirements and corporate policies.

Providing management assurance usually requires a determination of the facility's compliance status and an effective means of reporting that information to management. Such programs typically require a significant degree of independence on the part of the audit team from the facility being audited. In addition, programs established to serve the needs of top management generally demand a more rigorous and in-depth review of facility operations than do programs designed to serve the needs of facility management.

As corporate auditing programs mature, they are conducted against an evolving set of criteria. Often when a company is first establishing its environmental auditing program, the focus is on identifying problems so that they can be corrected (Fig. 3-1). As the audit program evolves, the emphasis moves from identifying problems only, to determining compliance status, and then assess-

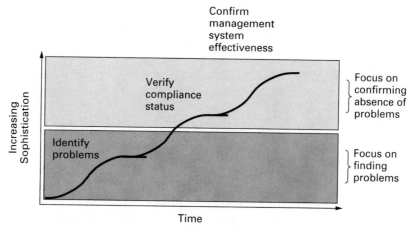

Figure 3-1. Natural evolution of audit programs.

ing the effectiveness of the environmental management control systems. The objective broadens to verify compliance at first against significant compliance regulations and, eventually, to verify full compliance. More sophisticated programs are designed to help the company control its environmental performance, confirm that there are no problems, and provide assurance that the environmental functions are operating satisfactorily.

Benefits

As the company becomes successful in meeting its objectives for the audit program, the most significant benefits it derives can be placed in two broad categories:

- *Increased environmental management effectiveness.*   The increased effectiveness results from identifying and reducing blind spots that may exist, clarifying issues that might otherwise be interpreted differently at different facilities, and developing a more uniform approach to managing environmental activities by sharing information with and learning from other facilities.

- *A feeling of increased comfort or security.*   Many environmental audit programs are established at the request of top management for the purpose of identifying and documenting the compliance status of individual facilities. The primary benefit of such programs is to provide senior corporate officers with a sense of increased comfort or security that environmental programs that they have approved are in place and environmental requirements are being met. The feeling of comfort is generally nonquantifiable and stems from the knowledge that operations are consistent with the organization's environmental policy and procedures and that legal and ethical responsibilities are being met.

The Role of the Audit Report

For more than a decade, environmental auditing has been a valuable tool for providing environmental, health, and safety assurance within a corporation. As a result, choices about how to make audit report findings and their implications readily understood by top management take on a new importance—for example, what report format to use and what type and level of detail of information to include.

Virtually all environmental audits involve gathering information, analyzing facts, making judgments about the status of the facility, and reporting the results to the appropriate levels of management. Information reported often includes an assessment of the strengths and weaknesses of EHS programs in place, as well as information regarding EHS compliance and other risk management issues requiring attention. These facility-specific findings can have value to management in and of themselves as an indication of how particular locations are carrying out their EHS responsibilities. When audit programs are staffed in a way that enables auditors to visit a number of facilities in the course of carrying out their duties, the auditors are in a unique position to provide an added perspective on how individual facilities compare with each other. They can help provide a barometer of how EHS matters are being managed on a day-to-day basis across the corporation. Even an observation about a lot of little things wrong in the facilities can be useful to management in suggesting that there is not enough attention being given to overall management of these EHS issues.

Thus, the environmental audit and its report present an opportunity to inform senior management about the company's EHS performance more broadly.

Some companies are going a step further by seeking an external, objective critique of their audit program (or environmental management systems more broadly). For example, companies that include ARCO Chemical, WMX Technologies, Texaco, Novacor, Chevron, and others have even included summary results of the outside critique in their annual financial or environmental reports—thus informing a broader public about their progress toward goals. Figure 3-2 provides such an example from ARCO Chemical's *1993 Environmental Progress Report*. The inclusion of the audit report summary is introduced in the report by Alan R. Hirsig, President and CEO of ARCO Chemical, as follows: "The Audit Report...is a crucial step for us in reconciling our environmental policies and programs with our actual performance."

Developing a Report to Meet Management's Needs

One key principle for conducting effective environmental audits is that the audit report be prepared with an "appropriate" form and content.[1] At a minimum, audit reports should present the purpose, scope, and results of the audit. However, audit reports that are highly valued within a corporation relative to their role in providing *assurance* go beyond the basics. They provide the right level and mix of information and contextual background to ensure that management receiving the report understands the implications of what is being reported. Moreover, these reports are written to draw the reader quickly into the importance of what is being discussed, and to identify the causes of underlying deficiencies in order to provide an effective trigger for implementing corrective actions, as appropriate.

As a result of the growing role of audits in the mix of corporate environmental management tools, together with the increasingly direct involvement of senior management in EHS issues, over the last decade we have seen an evolution of how companies report their environmental audit results. Although there are a variety of report formats used today, a look at this evolution provides a useful

---

[1]From a booklet entitled *Principles for Conducting Environmental, Health, and Safety Audits*, 2d ed., Arthur D. Little, Inc.

Following is an Executive Summary of an evaluation of ARCO Chemical's Manufacturing Audit Program by Arthur D. Little, Inc., an international consulting firm. This independent review found that, despite the fact that our formal audit program is young, it has many strengths and will continue to improve with some specific modifications.

We have reviewed the Manufacturing Audit Program in place at ARCO Chemical Company during the period of January 1, 1992, through April 30, 1993. Our evaluation is based on our extensive familiarity with the environmental, health, and safety auditing practices of several hundred companies, including virtually all companies that aim to have a leadership program. Our work did not include an evaluation of the compliance status of facilities included in the Manufacturing Audit program.

In our opinion, ARCO Chemical Company's Manufacturing Audit Program is designed and implemented in a manner that is generally consistent with, and in several instances exceeds, the prevailing practices in the chemical industry. In understanding our evaluation, it is important to note that the Manufacturing Audit Program is a relatively young program. If it can sustain the progress made over the past year, it is well on its way to becoming an industry leader.

We have identified several aspects of the program which are particularly noteworthy as well as those that need further improvement. Our evaluation of the Manufacturing Audit Program has been communicated to company officials. ARCO Chemical has accepted our opinions and we understand that action is under way to remedy the shortcomings.

STRENGTHS

ARCO Chemical has established a rigorous methodology to help ensure that facility practices are reviewed in a comprehensive manner. The methodology features the use of guidance documents that provide the structure for collecting information regarding plant practices. The program also fosters team consensus building and cooperation between the audit team and plant personnel in identifying problems and sharing best practices. In addition, the reporting process provides for clear and concise conveyance of audit results to the appropriate levels of management.

We found that the program is designed to tap the technical and managerial expertise within the company. Audits are led and conducted by staff members who collectively have relevant and appropriate technical skills, operations know-how, and environmental, health, and safety experience. In addition, team leaders consistently earn the respect of plant managers, and staff associated with the program are regarded as strong performers.

The scope of the program is comprehensive. Building on the Chemical Manufacturers Association's Responsible Care™ program, ARCO Chemical's program considers issues and operations that may have an impact on environmental, health, and safety performance. For example, the scope includes ARCO Chemical manufacturing standards that cover environmental, health, and safety issues as well as other standards related to operations, technology, maintenance, transportation, purchasing, and suppliers.

The stature of the Manufacturing Audit Program within the organization is impressive. ARCO Chemical's Chief Executive Officer not only is committed to the program, but has reinforced his commitment by allocating significant resources and management oversight to implement

Figure 3-2. Example of an audit report summary included in an environmental report. (*Excerpted from ARCO Chemical Company's "1993 Environmental Progress Report." Reprinted with permission from ARCO Chemical Company.*)

program activities. In addition, based on our interviews with staff at both the corporate and plant levels, we noted that the program is viewed as one of the two principal drivers of ARCO Chemical's Manufacturing Excellence Program, and ownership of the program extends beyond the Manufacturing Audit Staff.

AREAS FOR IMPROVEMENT

In reviewing on-site auditing practices we noted that the approach to data gathering was not always implemented and team resources were not always assigned in a manner to help achieve an in-depth review of facility practices. That is, the audit teams' approach to data gathering, in many instances, was more rigorous in the identification of problems than in the confirmation of sound or good facility practices. Also, the allocation of team resources did not consistently reflect an analysis of audit topics judged to have significance in terms of compliance or risk management. Thus, we pointed out to ARCO Chemical that by refocusing slightly the approach to data gathering, reallocating team resources where appropriate, and maintaining the skill level of the audit teams, the nature and depth of facility reviews could be enhanced.

In reviewing the preparation and content of ARCO Chemical's working papers or audit team field notes, we noted that they are not entirely consistent with general industry practices. Working papers typically contain a record of observations, interviews, data reviewed, and audit procedures followed so as to provide the basis for substantiating an audit team's conclusions regarding facility practices.

ARCO Chemical's working papers generally documented interviews and deficiencies well; however, the working papers frequently did not provide a clear "road map" of how the auditor accomplished each step of the audit process. We recommended to ARCO Chemical that additional efforts be placed on developing working papers so that they capture, in greater detail, information on what the team did, how and why.

We also recommended that ARCO Chemical consider establishing audit program performance measures or benchmarks to assess company performance with respect to compliance with governmental requirements and internal manufacturing standards. The purpose of this activity would be to provide the company with a vehicle to track progress made in implementing risk management and quality control measures, to identify compliance or manufacturing programs where resources are most needed, and to help articulate performance expectations.

A variety of audit program policies and procedures have been prepared by the Manufacturing Audit Staff. However, many of these documents are in the early stage of development and do not fully describe the performance expectations of program management. By further modifying and expanding these policies, procedures, and standards, ARCO Chemical will have necessary reference documents to help ensure consistency in auditing activities over time, and to provide for customers the context and parameters under which the program operates.

Arthur D Little

Figure 3-2. (*Continued*)

benchmark for evaluating whether or not your company's audit report is consistent with the needs of senior management and the culture of your organization.

In the early days of audit program implementation, many companies were concerned that the written word might come back to haunt them. Thus, textual discussion in these audit reports—particularly interpretive information—was kept to a minimum. Audit teams generally produced *audit memos* that provided a basic message that We visited the site and here are our general observations. As companies became more confident both in the sophistication of their EHS systems and of the value of their audit programs as a management tool, the audit teams used the reports to provide a message that We visited the site and here are the deficiencies noted. This *exception report* listed departures or exceptions from governmental requirements and company standards and provided observations related to the management of EHS programs reviewed during the audit.

With the increasing sophistication of EHS systems and audit programs, companies with procedures and checks and balances in place continued to list the exceptions and observations in audit reports. They also often included a generic statement about areas reviewed during the audit. The basic message of this *generic opinion report* was We reviewed site programs and practices following our standard audit guides and procedures, and we believe the site is generally in compliance except as noted in this report. This report implied a level of comfort to management, but because it always seemed to provide the same information, it continued to beg the question of just what the results meant for the company as a whole.

Taking the next step in providing assurance, today management at some companies are asking their audit teams to provide a *true opinion* of their overall analysis of the management systems in place and the level of compliance achieved with respect to a spectrum of relative performance, while also including a list of exceptions and observations ordered by their relative significance. The *true opinion report*, which is based on the professional judgment of the audit team, is gaining in popularity among top management who count on audit reports for three main purposes:

- To help them understand relative overall environmental, health, and safety performance and the existence of specific problems
- To assist them in understanding the significance of the audit results so as to focus resources on areas where improvement is most needed
- To measure improvement in performance from audit to audit

Companies that have been sharing their audit results with top management and the board of directors found in early years that providing a "comfort statement" about the general results of the audit program was sufficient to meet the board's needs. Increasingly, however, the board members wanted to know more and today are receiving entire presentations about the findings of the audit program and how those compare to past findings. Figure 3-3 provides an

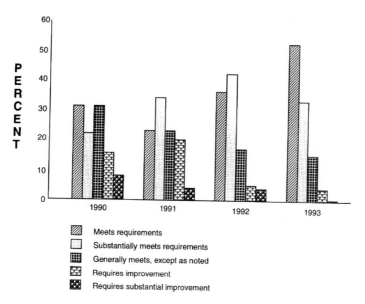

Figure 3-3. An example of company's overall EHS performance as reflected by the EHS audit results.

example of a summary sheet presented to a board (as part of a detailed audit-results presentation) of one company's environmental performance as reflected in the audit results.

Companies that are particularly successful in making the audit report valuable to senior management and the board understand both the changing role of the audit report in providing assurance and also the characteristics of an audit that complement this role. Some helpful rules of thumb for EHS management to consider in developing a report useful to senior management include:

- *Find out just what top management wants the audit report to accomplish*—the answer may trigger a complete rethinking of your audit program and audit report approach.

- Including comments on the effectiveness of EHS programs is part and parcel of a *rigorous and structured approach to auditing.* Consequently, the standards established for the audit and staff chosen to participate as audit team members become more critical—consistency in audit team makeup from audit to audit is important.

- In order to meet the needs of senior management effectively, a *clear writing style* is important in an audit report. Companies should examine the resources typically devoted to the writing of audit reports—time allotted and the skill level of the individual actually doing the writing—to make sure they are sufficient to fulfill the expectations of the management audience to whom the reports are being sent.

The Future of Stewardship

Until now, the corporate board's challenge has been to ensure that it has the right information to be confident that environmental matters do not threaten a company's well-being. Now, as senior management moves from reacting to environmental pressures and liability to making environmental considerations a key part of business processes, the board's focus will also shift. The choices companies make in how they manage EHS affairs can give them control over environmental forces in business. Or their choices could give a large measure of that control to regulatory bodies and competitors. As management addresses these choices, the board's role as environmental stewards will help them guide companies toward decisions that give them control over their environmental destiny.

Companies with a well-defined environmental stance, for example, have a basis on which to build communications within the company and outside of it and from which to act on regulatory and industry developments. Companies that do not define their environmental stance will find instead that the regulatory framework and industry standards do so by default—and that regulations increasingly drive their cost structure. Companies that effectively communicate environmental performance to stakeholders can benefit from increased levels of trust and a more positive public image, while those that do not may see the media and competitors frame that discussion instead. Companies that plan for the impact of environmental issues on competitive assumptions will be prepared to move ahead of competitors, while those that do not could face an increasingly uncertain hold on the marketplace.

To ensure that the company is doing what it must to meet its environmental responsibilities, directors can ask themselves and management critical questions about the company's environmental stance and performance. Are we getting the information we need to understand the company's approach to environmental management? Is the company protected from financial liabilities associated with environmental matters? What is the company doing to reduce its overall environmental impact? Will these initiatives meet stakeholders' needs? Is the company communicating effectively and appropriately to stakeholders? In finding the answers to these kinds of questions, the boards will find that there is a range of useful management tools to help provide the assurance, but the environmental audit (and audit report) is a particularly valuable tool in understanding the company's progress in meeting goals for environmental performance.

4

Environmental Auditing and Public Consent

James E. Lukaszewski

Chairman, The Lukaszewski Group Inc.

Introduction: Fundamental Concepts

No matter how well a corporation's overall environmental management system or its environmental health and safety auditing department function, there is no guarantee that management will be able to build that new plant, introduce that new product, or maintain existing permits. The key to survival and success is the corporation's relationship with the local community and a variety of regulatory and special publics. Unfortunately, these relationships are often either ignored or poorly maintained. Communication with the community and special publics is the crucial ingredient. No environmental audit is complete without an audit of community relationships and of communication techniques and approaches.

When it comes to talking about news that is unpleasant, controversial, or outright bad, corporations often revert to denial, deception, or disinterest. The following comments from management are not uncommon:

- The less we say, the better.
- Let's wait to see who really cares about this.
- Why do they want to know about that?
- They shouldn't be interested in this!

- That's not important. Don't tell them!
- Just tell them the important stuff. They'll just get nervous if you tell them everything.

The mindset exemplified by these comments can and will lead only to trouble. Throughout the world, public concern on an individual basis over the environment continues to grow. People, including neighbors, employees, area residents, customers, children, and government officials, have questions and concerns. They want to know about the:

- Nature of the risk
- Actions to be taken
- Environmental effects
- Management of response to serious or worst-case problems
- Effects on future generations
- Effects on very long term exposure to extremely low dosages of toxic compounds or energy fields

These concerns are legitimate and powerful because of the community's ability to stop, shut down, or hold up corporate projects. If a corporation's communication plan fails to address the community's questions, there will be no relationship with the community, and in all likelihood attempts to gain the community's trust, cooperation, and consent will be seriously jeopardized, if not impossible to achieve.

Today's business operating environment finds a public very sophisticated about science, environmental impact, and current issues. Contrary to industry and management assumptions, and despite this high level of knowledge, the public often sets rationality aside, making decisions using highly emotional approaches, ideas, and even hunches. Why is this so? Because community decision making is always values-driven. As corporate management plans to assess its relationship with the community, it should recognize at least 10 communication realities:

1. Public consent is required, continuously.
2. Public involvement is necessary, ongoing, and often government-mandated.
3. Public involvement can kill projects as well as permit them.
4. Public officials expect the business to win and maintain the public's support.
5. The news media will focus on the conflict, controversy, and opposition.
6. Personal self-interests, values, and needs take precedence over social values and needs.

7. Industrial and business facilities are often seen as threats to personal and self-interest values.

8. Business facilities have few inherent political constituencies and little political clout.

9. Personal fear is a factor.

10. Complex and scientific information about risk and probability—even when openly and clearly communicated to broad audiences—can, and often does, cause grave concern.

Business executives, scientists, bureaucrats, technologists, and government officials who believe there is a magical way to bring total rationality to environmental decision making are destined to live lives filled with Maalox moments. The irritations to the process caused by values-driven decision making are very real:

- Emotional communication has replaced reason.
- Activism has overtaken scientific investigation.
- Exaggeration often overwhelms precision.
- Grassroots manipulation is the new realism.

Science, data, and facts are important, but in the public decision making process their importance is only as *background* to building emotional comfort which allows the public to accept a proposed environmental change. If the public's concerns are not addressed, or worse, minimized, trivialized, ignored, or belittled, the publics—whether the community, neighbors, activists, elected officials, or some combination of all these—in a position to manage the destiny of your company, your product, or your environmental situation will take control and bring about defeat.

Management's principal linkage to these powerful interests is through communication, and much of the information communicated flows out of the environmental audit process. As such, in this chapter we will examine the critical communication concepts that, if correctly applied, can help businesses take advantage of information gained through environmental, health, and safety (EHS) audits to develop an effective strategy to gain and maintain consent from the community.

This chapter is constructed to meet several objectives:

- Audit checklists are included to help evaluate existing communication plans, strategies, policies, and tactics.
- Various planning document models are included to aid in constructing communication based relationships which will minimize the chances of failure and maximize the obtaining of community consent.
- Different communication plan formats are presented for use as both audit and operational models.

Major communication audit points appear throughout the introduction to this chapter. They call your attention to fundamentally sound approaches and will help you establish specific communication strategies and techniques which ought to be a part of the communication plan. These audit points are based on the attributes successful communication strategies have in common.

AUDIT POINT NUMBER 1:

Successful environmentally related communication programs and strategies have seven major attributes:

- Proactivity
- A focus on consequences over tactics
- Fundamentally ethical behavior
- Values-sensitive strategies
- Conversation-based relationships
- Prioritized action and decision plans designed from a community perspective
- Continuous, seamless communication operations

---

1. *Proactivity.*   Act now; talk now; listen now. This is what the community expects; what your employees, neighbors, government, and opponents expect. Those who lay back fail.

2. *Focus on consequences over tactics.*   Realistically look at the effects of what you propose to say and do. If what you propose will make individuals angry or more resistive, change what you plan to do. If what you plan to do will negatively impact community core values, change your plan. Advertising, meetings, face-to-face exchanges, and local do-goodism will not overcome bad, untimely, engineering-driven, or dumb ideas.

3. *Fundamentally ethical behavior.*   If your corporation behaves or talks in a way that is unacceptable to the community, success—based on the community's standards rather than your own—is impossible.

4. *Values-sensitive strategies.*   It is not possible to get the community to approve, reinstate, or continue actions that negatively impact community core values, property taxes, health and safety, the environment, personal comfort, freedom from fear, etc. To be successful, strategies must accommodate the value system of the community and of those most directly affected.

5. *Conversation-based relationships.*   Get out there and talk to people face-to-face, belly-button–to–belly-button. Small group and individual meetings are best. Humanize, be empathetic, use community oriented language.

6. *Prioritize actions and decisions from a community perspective.*   Put yourself in the community's shoes—those of a neighbor, opponent, those most directly

affected—and honestly project the impact of what you plan to say or do on those individuals. If you do this, you will not make mistakes which keep you from success.

7. *Continuous, seamless communication operations.*   Effective community relations is ongoing and done on a daily basis. The corporation that gears up at the last minute will have no base of support, no common connection with the community, and no help from public officials. Far too often we underrate the community's anger and overrate our preparation. We ignore the true feelings of the community, trivialize the risks we pose, and then disparage the community's opposition, representation, and science. Intermittent, timid, cutesy, insincere, and highly technical communication are ingredients in the recipe for losing support.

Emotions are running high, especially following a string of major environmental disasters that include Chernobyl, the *Exxon Valdez,* the Sandoz chemical spill in Switzerland, the well fires and oil spill off the coast of Kuwait during the Persian Gulf War, the Shetland Islands oil spill, and various plant explosions. Research shows that poor environmental behavior of countries and businesses worries people in very large numbers.

Another factor has come on the scene. Aggressive government intervention around the world is forcing businesses and individuals to be more conscious of how their actions affect the environment. In the United States, criminalization of environmental rule and law infractions is escalating. The U.S. Environmental Protection Agency has in place an aggressive criminal prosecution process to hit businesses with tough penalties quickly. Such actions are widely supported by the public. In fact, most polls demonstrate that the public thinks the government should go even further and be even tougher.

AUDIT POINT NUMBER 2:

Reduce the media's power by planning for and accommodating the media's behavior patterns and communicating directly with those most affected by your actions.

The news media, virtually worldwide, are emotionally committed on the issue of environment. Routine reporting has become *alarmist* in nature, irrespective of what the facts might indicate. *Allegations,* no matter what the source, are carried instantly, often without any verification. *Interpretation* of events, issues, and problems on the flimsiest of information is now the daily routine. Media behavior, driven by competition and deadline pressure, gives rise to the attitude of, If we make a mistake, we can fix it tomorrow…, no matter what the impact on reputation, market share, or company value.

The media have become *interventionist,* often cooperating with and even helping to stage environmental situations which gain enormous visibility, whether based on reality or simply the hunch of an environmental activist organization. Plaintiff's attorneys aggressively pursue media interest to create emotional situations and attract "victims" to class action lawsuits.

The media have become *speculators,* almost always devoting their interests to the worst-case scenario. We can complain about this all we like, but the fact is the public depends on the news media's *exaggerations* to help create a climate of fear which builds leverage against environmental decisions that make the public uncomfortable.

These media behaviors and attitudes are uncontrollable. Therefore, one important goal is to communicate as directly as possible with those most directly affected. What is crucial from a communication planning perspective is to recognize that these situations will occur and that they can be minimized through the structure, language, intent, and execution of your communication. Concepts and strategies in this chapter are all elements designed to reduce the media's power, influence, and interest in what you are doing.

AUDIT POINT NUMBER 3:

Build community trust and comfort continuously.

A communication plan that doesn't address the basic trust-building needs of the community—from the community's perspective—will fail.

When we analyze successful community communication programs—successful from the community's perspective—even if a siting is not immediately needed or a permit is not granted, we find that these programs focus relentlessly on building community trust. Community trust-building programs share at least seven common elements. They:

1. Provide advance information;
2. Seek community input;
3. Really listen (respond to community concerns);
4. Demonstrate that community ideas have had impact;
5. Keep in touch (through aggressive question answering);
6. Speak in community language; and
7. Bring the community into the decision-making process.

We will talk about each of these specifically when we discuss the model consent-building process later in this chapter.

AUDIT POINT NUMBER 4:

Good relationships are based on ethical behavior.

   Community relations which lead to public permission to move ahead work better when based on fundamentally sound, ethical concepts and openness.
   Levi Strauss and Co. has developed an excellent model of ethical principles against which any behavior model can be taught, soundly analyzed or planned:

- Honesty
- Promise keeping
- Fairness
- Respect from others
- Compassion
- Integrity

   Each of these elements can be defined according to the culture in which they are applied. It must be done visibly and without reservation. The truth is, the public as a whole is raising the bar of expectation for the behavior of individuals and organizations who have control over environmental decision making. In addition, regulatory agencies are increasingly imposing new standards of integrity and compliance on organizations that have difficulty in these areas, often very harshly and publicly.
   This aggressive, ethical approach leads to a simple and direct environmental communication policy involving four internal and external concepts:

1. A willingness to talk
2. Relentless truthfulness (from the audience/public perspective)
3. Willingness to answer any and all questions from any and all sources
4. Recognition that there are no secrets, that everything comes out eventually

---

AUDIT POINT NUMBER 5:

Build or analyze your plan using successful models.

The Master Public Consent Model

The *master public consent model* explained here in detail provides an excellent series of audit checklists to analyze and evaluate your existing plans and processes.

Each section can be used as a benchmark checklist for action and criteria for comparison and modification.

Figure 4-1 demonstrates in a single master diagram the communication and behavior structure upon which public consent rests. While the resulting structure looks sturdy, it is in fact incredibly fragile and ready to crumble virtually at any time.

Section A: Organizational Faults (Behaviors and Assumptions to Avoid).   At the bottom of the model are illustrated the behaviors, notions, assumptions (even delusions), which tend to destroy effective communication and public consent building. These are the cracks, the fissures, the faults that lie below the best laid foundational concepts for effective communication.

Avoiding these faults is a conscious, nonstop auditing and evaluative process in effective environmental communication. Here are some examples as illustrated in Fig. 4-1.

- False assumptions: "The public cares more about jobs and the economy than the environment."
- The media is to blame: Unfortunately irrelevant; the media doesn't sign permits.
- Erroneous data: Usually means we did not do enough homework.
- "We've got the connections": A delusion quickly made real when public officials stop talking to us or oppose us.
- "It's obviously needed": Only to those whose bonuses or careers depend on the outcome.
- Overrate your preparation: Because the telephone is not ringing does not mean no one cares. Remember, being an opponent is not a full-time job.
- Denial/delay: "If we don't talk about it, maybe no one will find out."
- Empathy/arrogance: If we talk in scientific, technical language without responding to the emotional concerns, we will not be credible.
- No respect for opposition: Opponents, media, and citizens without credentials will always have far more credibility than we do.
- Underrate negative community emotions: Neighbors and opponents don't get angry until something actually starts happening that affects them (that's why companies so often feel they are blind-sided).
- No Plan B or Plan C: There is an arrogant, usually mistaken belief that the first site/approach/technology proposed is the one that will ultimately be approved (it almost never is).
- Self-talk/self-delusion: If we tell ourselves often enough that something is correct, we believe it.

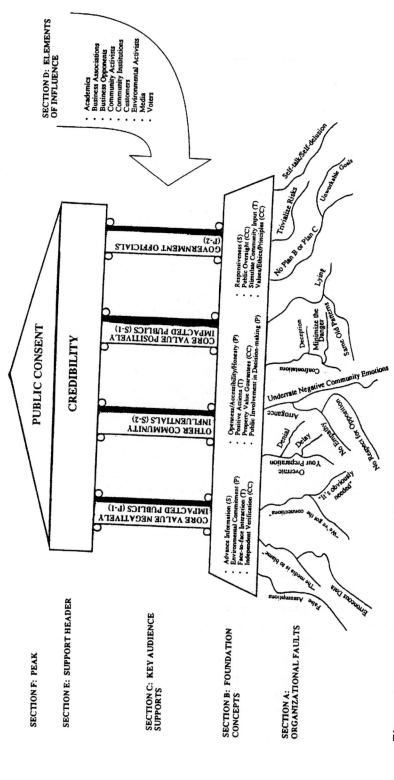

Figure 4-1. Public consent model. *(Copyright © 1993, James E. Lukaszewski. All rights reserved.)*

Section B: Foundation Concepts (Every Concept Missing Makes Your Structure Even More Fragile). A place to begin planning your environmental communication program is around a core group of positive foundational ideas and behaviors which meet or exceed community expectations:

- *Advance information.* Be prepared to get information out to those most directly affected early, often, and whenever they feel they need more.

- *Environmental commitment.* At every step avoid arguing, haggling, or negotiating; simply do more than is expected.

- *Face-to-face interaction.* Large meetings, although necessary, can also turn into lynchings. Focus most of your efforts on smaller meetings where you can be face-to-face with neighbors, friends, supporters, and even opponents.

- *Independent verification.* Early on in your communication process, structure either outside advisor groups or experts acceptable to all sides who can look at the facts, look at the data, apply some rationality to the emotionally charged atmosphere, and bring wisdom to the process.

- *Openness/accessibility/honesty.* Be accessible, follow the four-point communication policy outline at the beginning of this chapter. Openness and honesty undermine the power of negative opposing forces.

- *Positive actions.* As you examine the time line of decisions, engineering questions, government action, and citizen involvement, plan positive, favorable responses at every available opportunity. Positive attitudes and actions also undermine the power of activism and negative opposition.

- *Property value guarantees.* If something you do will affect the value of the possessions or property of those who live around what you have done or plan to do, immediately initiate some mechanism for protecting the value of those possessions and property. This is a common practice across the United States. Many models are available. Avoid being forced to guarantee values as a last ditch negotiating nugget. Volunteer it; get started; be ahead of the opposition.

- *Public involvement in decision making.* Set up credible advisory committees. Credibility is determined primarily by the presence or representation on your advisory groups of those who are either unconvinceable or who are unalterably opposed. Avoid setting up cheerleading squads. They have no credibility and are likely to say and do things which will embarrass you and cause irreparable damage.

- *Responsiveness.* Stay ahead of the communication needs of your audiences and those most directly affected. There is not a question you will be asked that will be a surprise. Prepare to answer all questions as early as possible. Raise questions out-of-sequence, well before the time the opposition would normally bring them up, and answer them. This tactic calms public officials, neighbors, employees, and others who are concerned about how you will behave in environmental-related situations.

- *Public oversight.* Invite public officials and the media in early, ahead of regular deadline requirements. Openness, although occasionally painful, undermines the power of opposition and the position of unsympathetic policy makers.

- *Stimulate community input.* Hold your own meetings—usually small, but occasionally large group meetings. Stand up and say it out loud. Ask for input from the community; then demonstrate that you have listened to the input.

- *Values/ethics/principles.* Recognize that all decisions in the community are values-driven. Those values include powerful personal issues like health and safety, valuable possessions and property, peace of mind, pride in the community, absence of conflict, freedom from fear, and economic security. They involve gut-level emotions that are more powerful than all the facts you can possibly muster. A communication plan that does not recognize the values-driven nature of public decision making is a communication program that will not succeed.

AUDIT POINT NUMBER 6:

Workable plans depend on good community assessments, including socio-economic information.

Community assessment involves gathering useful intelligence and information about the community including its political structure, demographics, business, environmental, community and other organizations, and the local news media. In addition to gaining an understanding of the community structure, the assessment also identifies issues and concerns in the community that may affect your operations. The information gathered during the assessment process forms the basis for the community relations program and approaches appropriate for the community and what you are trying to accomplish.

Model Form 1 shown in Fig. 4-2 is a sample community assessment survey.

Section C: Key Audience Supports. Figure 4-3 reflects the grid priority approach to understanding relationships based on how a given segment of the public is affected by environmental action.

AUDIT POINT NUMBER 7:

Successful strategies involve early and continuous audience study and prioritization based on core value impacts.

COMMUNITY ASSESSMENT SURVEY

This Community Assessment Survey process provides a perspective on where a proposed site stands with key audiences including employees, local government officials, the media, and the community-at-large. It will help pinpoint community needs as the community relations planning process begins.

Name: _____

Branch: _____

Date Completed: _____

Part 1: Evaluation of Community Position

1. How many people are employed at your branch? _____
    What percentage of employees live in the community? _____

2. Within the next five years, do you anticipate the employment level to:
    _____ Increase _____ Decrease _____ Remain the same
    Comments: _____
    _____
    _____
    _____

3. What is the annual value of the local goods and services your branch buys from _____
    vendors in the:
    $ _____ Community $ _____ State
    $ _____ Other $ _____ Total

4. What is the annual payroll for your branch? $ _____

5. How would you rate the economic outlook for your community?
    _____ Poor _____ Fair _____ Average _____ Good
    _____ Excellent

6. What current issues or activities (i.e., permitting violations) do you have under way, which could affect your relationship with the local community?

Figure 4-2. Model Form 1.

Part II: Community Relations

1. How would you define your branch's community in terms of its geographical boundaries?
    _____
    _____

    Population: _____
    _____
    _____

2. How close is the branch to:
    The nearest residential area? _____
    School? _____
    Play Area? _____
    Church? _____
    Hospital? _____

3. What is the frequency of the branch's formal and informal communications with the community? _____
    _____
    _____

4. What is the nature of your communications with the community? _____
    _____
    _____
    _____

5. What is the tone of the branch's contracts with the community?
    _____ Friendly _____ Neutral _____ Hostile _____

6. Who in your branch does community relations work on behalf of our company?
    _____
    _____
    _____

7. Our company's image in the community is generally: (Poor, Fair, Good, Excellent, ___ Don't Know) for:
    Your Branch: _____
    Corporate: _____

8. What makes you say that? _____
    _____
    _____

Figure 4-2. (*Continued*)

9.   Please rate our company's image on:
     Environmental Stewardship: _____
     Safety: _____
     Community Service: _____
     As a corporate Citizen: _____

10. Is your branch generally considered by the community as a good employer?
     _____ Yes _____No
     Why? _____
     _____
     _____

11. What is the current status of your branch's relationship with the community?
     _____ Little or no communication.
     _____ Frequently hostile communication.
     _____ Frequently friendly communication.
     _____ Some communication on certain topics.
     _____ Excellent two-way dialogue.

12. What do you consider your branch's greatest strength in its relationship with the
     community? _____

13. What is the branch's greatest weakness? _____
     _____
     _____
     _____

14. Do you believe the business leaders in your community understand your branch's goals
     and needs? _____ Yes _____ No

15. Does the branch manager (or other personnel) maintain contact with business and other
     community leaders? _____ Yes _____No

16. How many contracts have you had in the last 12 months and with which group?
     _____
     _____
     _____

17. Describe the contracts: _____
     _____
     _____
     _____

Figure 4-2. (*Continued*)

18. In the next six months, do you know of any public hearings, meetings, visits by key government officials or other key events at your branch?
    _____ Yes _____ No

19. Please list them: _____
    _____
    _____

20. How often (Rarely, Occasionally, Often, Regularly) does your branch receive complaints from the community about:
    _____ Environmental problems
    _____ Odors
    _____ Traffic
    _____ Other operational aspects

21. How would you rate your branch's involvement or participation in activities in the community?
    _____ Low _____ Average _____ High _____

22. Please detail your activities: _____
    _____
    _____

23. Does your branch have a formal management-supported program for community contracts?
    _____ Yes _____ No

24. If formal, please describe them: _____
    _____
    _____
    _____

25. Are the volunteer efforts of your employees the result of a formal program or informal, independent activities, or both? _____
    _____

26. What community clubs, groups, or associations are employees currently involved in?
    _____
    _____
    _____

27. Does the branch's management encourage employee participation in these activities?
    _____ Yes _____ No

Figure 4-2. (*Continued*)

28. Have you had any contact with environmental groups or organizations in your community? If so, which ones? _____
   _____
   _____
   _____

29. Is the branch's management actively involved in community organizations? Which ones? _____
   _____
   _____
   _____

30. List local environmental or other activist organizations: _____
   _____
   _____
   _____
   _____

31. What contract have you had directly or indirectly with local environmental or community organizations? _____
   _____
   _____
   _____

32. How often does your branch receive requests for support from community organizations, schools, governmental groups and professional organizations?
   _____ Rarely _____ Occasionally _____ Often

33. How many speeches/presentations have facility representatives made in the last 12 months?
   _____
   _____
   _____

34. What is your current level of contributions to community organizations?
   $_____

35. To which organizations? _____
   _____
   _____
   _____
   _____
   _____

Figure 4-2. (*Continued*)

36. Do you believe these contributions benefit your branch as well as the recipients?
    _____ Yes _____ No

37. Describe the community's attitude toward our company over the past five years:
    _____ Changed substantially
    _____ About the same
    _____ Worsened

38. Does your branch have any direct involvement with the local school system?
    _____ Yes _____ No

39. Describe the involvement: _____
    _____
    _____
    _____

40. Have you ever had an "Open House"? _____ Yes _____ No

41. If yes, when and how many people attended? Was the event successful?
    _____
    _____
    _____

42. How do you measure the level of trust, respect and appreciation for our company's
    presence in the community? _____
    _____
    _____
    _____

Part III: Employee Communications

1. Do you believe that your facility thoroughly communicates its positions, objectives and
   programs to its employees?
   _____ Yes _____ No
   Comments: _____
   _____
   _____

2. In what way? _____
   _____
   _____
   _____

Figure 4-2. (*Continued*)

3. Through what avenues do you communicate to your employees? _____
   _____
   _____
   _____

4. Your employees' attitude toward our company is generally (Unfavorable, Neutral, _____
   Favorable) for::
   _____ Your Branch
   _____ Corporate

5. How do employees communicate their feelings/attitude to management?
   _____
   _____
   _____
   _____

6. What is the current status of the relationship between the branch's management and the
   employees?
   _____ Poor communications and bad feelings
   _____ Some problems, but a desire to work together
   _____ Good communications and relationships
   _____ Other (explain) _____

7. What do you consider your branch's greatest strength in its relations with employees?
   _____
   _____

   What is its greatest weakness? _____
   _____
   _____
   _____

8. How do your employees usually receive information about your branch and the company?
   _____
   _____
   _____

9. Do you have any special awards programs for employees?
   _____ Yes _____ No

Figure 4-2. (*Continued*)

10.   What are they? _____
_____
_____

11.   Do you have an annual branch picnic or similar social event for employees?
_____ Yes _____ No

12.   What is the event? _____

Part IV: government Relations

1.   What is the current overall status of your branch's relationship with local government _
officials?
_____ Non-existent _____ Poor _____ Fair
_____ Good _____ Excellent

2.   What do you thick is the local and state government's attitude (Unfavorable, Neutral, _
Favorable, Don't Know) toward our company:
_____ Your Branch
_____ Corporate

3.   How is local government viewed by the community at large?
_____ Favorably _____ Neutral _____ Unfavorably

4.   What do you consider your branch's greatest strength in its relations with local
government? _____
_____
_____

What is the greatest weakness? _____
_____
_____

5.   Is your branch a member of:
_____ Local Chamber of Commerce
_____ Industry Group (specify)
_____ Other

6.   From what source(s) does your branch receive information on local and state government
activities related to your operations? _____
_____
_____

Figure 4-2. (*Continued*)

7.   Do local or state politics have any effects on the operations at your branch?
     _____ Yes _____ No

8.   What has been the history of local government's participation and support or non-support of issues affecting your branch? _____
     _____
     _____
     _____

9.   Do the branch manager or other personnel maintain regular contact with the following?
     _____ Elected City/County _____ State representatives
     _____ State Senators _____ Federal Officials
     On what issues? _____
     _____
     _____
     _____

10. Do government officials visit your branch? _____ Yes _____ No

11. In what context and when? _____
     _____
     _____

12. How many such visits per year? _____
     _____

13. Are any employees at your branch elected officials?
     _____ Yes _____ No

14. What positions do they hold? _____
     _____
     _____

15. Describe your branch's relationships with local, state or regional environmental officials:
     _____
     _____
     _____
     _____

Figure 4-2. (*Continued*)

Part V: Media Relations

1. How many of each news media cover your community?
   _____ Newspaper _____ Radio _____ Television
   _____ Other (specify) _____
   _____

2. What is the current status of your branch's relationship with the media?
   _____ Little contact
   _____ Some contact during news events
   _____ Frequent contact

3. Please describe the most recent contact with the media and the result:
   _____
   _____
   _____
   _____

4. What kind of coverage (Unfavorable, Neutral, Favorable, Don't Know) do you think the media gives our company for:
   _____ Your Branch
   _____ Corporate

5. What do you consider your branch's greatest strength in its relations with the media?
   _____
   _____

   What is the greatest weakness? _____
   _____
   _____
   _____

6. How often do the local media contact your branch?
   _____ Never _____ Rarely _____ Sometimes
   _____ Frequently

7. How often do you contact the local media?
   _____ Never _____ Rarely _____ Sometimes
   _____ Frequently

8. Are news reports about your branch generally complete and accurate?
   _____ Yes _____ No

Figure 4-2. (*Continued*)

9. Do you have a system for receiving and handling routine and emergency news media inquiries? _____ Yes _____ No

10. Describe the system: _____

_____

_____

_____

11. How would you describe the relationship between the branch and the media?
_____ High level of trust _____ Moderate trust
_____ minimal trust _____ No trust

12. Do you or someone else from your branch maintain regular contact with local reporters?
_____ Yes _____ No

13. Who? _____

14. Do you have a designated branch spokesperson(s)?
_____ Yes _____ No

15. Who? _____

16. Have your spokesperson received training for dealing with the news media?
_____ Yes _____ No

Figure 4-2. (*Continued*)

The strategy for prioritizing relationships is governed by core value impact. Here is how the prioritization process works:

1. Identify and classify all publics into primary and secondary categories, depending on their core value impact. Within each category, every public is placed into one of two grid sectors.

2. Publics that usually have the most influence (the P1 sector) are those whose core values are negatively affected by a proposed facility, action, remediation, or legislation. Publics whose core values are negatively impacted should receive the highest priority for building and maintaining relationships. (Unfortunately, it is these publics who are often ignored, downplayed, or labeled as crazies by management.)

3. Early identification of publics who are severely impacted is essential, as is the development of an effective relationship with these publics throughout the process.

	Secondary (S)	Primary (P)	
S1 **Secondary** **influentials**	**Core values positively affected by project** A. Project users B. Community, business and civic groups	**Core values negatively affected by project** A. Site neighbors B. Pathway neighbors C. Activists	**P1** **Primary** **influentials**
S2 **Secondary** **gatekeepers**	**Other publics** A. Media B. Business/civic groups C. Educational institutions	**Governmental decision makers** A. Elected officials B. Appointed officials C. Staff D. Hired technical consultants	**P2** **Primary** **gatekeepers**

- The relationships most important to a project are those classified in squares P1 and P2.
- Individuals/publics in Sector P1 have the most influence with environmental decision making.

Figure 4-3. Priority relationships grid.

4. P1 audiences must receive the highest priority in the communication strategy for staff time and budget.

5. The goal of "effective" relationships is to find ways to mitigate, negotiate, or eliminate negative impact.

6. Publics whose core values are negatively impacted will react to you with suspicion, caution, anger, and hostility. While not a very inviting set of attitudes, it is precisely these emotions that must be successfully addressed or public permission will not be retained or obtained.

There always are unconvinceables. They are influential because:

1. An organized, vocal, uncontrollable minority whose core value are negatively affected will usually be more successful with locally elected officials and opponents than a majority of the community that may be neutral or supportive.

2. While the ultimate power is held by a majority of the public in general, the dynamics of public politics often gives a disproportionate and controlling amount of power to a minority that is organized and vocal when the majority is passive and unorganized.

3. Governmental decision makers will usually not take an action that is opposed by an organized and vocal group of citizens whose core values are negatively impacted.

4. Negative impact on individuals and publics tends to have greater weight than positive impact.

5. A real negative impact is given greater weight and has more influence than a prospective future positive benefit.

The message of Fig. 4-3 tends to fly in the face of current conventional communication wisdom. Time and time again, when vocal negatively affected influentials are ignored or contact is delayed until the end of the process, the potential for victory is diminished.

Section D: Elements of Influence.   Enormous pressure is placed on the *master public consent model* when sectors of influence focus their power against it. In Fig. 4-1 we show nine of the many sectors of influence which can alter the outcome of your communication strategy. Here are some typical sector of influence examples:

1. *Academics.*   No matter what the data or facts tend to demonstrate from the perspective of your academic experts, every side in the discussion can bring in its own academic experts. The result is the "battling Ph.D.s" phenomenon, where different conclusions are drawn from the same data. The consequence is a confused public and elected officials reluctant to make decisions.

2. *Business associations.*   Often visible and supportive in the early going, but when government regulators come around or activists decide to demonstrate outside their doors, business associates quickly disappear from the scene.

3. *Business opponents.*   Often your environmental difficulty is someone else's marketing opportunity. Rumors, false information, and sloppy communication can occur and do substantial damage, even if it merely distracts from what you must accomplish.

4. *Community activists.*   Activists are energized by the increasingly emotional way in which you react to their statements, ideas, actions, or threats. If your communication follows the guidelines described in this chapter, you will be unassailable and, therefore, can operate without fear of these individuals and organizations.

5. *Community institutions.*   These could be local churches, mosques, synagogues, or temples. If your employees are concerned about your behavior they may not talk to you, but they may go to their places of worship and talk to ministers, priests, rabbis, and other spiritual leaders. These groups will then bring pressure on you and on government to resolve the situation or to negotiate or mediate a solution.

6. *Customers.* If what you do is bad enough, silly enough, or stupid enough, customers will think twice about buying from you—at least until the situation is resolved.

7. *Environmental activists.* While they may well seem unconvinceable, your interaction with these individuals is crucial to your credibility with key audiences, most notably public officials and perhaps even your employees.

8. *Media.* A lot of time is spent worrying about and attempting to control the news media. Instead, do the right things to begin with. Put the media in a position where they *can* report on your organization and its activities as progressive, rational, and environmentally oriented.

9. *Voters.* Public officials have learned that if they appear to cave in to the wishes of business, or make the decisions too easy, or impose too few restrictions, they will be tossed out of office. Besides, almost every player is a voter.

Section E: Support Header—Credibility. A successful relationship with the public is based on credibility, but what is credibility?
Credibility has four attributes:

■ It is conferred by outsiders on the organization, individual product, or issue. It cannot be built or created. It is always externally conferred.

■ Credibility is often a function of a reputation for openness, honesty, accessibility, and promise keeping—the attributes which make up what we normally call ethical behavior.

■ Credibility in the future is predicated principally on past behavior.

■ The credibility necessary to attract and support public consent in the long term is built on a behavior framework which is unassailable.

AUDIT POINT NUMBER 8:

The public expects generally unassailable behavior, even when mistakes are made.

Figure 4-4 demonstrates a contrast between the unassailable approach and the credibility-destroying approach.
Unassailable behavior leads to trust and community comfort—or at least neutrality. Often, winning is a function of lack of opposition rather than overwhelming votes or public action. Trust is a very fragile commodity.
Figure 4-5 demonstrates visibly how behaviors and actions either build or lose trust.

Unassailable approach	Credibility-destroying approach
1. *Responsiveness:* When problems occur we will be prepared to talk about them internally and externally as aggressively as we respond to them operationally.	1. *Aloofness:* Wait to respond—"No one may notice." Develop our own story.
2. *Openness:* If the public should know about a problem we are having, or about to have, which could affect them or our credibility, we will voluntarily talk about it as quickly and as completely as we can.	2. *No Commitment:* Refuse to talk; volunteer nothing. Answer only if they get the question right.
3. *Concern:* When business problems occur, we will keep the community and those most directly affected posted on a schedule *they set* until the problem is thoroughly explained or resolved.	3. *Delay:* Stall responses. Hire big-time outside expert to study; report something next year (maybe). We can't talk until we know all the facts.
4. *Respect:* We will answer any questions the community may have and suggest and volunteer additional information in the event the community does not ask enough questions. We will respect and seek to work with those who oppose us.	4. *Disdain:* Avoid opponents; disparage them. Belittle uneducated questions and people.
5. *Cooperation:* We will be cooperative with the news media as far as possible, but our major responsibility is to communicate compassionately, completely, and directly with those most directly affected by our problems, as soon as possible.	5. *Umbrage:* "They have no business being involved in this." "There is no news here, why do they care?" "Be careful not to appear *responsible*."
6. *Responsibility:* Unless incapacitated or inappropriate, the senior executive on site is the spokesperson during an emergency.	6. *Stonewall:* "Not to my knowledge." The lawyers will convey our "no comment."
7. *Sensitivity:* At the earliest possible moment we will step back and analyze the impact of the problems we are having or causing, with the intention to communicate with all appropriate audiences to inform and to alert.	7. *Hunker down:* Anything we learn will be saved for litigation. We'll talk only as a litigation prevention strategy. "If they can't get it right, we don't and won't have to talk to them."
8. *Ethics:* If we are at fault, we will admit, apologize for, and explain our mistakes as quickly as possible.	8. *Arrogance:* No apology; no admission; no empathy. "Up yours."

Figure 4-4. Principles of unassailable behavior.

9. *Compassion:* We will always show concern, empathy, sympathy, and remorse or contrition.

9. *Reticence:*
"We can't set a precedent."
Do nothing which can be interpreted as taking responsibility.

10. *Generosity:* We will find a way to go beyond what is expected or required, even to "do penance" where appropriate.

10. *Avoidance:*
"Offer them ten percent less than they need."
Let them sue; we'll investigate, stall, and pay as little as possible as far from now in time as possible.

11. *Commitment:* We will learn from our mistakes, talk publicly about what we've learned, and renew our commitment to keeping errors, mistakes, and problems from recurring. Our goal is zero errors, zero defects, zero mistakes, zero crises.

11. *Abstention:*
Our mistakes are our business. Accidents happen; everything in life carries some risk.
Zero is impossible.
We'll do the best we can and that will just have to do.

Figure 4-4. (*Continued*)

Trust retaining:

The other party feels that you have listened, *actually heard,* and *have accepted* the value of their positions and feelings.

Visible sacrifice, accommodation, compromise, or some of each.

Acceptance of community ideas, values and concerns is directly reflected in what you do and how you do it, what you say and how you say it.

Some portions of your self-interest are clearly subordinate to the needs and wants of the other party.

Trust subtracting:

"Just go tell them our story." "They'd better believe it."

"Our plan is the plan." "The deadlines are final and can't be changed." "We have a schedule to meet."

"They just don't understand how costly this is; we can't just change on a whim." "Rewrite the original proposal with more reasons why we can't give in to them." "Say it louder and more often."

"If we say this often enough, they know they'll have to believe it, they'll know it must be true." "It's not really any of their business anyway." "Butt out."

Figure 4-5. Trust retention contrast analysis.

1. Don't involve people in decisions.
2. Hold on to information.
3. Ignore people's feelings.
4. Don't follow up.
5. If you make a mistake, deny it.
6. If you don't know the answer, fake it.
7. Don't speak plain English.
8. Be a bureaucrat.
9. Delay talking to other organizations.
10. Send your introverted scientists.

Figure 4-6. Ten ways to lose trust and credibility.

Figure 4-6 demonstrates vividly, and without the need for further explanation, how easy it is to damage relationships, audience support, and alliance.

AUDIT POINT NUMBER 9:

Trust-building communication programs address or respond to key audience concerns and values, especially during high-profile environmental situations.

There is a pattern of public interest and concern which requires extensive communication-planning execution. The most frequently recurring areas of public interest are:

- Health and safety
- Natural environment
- Social environment
- Cultural environment
- Technical considerations
- Financial considerations
- Economic considerations

Checklist 1 shown in Fig. 4-7 reflects seven major public interest areas of environmental concern requiring specific response management, especially dur-

Checklist #1

CHECKLIST OF PUBLIC INTEREST, CONCERN, AND INQUIRY DURING HIGH PROFILE ENVIRONMENTAL SITUATIONS

Category A – Public Health and Safety

1. **Groundwater Contamination:** The potential for contamination of groundwater resources should be minimized.
2. **Surface Water Contamination:** The potential for contamination of streams and other surface water should be minimized.
3. **Gas Migration:** The potential for undetected sub-surface migration of landfill gases should be minimized.
4. **Odors:** the number of people potentially affected by odor problems should be minimized.
5. **Noise and Dust:** The number of people potentially affected by noise and dust should be minimized.
6. **Birds:** The potential effect of birds on air traffic safety and nearby land uses should be minimized.
7. **Rodents, Insects and Litter:** The potential health and nuisance effects to people and nearby land uses as a result of rodents, insects and litter should be minimized.

Category B – Natural Environment

1. **Mineral Resources:** The loss of mineral resources should be minimized.
2. **Agricultural Soils:** The quantity and quality of agricultural soils lost should be minimized.
3. **Forest Resources:** Impacts on terrestrial flora and fauna should be minimized.
4. **Terrestrial Ecology:** Impacts on terrestrial flora and fauna should be minimized.
5. **Aquatic Ecology:** Impacts on fish-bearing streams should be minimized.

Category C – Social Environment

1. **Future Land Use:** Impacts on planned future development should be minimized.
2. **Existing Land Use:** Impacts on people and existing land use should be minimized.
3. **Agricultural Land Use:** Impacts on agricultural land use should be minimized.
4. **Community Characteristics:** Changes to the character and stability of the local community should be minimized.

Category D – Cultural Environment

1. **Heritage Resources:** Impact on significant heritage resources should be avoided.
2. **Archaeological resources:** Loss of significant archaeological resources should be avoided.
3. **Visual Aesthetics:** Impact on local visual aesthetic characteristics should be minimized.

Figure 4-7.

4. **Cultural Communities and Facilities:** Impacts on distinctive cultural communities and on community facilities should be minimized.

Category E – technical Considerations

1. **Geotechnical Factors:** The site should be able to be developed using proven engineering practices and with a minimal requirement for import or export of earth materials.
2. **Capacity and Flexibility:** The site should have sufficient capability and flexibility to meet the waste disposal needs of the Master Plan Area over the planning period..
3. **Servicing:** The work required to provide necessary site servicing including water, leachate disposal, electricity and road access should be minimized.

Category F – Financial Considerations

1. **Overall Facility Cost:** These costs, including site acquisition, development, operating, financing, closure and long term care costs, should be minimized..
2. **Haul Costs:** The costs of transporting waste to the site should be minimized.
3. **Affordability:** The facility should be affordable as defined in OMB guidelines and the financial impacts on the Master Plan Area associated with the development of the facility should be minimized.

Category G – Economic Considerations

1. **Property Taxes:** Site development should not significantly affect property tax rates.
2. **Resource Utilization:** reductions in revenues generated by agricultural land and other natural resources should be minimized.
3. **Employment and Income:** Net losses of local employment and income should be avoided.
4. **Property Values:** Impacts on local property values should be minimized.

Figure 4-7. (*Continued*)

ing high profile situations such as lawsuits or repermitting. Use the checklist to audit your project for problems or issues that could be raised in relation to these issue categories.

AUDIT POINT NUMBER 10:

Build in broadly based, environmentally sensitive concepts and principles.

Figure 4-8 describes one of the most generally accepted operating approaches, the CERES Principles. Originally called the "Valdez Principles," the principles were redefined and renamed after much discussion, analysis, and some controversy. The importance of the principled approach, using CERES as an example, is that increasingly, even at the local level, environmental responsibility is viewed as part of the global biosystem rather than as just a local issue or problem. The exercise is to examine the fundamental principles that govern

THE CERES PRINCIPLES

By adopting these Principles, we publicly affirm our belief that corporations have a responsibility for the environment, and must conduct all aspects of their business as responsible stewards of the environment by operating in a manner that protects the Earth. We believe that corporations must not compromise the ability of future generations to sustain themselves.

Protection of the Biosphere

- Reduction of emissions of substances which may cause environmental damage.
- Safeguarding of ecosystems affected by our operations.

Sustainable Use of Natural Resources

- Careful use of renewable resources.
- Conservation of nonrenewable resources.

Reduction and Disposal of Wastes

- Waste reduction through source reduction and recycling.
- Safe and responsible disposal.

Energy Conservation

- Improvement of energy efficiency in our operations, goods, and services.
- Use of safe and sustainable energy sources.

Risk Reduction

- Minimization of environmental health and safety risks to our employees and surrounding communities.

Safe Products and Services

- Reduction of products or services that pose environmental, health, and safety hazards.
- Informing our customers of products' environmental impacts.

Environmental Restoration

- Correcting of damaging conditions we have caused to health, safety, or the environment.
- Redressment of injuries.

Informing the Public

- Informing those who may be affected by the conditions we cause.
- Dialogue with neighboring communities.

Management Commitment

- Involvement of upper-level management in environmental issues.

Audits and Reports

- Annual self-evaluations.
- Completion of the CERES Report.

Figure 4-8.

how you manage your environmental affairs and compare them with an appropriate model such as the CERES Principles. The more you can reflect this approach, the more comfortable the community and government will be with your corporation.

Other Communication Principles Formats

Communication principles are statements of how you and your employees will operate on a day-to-day basis in key areas of community concern. Figure 4-9 is a model from a company which operates one of the largest landfills in America.

AUDIT POINT NUMBER 11:

Effective community relationships = Early, honest, empathetic communication + public involvement.

To truly develop a relationship with a community or set of audiences to obtain public consent requires more than putting out news releases, press kits, and videotapes. Face-to-face as well as large and small group interactivity is a prerequisite from the perspective of both audiences and public officials (remember, they are the ones who have the power to grant or give permission). Developing public involvement begins to build new or enhances past behaviors that are the source of trust and credibility.

Figure 4-10 demonstrates a wide variety of direct and indirect information gathering and dissemination techniques. The lesson is that the higher the profile of your problem, the greater the variety and intensity of public involvement techniques you will need to use to stabilize the public attitudes, keep your key audiences in position, and maintain your credibility.

AUDIT POINT NUMBER 12:

Managing communication with government and planning for governmental needs is a crucial success factor in environmental communication.

Relationships between government officials and various constituencies in a democracy are governed by a unique set of core values that public officials embrace as a part of holding office. These public official core values include:

- Protection of individual rights and core values
- Protection of minority rights and core values

- Protection of the public from negative events
- Due process for all
- Consensus motivation
- System for integrating all available science, technology, data, and other important factual information into the process as background to the emotions and values of the public

An effective environmental communication plan also anticipates public official expectations. These expectations include:

- Early, frequent communication
- Specific answers to questions
- Owning up to mistakes and stupidities
- Covering for fears and misgivings
- Going beyond what is required
- No surprises
- Direct cooperation and contact with the opposition

Typical behavior of government environmental protection and pollution control agencies means that only parts of the huge monoliths move at any given time. Frequently, the remainder of the body has no idea which parts are moving, at what velocity, or in what direction. Once movement by one part is detected by another part, countermeasures are often initiated and the original movement is altered, canceled, or redirected. You must be cautious or your company could become a victim. Yet, to conserve time and resources, and to implement the best solution, there is a real need to manage the relationship with these agencies much more aggressively.

You have probably already seen parts of the pattern you will be subject to: extremely slow action; people who do not do their homework until absolutely forced to; key government personnel changes at critical times which cause delay after delay; and very little coordination except when driven by external forces, events, or publicity.

There are five steps you can take toward accomplishing two important objectives:

Objective 1: Create a process that will help lead these governmental agencies toward making decisions and completing actions according to an agreed-on schedule.

Objective 2: Set a public record which demonstrates your company's consistent, aggressive, and positive efforts to move the government process along and solve the problem.

THE PRINCIPLES OF OUR ENVIRONMENTAL COMMITMENT

1. We will operate in a manner which protects the environment, health, and safety of the citizens of the communities where we operate as well as our employees.

2. We will comply with all federal, state, and local environmental laws, regulations, and permits.

3. We will anticipate environmental regulations and take appropriate actions which may precede laws or regulations.

4. Internal and external specialists will be available to address environmental issues at all times.

5. Environmental assessments will be conducted for all real estate we own or plan to buy or sell.

6. We, in addition to state protection agencies, will audit our operations routinely for conformance to existing environmental standards.

7. Our personnel will participate in continuing education, studies, programs, and other activities to help develop long term solutions to environmental issues.

8. We will communicate our environmental policy to all employees and to all others involved in or affected by our operations. The company will be responsible for environmental performance and results. Facility and area managers will:

   - Monitor and certify compliance.

   - Promptly report noncompliance conditions to appropriate regulatory authorities.

   - Take direct action, including curtailment of operations, if necessary, to prevent serious harm.

9. We will respond openly and promptly to public inquiries about environmental issues our operations may create, and initiate communications with others who might be affected.

10. Managers and employees will promptly communicate to management significant environmental developments which may have an impact on employees, communities, or the public.

11. We will promote the development and adoption of scientifically sound and balanced environmental policies, laws, and regulations through active support of and participation in governmental legislative and rule-making processes and other forums dedicated to providing public officials with technical information and advice.

12. We will factor aesthetics into all future siting decisions and maintenance of existing sites.

13. We will develop and communicate to appropriate local authorities environmental incident plans for any operations that potentially impact a community.

Figure 4-9.

PUBLIC INVOLVEMENT TECHNIQUES

Direct Methods of Information Dissemination
Techniques

Briefings	Guest speaking	Open houses
Brochures	Handbills	Personalized letters
Direct mailings	Information fairs	Purchased advertising
Door-to-door visits	Information hotline	Slide shows
Drop-in center	Mobile office	Telephone
Fact sheets	Newsletters	Videos
Flyers	Newspaper inserts	Volunteers

Purpose

To provide detailed information to a targeted audience in your own words and on your schedule.

Indirect Methods of Information Dissemination
Techniques

Feature stories	Press conferences	Public service announcements
Guest editorials	Press interviews	
News releases	Press kits	

Purpose

To provide information to the media and the general public.

Information Gathering
Techniques—Purposes

Information contact person—Identify a point of contact where the public can place a single call and receive either an answer or be called back with information.

Interviews of community leaders, key individuals—To identify reactions to, and knowledge of, project. To identify issues of concern and historical controversies. To identify other groups or individuals to be contacts or added to the mailing list. To assess the political climate and relationships among various interest groups.

Mailed surveys or questionnaires—To assess public awareness of project actions, public issues, and concerns. To assess values and issues of concern to the public.

Telephone survey—To assess public awareness of meetings, project actions, public hearings, etc. To track the movement of public opinion to the project.

Figure 4-10.

Focus groups—To gather emotional/intellectual reaction to possible activities.

Door-to-door—Give site neighbors the opportunity to directly express opinions.

Open forums—For the public to have an opportunity to ask questions and express views.

Brainstorming sessions—Give diverse group of public opportunity to define problems and develop alternatives.

Citizen and Agency Involvement
Techniques—Purposes

Advisory groups of key publics—To advise on policy and technical matters, critically review results, help find compromises between competing local interests, advise on public involvement approaches, and promote consensus with constituents.

Public workshops/task forces—Small diverse groups to explore specific topics solutions to particular problems.

Project liaison—Contact person in key public groups and agencies that is kept fully informed of project activities.

Conflict Resolution/Consensus Building
Techniques—Purposes

Facilitation leader—To impartially lead discussions.

Mediation process—To re-establish communication when all positions are polarized and move parties to mutual understandings and agreement.

Nominal group workshop—To build consensus on project actions, issues, or mitigation plans.

Delphi technique—To identify options using independent experts.

Public values assessment—To combine public values with technical facts to identify alternatives that most closely meet what the public has said is important to them.

Analysis and Documentation
Techniques—Purposes

Computerized comment storage and retrieval system—To objectively summarize and make available public comments.

Summary and evaluation reports—To provide written documentation of activity, attendees, issues, and comments, and to evaluate the public involvement program.

Figure 4-10. (*Continued*)

These five steps are:

1. Request and hold a monthly review meeting with appropriate government officials. Use a similar agenda for each meeting:
   a. Review of agency progress against its own timelines
   b. Your concerns about what the agency is doing
   c. Correction or explanation of previous comments, actions, or upcoming decisions
   d. Review of your community relations plans and actions
2. Brief local officials, specific thought and opinion leaders, and other interested parties on progress (or lack thereof) monthly.
3. Develop a monthly letter to the EPA's regional administrator which raises issues, clarifies concerns, and generally prods the agency to move things ahead on a variety of fronts.
4. Support the governmental agency's own vested interests:
   a. Help it show progress to its own publics (for the federal EPA these are typically Congress, other federal agencies, and the administrator and other senior officials of the EPA in Washington).
   b. Help environmental protection agencies use your company as the potential success story you are.
5. Assist the bureaucracy:
   a. Predevelop documents you know these governmental agencies need:
      (1) Presubmit your company's own version of an order which can trigger an early start to the discussion
      (2) Resubmit engineering or other related data that governmental agencies may not have seen.
   b. Preapprove processes and plans, where possible.
   c. Prioritize issues and problem areas, preagree, or preapprove as many as possible leaving only the crucial issues to discuss and negotiate.
   d. Conduct seminars or briefings (open to others in EPA, local officials, community leaders, other audiences or publics) on:
      (1) The community relations and legal process you intend to follow
      (2) Technical/scientific issues related to the site and other aspects of the process
      (3) Other issues, questions, or problems whose explanation will accelerate governmental agency knowledge and help move the process forward

CAUTION: Even with this level of effort, your urgent, consistent, positive action will move the process along in a fashion that may much of the time be only barely noticeable. Without it, there may be no movement for great periods of time.

Model Approach and Documents

One genericized model approach is shown in Fig. 4-11 and five model documents based on real situations are shown in Fig. 4-12. They are all quite instructive to consider.

Model Approach 1	Public Affairs Priorities in High Profile Environmental Situations
Model Document 1	Superfund Communications Strategy
Model Document 2	A Chemical Plant Greenfield Siting
Model Document 3	Communication Plan for Siting a Municipal Solid Waste Facility
Model Document 4	Communication Plan for Siting a Medical Waste Incinerator
Model Document 5	Sample Format for a Superfund Community Relations Plan including tables of contents for a Communications Control Book and a Contacts list Data Base

MODEL APPROACH #1:

PUBLIC AFFAIRS PRIORITIES IN
HIGH PROFILE ENVIRONMENTAL SITUATIONS

Note: Two essential tasks in managing high profile environmental situations are providing vision and leadership, and focusing on the critical goals and key strategies. Here are seven key goals with objectives, strategies, and tactics. Adapt them to the needs of your program and process.

Goal 1:   Aggressively Manage the Process:

- Your role is to be the chief explainer of the entire effort.
- Your role is to be the manager of the people and the people-related components of the project.
- Your role is to generate and then manage the strategy.
- Your role is to create and carry the principal messages and to "animate" the process to facilitate understanding – internally and externally.
- Your role is to be the creator of the appropriate perceptions among critical audiences.

Objective: To keep the process moving forward by always having a Plan B, C, and D in place.

Strategy:

- Explain issues, techniques, and tactics before the questions arise.
- Focus on the key messages.
- Mold perceptions rather than educate the public.
- Communicate directly with key audiences using vehicles you control, like video, direct mail, and newspaper advertising.
- Choose tactics which focus on the goal – eliminate distractions.
- Escalate your decisiveness to reduce instability and lack of direction.

Tactics:

- Communicate through the daily newsletter.
- Be powerfully visible. Set the course and the policy; let others chair the committees.
- Create and use an effective, doable timeline.

Goal 2:   Maintain and Build Critical Audiences:

- Old audiences:
  - Area residents;
  - Township residents;
  - Township residents within the two-mile radius of the site;
  - Business leaders; and
  - Local and state political officials (appointed and elected).
- Build your base audience of company employees and retirees.
- Expand and develop other natural audiences:
  - Senior clubs/civic groups;

Figure 4-11. A genericized model approach.

- Religious groups;
- Senior citizen organizations; and
- Fraternal organizations.

Objective: To keep audiences most directly affected in tune with our concerns and theirs, and to be able to move them to action when necessary.

Strategy:

- Regular direct contact by mail and phone.
- Regular indirect contact through paid advertising, public appearances, the grapevine and third party conversations.
- Provide feedback quickly to what is heard, said, and done.

Tactics:

- Build your base audiences:
  - Employees.
  - Retirees.
    - Create an organized group of retirees who would recruit and carry the message to the community through their activities. They would also influence the behavior of those in the plant on these issues.
    - Create events where these audiences can mingle and develop relationships.
- Target old audiences:
  - Develop monthly letter (more frequently when decisions approach) describing the progress made.
  - Continue promoting applications for the Property Value Guarantee Program.
  - Maintain personal contact with key groups like the Chamber of Commerce.
- Develop new, natural audiences:
  - Recognize through announcements, financial support, or awards the work of key groups in the community.
  - Establish a benchmark survey in the awareness groups.

Goal 3: **To Grow Personal and Corporate Relationships with Public Officials** (local, regional, and state)

- Build personal relationships through:
  - In-person contact, telephone, and personal letter/clipping-type mailings; and
  - Tools that are developed to fit the needs of these public officials.
- Use surrogates where appropriate, but generally as a secondary source of contact or to open doors.
- Use existing pathways:
  - Technical contacts from within with state and federal environmental and other regulatory agencies;
  - Union leadership; and
  - Encourage and foster contacts to government and political organizations.

Objective: To ensure that the right public officials know who your company is, what your company does, and what your company needs.

Figure 4-11. (*Continued*)

Strategy:

- Use the "Adopt-A-Politician" concept as the basis for a permanent plant-based program.

Tactics:

- Tie in corporate public affairs.
- Make regular direct contact with key county/major state officials.
- Assign officials and politicians to other members of the plant team.
- Work to build key executive branch relationships, open doors, and generate feedback from state Capitol.
- Stay in touch using periodic (i.e., monthly) brief bulletins on issues, questions, and concerns.
- Respond to feedback immediately.
- Use natural contacts which exist between the plant, retirees, and business leaders.

Goal 4: Manage the Legal Process:

This process includes:

- Township counteraction strategy in the event they cite us for zoning violations.
- Grandfathering under the new zoning plan.
- Site-related issues such as graves and historical concerns.

Objective: Daily progress through control which reduces surprises, eliminates duplication, facilitates forward movement, and keeps individual legal efforts on their appropriate tracks.

Strategies:

- Use local counsel for local work.
- Use environmental counsel for environmental work.
- Use corporate counsel for overview, advice, and coordination with public affairs strategy.
- Set goals for each facet of the legal process.
- Encourage creative ideas, but control the focus of each legal team.

Tactics:

- Develop work plan for each team.
- Make daily contact with each team to review work plan and assess programs.
- Require written opinions and comments.
- Only hold meetings when absolutely necessary.
- Build consensus – decide the outcome of meetings before they begin.

Goal 5: Recruit Department Managers to the Fight:

- Exercise positive leadership.
- Help managers look beyond the plant's problems. Victory lies outside the plant.
- Help managers simplify their own priorities, one of which *must* be the landfill process.

Objective: To put your plant management group on a war footing (designate "foxholes" for everyone because war *is not* a participative process).

Figure 4-11. (*Continued*)

Strategies:

- Through positive leadership and effective delegation, help department heads assume an appropriate role in this struggle.
- Raise their sights.
- Move beyond the past.
- Help them simplify their own priorities.

Tactics:

- Pick the tough issues and ask for volunteers.
- Assign those who don't volunteer to the issues that are left.
- Participation in "Adopt-A-Politician" should be mandatory.
- Recognize achievement and participation.
- Take the managers into your confidence because they can't come in on their own.
- Ask each manager for one constructive suggestion or alternative a week from his area for conduct in the war.
- Forecast outcomes, i.e., "Yes, we're going to lose this election, but let's do so as narrowly as possible."
- Play the "what if" game constantly.

Goal 6:   Manage the Most Urgent Issues First and, Where Possible, One at a Time:

- Continue explanations of why a landfill is needed and how the process is progressing.
- Continue promotion of the Property Value Guarantee Program.
- Continue the search for technical truth by asking the tough questions and not tolerating surprises or resentment because of the way things were done in the past. Win by moving forward. If the technical team won't or can't work, replace it quickly.

Objective: To maintain control and direct the evolution of issues.

Strategies:

- Anticipate the questions and answers early.
- Challenge all proposed technical solutions and alternatives.
- Ask the unaskable.
- Talk in terms of rationally evolving solutions rather than rigid, ultimate answers.

Tactics:

- Continue aggressive communication regularly and directly with key audiences on landfill issues.
- Communicate first with your plant audiences.
- Continue progress on the Property Value Guarantee Program plan.
  – Make a deal with the bank.
  – Continue advertising for participation.

Goal 7: Continue Building a Community "Relationship":

- Continue to request meetings with the local officials.
- Implement "Adopt-A-Politician."
- Contact major audiences directly at least once a month.

Figure 4-11. (*Continued*)

- Use paid advertising to explain, clarify, or set the public debate. Keep the other side honest.

Objective: To continue generating overt and explicit communication and support on our behalf.

Strategies:

- Focus on participation when you ask members of the community to do something.
- Always ask for a response in writing or in person.
- Emphasize simplicity, concern, caring, understanding, feelings, common sense, fairness, and the right things to do.
- Stress the need for the community to talk back to your company. Communicate that you want to hear community needs and concerns.
- Use "please" and "thank you" a lot.

Tactics:

- Use "Dear _____" cards wherever the plant manager goes . . . in speeches, in letters, etc.
- All paid advertising should contain the "Dear _____" coupon.
- Be visible in powerful settings.
- Affect rather than set the agenda of others.
- Begin plant tours to show people what you do.
- Take community leaders and base audience members to similar sites elsewhere to further demonstrate "what's in it for them."

Figure 4-11. (*Continued*)

MODEL DOCUMENT #1:

SUPERFUND COMMUNICATIONS STRATEGY

CORPORATE STRATEGY

Five philosophies/strategies form the basis for the specific optional activities suggested in the XYZ Company Plan. The objective is to prepare for, participate with, and respond to EPA-mandated and optional community relations activities:

I.   Give the Public an XYZ Company Community Relations Process

II. Remember the Record

III. Forecast the Process

IV. Test-run the EPA Process

V.   Monitor/Evaluate the EPA's Community Relations Activities

DISCUSSION OF THE FIVE STRATEGIES

I.   Give the Public an XYZ Company Community Relations Process

Give the public name(s)/telephone number(s) of XYZ Company contact person(s) and demonstrate through frequent action that the public can contact, talk to, and be comfortable with the process of working with XYZ Company.

II. Remember the Record

An administrative record[1] will be established by the EPA – therefore, we should have both a pro-active and a defensive strategy regarding this record, and we need to monitor the record.

A. Pro-active Actions

XYZ Company on a frequent basis will submit letters, memorandum, reports, etc. to the EPA so that the file shows that XYZ Company initiated actions and responded rapidly and completely to community concerns. Submit letters frequently to outline

---

[1]   Administrative Record: A file which is maintained and contains all information used by the lead agency to make its decision on the selection of a response action under CERCLA. This file is to be available for public review and a copy is to be established at or near the site, usually at one of the informational repositories. Also a duplicate file is to be held in a central location, such as a Regional or State office.

Figure 4-12. Model documents.

our ideas and suggestions, recommendations and offers of assistance to the EPA and community.

Frequent and systematic letters to the Administrative record can establish a favorable analysis of XYZ Company based solely on documents in the record.

B. Defensive Actions

We need to remember that all correspondence from the EPA and other agencies as well as our response – or lack of response – will also be part of the record. All letters and documents to the EPA and other related agencies should be reviewed with the thought in mind that it will definitely become a public document and will be reviewed by someone in the public realm (i.e., area resident, news media, local official, opposition, etc.).

C. Check Record Frequently

On a frequent and regular basis check the administrative record for new documents, reports, letters, etc. – perhaps a written request for new information should be filed every two weeks.

During a key public notice/public hearing process – the record should be checked daily.

III. Forecast the Process

CERCLA and related laws/regulations require a very visible and highly interactive government-directed public relations program aimed at local residents.

XYZ Company should show its leadership by forecasting to base audiences, residents, and area public officials the required and likely EPA community relations activities.

Forecasting will present XYZ Company as a responsible corporate citizen and will reduce the "shock" or "prominence" of EPA lead activities.

IV. Test-run the EPA Process

XYZ Company will scope out the EPA community relations activities and on a model-basis conduct some of the key activities by itself in advance of EPA action (interviews, small meetings, etc.).

This will allow XYZ Company to both forecast the EPA process and collect information, in advance of the EPA process, that will allow XYZ Company to assess its overall community relations plan and determine what kind of community concerns the EPA will likely identify in its community interviews and community relations plan.

V. Monitor/Evaluate the EPA's Community Relations Activities

On a limited basis, XYZ Company will conduct follow-up interviews with public officials and area residents after they have been interviewed by the EPA. This action allows XYZ Company to closely monitor the actions the EPA is considering, the type of information it is sharing with the public, and the information the public is providing the EPA.

Figure 4-12. (*Continued*)

UNINTENDED CONSEQUENCES/EVENTS

XYZ Company's objectives throughout this process are to be responsive to community concerns and issues as they affect the company, show concern and understanding of community attitudes beyond simply paying for site remediation, and help keep the entire process in perspective until the site is delisted.

Another purpose for the XYZ Company's community relations program is to prepare for unintended events and consequences that very commonly occur during this process. For example:

- New information from unrelated sources cause unanticipated concerns;
- Anti-company activist actions;
- Significantly greater contamination than anticipated;
- Exceptionally antagonistic or distorted media coverage;
- High profile law suits; and
- Aggressive political attacks.

The lead agency community relations contractor will have response plans in place for the government components of these unanticipated developments – but so will XYZ Company.

DEALING WITH UNANTICIPATED EVENTS

Developing, practicing, and managing our own responses to unanticipated consequences and events will make the difference between a community relations program that is in the public interest and under control and one that isn't.

We will:

- Respond quickly;
- Take appropriate responsibility;
- Ask for help and understanding;
- Inform company employees immediately;
- Show concern;
- Be open to suggestions;
- Rehearse all statements and messages;
- Explain to the community as soon as possible;
- Invite in local officials to help with the explanations, where appropriate;
- Talk about prevention of future occurrences;
- Seek out and talk to affected groups;
- Seek out and talk to affected agencies;
- Use simple, direct, positive messages;
- Stick to the facts and company policy; and
- Use common sense.

Figure 4-12. (*Continued*)

MODEL DOCUMENT #2:

A CHEMICAL PLANT GREENFIELD SITING

PROJECT PREREQUISITES/REALITIES

Siting manufacturing/chemical plants is a multi-year, multi-disciplinary process which is costly at all levels, time consuming in every phase, and subject to many fractious and irritating delays and distractions. There are some important realities and prerequisites that need to be dealt with as these projects move forward.

1. *Long term commitment.* There will be periods of tremendously intense activity and lengthy lulls in this process. There will be times of enormous expense which seem to produce nothing but frustration, anger and irritation. At a minimum the process will take several years, probably twice the length of time you optimistically forecast. No one will be grateful for this project once it is formally announced and gets under way until the gates officially open and operations begin.

2. *Community relations budget.* Currently, the cost ratio used in budgeting community relations efforts is a minimum of 15 percent of the gross siting budget.

3. *Flexibility.* It is likely that the initial site chosen will not be the ultimate site that is permitted. In fact, two or three different locations may be considered before siting is accomplished. It is advisable to have a Plan B and Plan C moving forward in parallel with Plan A, just in case.

4. *Focus on the goal.* A lot of unpleasant things will be said about your company and people, mostly near the site. Just remember sticks and stones are what need to be feared. Words and attitudes can be overcome and hurt mostly your ego and pride. To win, we must stay focused on the process of getting permits and permissions.

5. *Communication drives construction.* One crucial reason for siting failure is the lack of commitment to communicating and involving the community. Another is bullheaded, technical and management decision-making which ignores or attempts to override public concerns. If a decision represents a shortcut, an extraordinary procedure, or an unusual approach, the communication implications on the various sectors of influence must be considered and *will likely drive* final decisions.

6. *Quick reaction.* Often the quickest way to diffuse community concerns, outrage and negative action is through instant response to questions and problems which arise. Most companies, even in good times, don't behave this way. In community relations communications involving environmental issues, the ability to respond accurately and quickly, as well as directly to those raising the issue or who are directly affected by something we are doing, is a critical control factor in managing unplanned visibility and the unintended consequences of our actions.

We must make and put in place ahead of time policy decisions to facilitate instant response.

Figure 4-12. (*Continued*)

PROJECT PHASES

Here is a generalized timeline for a model community communications program, again assuming ideal circumstances – without a single glitch, without a single complication, and without a single legal, administrative, or procedural detour. Please remember that the events or tasks outlined here may not necessarily occur in linear order. Many of them may occur at different velocities, but at the same time.

Year One

Phase I – Communication Assessment and Planning

- Community assessment
- Key issues identified
- Key audiences identified
- Company spokesperson training
- Benchmark attitude survey
- Spokespersons selected and trained
- Early messages/themes identified
- Base audience programs created
- Opponents identified
- Timelines for legal, engineering, public policy, and local government developed
- Analyze engineering design assumptions
- Analyze all steps in the process
- Conduct in-depth need for a facility analysis
- Devise hunch reduction process
- Likely local scenarios developed

Phase II – Message Development and Targeting

- Move messages out to key sectors of influence
    - Media
    - Activists
    - Public officials
    - Academicians
    - Business groups
    - Employees
    - Neighbors
    - Children
- Consider doing community needs assessment
- Do models and open houses
- Measure interest, attitudes
- Monitor, re-assess community interest and attitudes

Phase III – Strategic Communication Implementation

- Public communication begins
- Base audience programs implemented
    - Employee meetings/letters
    - Supervisory training
    - Coaching other support groups
    - Finding sympathizers
    - School programs developed
- Government affairs communication process initiated

Figure 4-12. (*Continued*)

- Door-to-door strategy implemented
- Public meetings announced
- Quick response mechanisms established
- Support center set up
  - Recruit employees/families to volunteer

Phase IV – Communication Timeline Management

- Focus on working the communications aspects of the:
  - Engineering timeline
  - Public policy timeline
  - Community involvement timeline
  - Environmental/regulatory process timeline
  - Legal timeline

Phase V – Contingency Management

- Develop scenarios to anticipate reaction from and participation by various sectors of influence:
  - Activists
  - Academics
  - Business opponents
  - Children
  - Competitors
  - Corporate campaigns
  - The media
  - Neighbors
  - Public officials
  - Public interest groups
  - Whistle blowers
- Develop scenarios to anticipate other ongoing facets of the plan:
  - Litigation visibility
  - Regulatory disruptions
  - Political intervention
  - Intense local activism
  - The well-meaning acts of our "friends" and family

Figure 4-12. (*Continued*)

MODEL DOCUMENT #3:

COMMUNICATION PLAN FOR
SITING A MUNICIPAL SOLID WASTE FACILITY

PROJECT SUMMARY

Our Goal: Create a public communication environment which facilitates the siting of a municipal solid waste facility (MSWF) for an 11-municipality consortium within the next 24 months.

Obtain staff and execute a coordinated public and private communication strategy which coordinates all appropriate elements including technical, legal, environmental, governmental, media, special publics, neighbors and opponents.

- Anticipate, prepare for and accommodate organized opposition
- Contain and control the four critical public issues in landfill siting:
  - Health and safety concerns;
  - Environmental concerns;
  - Personal and real property value concerns; and
  - The "what's in it for me" concern.
- Work through key local political leaders.
- Constantly assess, evaluate and respond to political environment.
- Win the active and open support of local thought and opinion leaders including local business people.
- Build support based on political and public realities.
- Thoroughly timeline all aspects of the project which can cause planned and unplanned visibility:
  - Engineering (hydrological testing and other technical procedures);
  - Regulatory procedures;
  - Public communication opportunities;
  - Political communication processes;
  - Fact-gathering and listening to various audiences;
  - Legal tactics; and
  - Litigation.
- Use personal meetings and public events to foster face-to-face discussion and involvement.
- Focus activities and interest within the municipality involved and as close to the site as possible.

Minimize media participation in negotiations and meetings by maximizing positive, yet simple messages communicated directly to affected audiences.

Talk in simple terms of solutions and the future.

SITUATION ANALYSIS

Public officials are increasingly reluctant to act and use the powers of condemnation or expropriation if those actions are not well supported by overt public action. The old days of back door arrangements are gone forever. Public support must be brought forward to get the desired result.

Figure 4-12. (*Continued*)

Public communication programs surrounding landfill sitings can only succeed if they also take into account the reality of public perception rather than assuming that good technical data and rational proposals will win. We must identify and repeatedly address the key questions of those into whose back yard a landfill will go – questions which revolve around four major issue areas: health, safety, property values and the environment.

We can anticipate some of the questions now:

- How does the surrounding area benefit from the landfill?
- What's in it for me; for the community?
- Since there has already been environmental damage from existing landfills, how do we know that you won't pollute at this site?
- What guarantees can you give that this new landfill will be properly inspected and regulations enforced?
- Isn't it possible for liners and compacted materials to be improperly installed, or accidentally broken, permitting leachate to pollute our ground water?
- Will you do full, ongoing, health monitoring studies of those who live in the vicinity of the site?
- Can you guarantee that there will be no adverse effect on property values or health?
- What happens when the landfill is full?
- What will it look like?
- What's to keep this site from becoming a regional facility at some later date?

Listening to the real concerns of the people in the local communities and the public officials who serve them is the only way a public communications program can begin to anticipate the questions, misinformation and areas of public ignorance that create opposition to landfill sitings. Anticipating the issues and the emotion enables us to understand, accommodate and preempt the tactics of organized or unorganized opposition.

PROJECT COMMUNICATION PHASES

Projects of this nature are constrained by the chronologic nature of the public process and distorted by the emotional nature of the argument. They often seem to take on lives of their own. Looking at the process in terms of four major phases allows us to do a certain amount of pre-planning and key issue identification which in turn allows us to have more control over message flow, and therefore a more realistic and publicly useful communications program. Here is a brief targeted list of the communications tasks and opportunities as the siting process proceeds:

Phase I: Pre-Public Announcement

- Political contact and research about the political environment – who to talk to, what to talk about and when to talk
- Site selections/obtaining land options
- Technical studies/need established
- Legal positioning/alternatives identified
- Communication planning
- Limited visibility/focus on local audiences
- Identify/activate base audience components/feedback channels
- Vulnerability analysis/ask key questions/anticipate opposition actions
- Identify/estimate breadth and depth of opposition
- Clarify competitive impact on sites (if any); competitors' status
- Assess community perception of XYZ Company

Figure 4-12. (*Continued*)

- Assess community perception of the issue/research
- Develop initial themes, audience and message priorities
- Complete time lines for:
  - Environmental assessment process
  - Engineering/technical milestones
  - Public communication process
  - Political communication process
  - Lobbying communication process
  - Legal/litigation (zoning/siting)
  - Political

Phase II: Early Disclosure

- Background contact with political and civic leaders
- Disclosure to local employees
- Public disclosure of intentions and plans
- Listening/feedback procedures implemented
- Filing of appropriate permits, technical documents and required information
- Initiation of public communication program to key audiences
- Framing of issues and themes for coming public debate
- Technical information and data translated into usable public messages
- Readying of technical and environmental communication tools for the intense public discussion phase

Phase III: Public Discussion/Debate

- Management of media-related involvement in the debate
- Intense face-to-face citizen discussion through public and private meetings
- Intense local public government contact
- Intense county government contact
- Some provincial political contact and discussion

Phase IV: Resolution

- Environmental information/messages conveyed intensively/conflict resolution
- Major community issues (health/safety, environment, property values, "what's in it for me") resolved
- Media support mobilized
- Community acceptance/rejection acknowledged

Figure 4-12. (*Continued*)

MODEL DOCUMENT #4:

COMMUNICATION PLAN FOR SITING A
MEDICAL WASTE INCINERATOR
(in the face of growing public opposition)

Gaining court decisions and public policy decisions take time. That knowledge, combined with an analysis of the current situation and our messages, leads to three very critical questions.

1. Can an incinerator be successfully sited given the current state of affairs and how this set of circumstances plays against the pattern we have come to recognize in the siting process?

2. Does XYZ Company have a secondary or fall-back position in the event it becomes clear that an incinerator cannot be sited as proposed?

3. What pollution control, engineering and operational alterations is XYZ Company prepared to offer to help the community accept and endorse a decision to site the incinerator?

This proposed communication plan is designed to create a turn-around in public attitudes by the end of eight or nine weeks. If attitudes cannot be neutralized or shifted in a comfortably recognizable way, XYZ Company should consider an alternative to course of action.

ANALYSIS OF CURRENT SITUATION AND OUR MESSAGES

Based on review of the public meetings and the flow of news and information from the area, here is our current analysis:

1. The public does not perceive or accept that there is a local need for such a facility. (Questions 4, 7, 38 from the public meeting.)

2. Even if there will be a future local need for such a facility, the public feels that need should be met at existing incinerators – or at other locations or using other processes. (Questions 17, 30, 32, 36 from the public meeting.)

3. The public does not accept the premise that they should support a new facility in the area even if other areas of the State have a critical need for such a facility. (Questions 5, 16 from the public meeting.)

4. The public does not accept the premise that the proposed site is a logical one simply because of existing compatible use.

5. The public resents the proposed site because it appears to primarily serve the needs/goals of XYZ Company.

6. The public feels that incineration, generally, and at XYZ Company's medical waste incinerators, specifically, will present grave threats and risks to their most basic and cherished values – those being:

   • The health of their families;

Figure 4-12. (*Continued 15*)

- The property value (and major life savings);
- The environment – air and ground water (therefore a threat to health and property value); and
- Their quality of life including peace of mind, pride in community, and absence of conflict.

Because the public does not accept the need, they challenge the credibility of both XYZ Company and the process for obtaining the incinerator itself.

- The public does not trust the permitting process.
- The public does not trust XYZ Company.

USE A DIRECT APPROACH

We must deal directly with the opposing groups now because:

- They represent the true feelings of the public – 45-to-49 percent of the public oppose incineration.
- Only 15-to-20 percent of the public support incineration of any waste.
- The vocal opposition is truly reflecting the majority viewpoint.
- The State will most likely go with majority public opinion when it comes to permitting.
- The number of residents who support incineration of any kind is a definite minority.
- The area municipal waste incinerator was successful because it had a significant constituency that wanted it, needed it and supported it . . . , i.e., locally elected officials. We will not have these same allies.

If this analysis is sound for future action, our communication goals then become relatively clear:

1. To reduce, contain and moderate the extent of the opposition;

2. To persuade the undecided, moderate residents to support the project; and

3. To provide a winning strategy so that allied groups will openly support the project.

THE PLAN

Based on these goals, here are the elements of the communication plan.

Strategies:

1. Directly address hard core opposition.

2. Directly address soft opposition.

3. Build base audiences.

4. Address the visible issues.

5. Develop support for the political process at the State level.

Figure 4-12. (*Continued*)

Assumptions About Permitting

1. Although the State has ultimate permit authority, officials will not issue a permit if extensive and vocal public opposition exists, or if local township and county officials vigorously oppose the project.

2. XYZ Company needs to win the hearts and minds of the public as if we were seeking the permit directly from area residents or the township board.

Communication Philosophy

1. This facility is needed, worthwhile and will be safe.

2. XYZ Company will respond to the gut questions as well as the technical issues.

3. The neighbors have every right to be concerned. XYZ Company has the obligation to answer all their questions.

4. To succeed, XYZ Company must focus on the real concerns of those affected by the incinerator.

Important Message Goals

1. The credibility of the company and the process are perhaps more important to gaining public support than the engineering facts.

2. It is unlikely that a local need can be proven or demonstrated for this facility.

3. XYZ Company needs to establish credibility and reduce the threat this project poses to the core values of area property owners.

4. XYZ Company needs to treat the public as a "host community" and go through the process of securing public acceptance of this project.

By viewing the public as a "host community" we:

- Accept the premise that local support is essential to obtaining a State permit.

- Find ways to eliminate, reduce and contain any threat – real or imagined – to the public's core values.

- Find a process the public accepts as credible.

- Find trade-offs the public views as beneficial, although this is of lesser importance than reducing the perceived threat to health, property, etc.

FOUR SPECIFIC STRATEGIES

Strategy #1:   Directly address "Hard Core Opposition"

1. Identify area residents who are openly opposed and critical and, where possible, their objections to the project.

Figure 4-12. (*Continued*)

2. Meet with these residents in small groups (through coffee parties or focus groups) of six-to-ten people to:

   - Identify specific concerns, attitudes, opinions, beliefs, values they have about the project, their neighborhood and the community.
   - On a "soft sell" basis, test our best arguments, information and data.

3. Take the opposition to a model site to:

   - Get their specific reactions to all phases of an existing site, including visits with local officials, neighbors and others affected by the existing site.
   - Continue "soft sells" of our best messages by showing the complete operation – all the time we are monitoring and seeking feedback.
   - Identify area/phases of the actual site that generated negative or positive reaction, or which tend to raise new questions.
   - Evaluate if an on-site visit changed pre-existing attitudes and opinions, and why and how.

4. Conduct a follow-up meeting with these opponents within a week of visiting the model site to:

   - Monitor any swing in position and evaluate, if any, positive images from the site visit which remain.

5. Join the predominant activist group, or at least find out when meetings are held and attend . . . then offer to answer the questions that are raised.

Strategy #2:   Directly Address "Soft Opposition"

1. Identify groups of area residents who are neutral, undecided or express willingness to be convinced.

   - Repeat the process followed with hard core opponents.
   - "Soft" opponents can be identified and initial meetings carried out concurrently near the end of the process of meeting with hard core opponents.

2. Directly plug these residents into the ongoing direct and indirect communication programs.

Strategy #3:   Build Our Base Audiences (to run currently with meeting the opposition)

1. Invite and involve key components of our base audiences in the same ways we are dealing with opponents. Hold "smaller" group meetings and get them to an operational site. Listen to what they share about what is being said in the community.

2. Identify where the community really is –philosophically and emotionally – at the present time through a benchmark public attitudes survey.

3. Develop updated letter, Q&A attachment and site drawing for those base audiences identified as neutral, if not somewhat supportive. They include:

   - Our own employees;
   - Vendor company employees;

Figure 4-12. (*Continued*)

- Physicians;
- Dentists;
- Clinics;
- Hospitals;
- Veterinarians;
- Morticians;
- Nursing homes; and
- Larger industries with medical facilities.

4. Complete an incinerator video to be used during small group and one-on-one meetings with members of base audience groups for the purpose of discussing the process and building support, or eliminating reasons for opposition.

5. Install the call-in telephone number for messages from the site manager or other noted authorities.

6. Address visibility issues:

- Through letters to the editor or controlled space essays; and
- Through a Q&A document prepared for public distribution by request using newspaper coupons.

Strategy #4:   Develop Support for the Political Process

1. Get elected officials to once again agree to visit both the proposed site and the model site.

2. Offer to have the key opponents travel with elected officials so both can hear our story again.

3. Attempt to identify individuals and organizations who can lend rational voices to the discussion – whether or not they can publicly support our project or any specific incinerator.

4. Identify ways meaningful to public policy makers through which community members can comfortably show their support.

5. Develop some level of open and active public support.

THE TIMELINE

This nine-week timeline utilizes a definite sequencing process. Sequencing, as opposed to running everything at once, keeps the process relatively low key. In addition, should the company decide to change course or should some unalterably adverse situation or event occur, sequencing allows the company to shut down the project without much cost. But the most important reasons for sequencing actions, besides manageability, are the cumulative effect of chipping away at the opposition and the absence, at least initially, of large-scale, company-sponsored efforts which cause emotionally-charged responses by the opposition.

Phase I:

01/19 - 02/02   1.   Hard Opposition

A.   Identify area residents openly opposed
B.   Schedule small group meetings

Figure 4-12. (*Continued*)

       C. Transcribe meeting audiotape

     2. Soft Opposition

       A. Identify soft opposition from meeting audiotape and news reports
       B. Set timetable for meetings with them

01/23 - 02/09   3. Base Audiences

       A. Conduct survey of public opinion in the area
       B. Re-initiate contact with base audiences by letter
       C. Arrange for "800" number and equipment
       D. Establish mailing list development process
       E. Develop newsletter format
       F. Draft "editorial" ad concepts
       G. Complete video program

     4. General Communication Programs

       A. Limited to requested interviews
       B. Perhaps draft a letter to the editor
       C. Keep very low key
       D. Prepare to react to developments in the legal strategy

<u>Phase II:</u>

02/02 - 02/16   1. Hard Opposition

       A. Arrange for guided visits to model site
       B. Follow-up in one-to-two weeks face-to-face
       C. Attend opposition group meetings to answer questions

02/09 - 02/23   2. Base Audience/Soft Opposition

       A. Arrange for guided visits to model site
       B. Do follow-up calls seven-to-ten days after the visit
       C. Refine Q&A communications process

02/16 - 03/09   3. Undecideds (neighbors/nearbys who could be convinced)

       A. Conduct small group meetings
       B. Arrange for guided visits to model site
       C. Follow-up to gauge and adjust attitudes

     4. General Communication Programs

       A. "800" phone line in operation
       B. First advertorial essay appears
       C. Videotapes made available to community – individuals and audiences
       D. Prepare to react to developments in the legal strategy
       E. Link "supporters" with public policy process in state Capitol

Figure 4-12. (*Continued*)

Phase III:

03/09 forward   1.   Hard Opposition

        A.   Continue face-to-face visits as appropriate
        B.   Continue visits to model site as appropriate

    2.   Soft Opponents

        A.   Continue face-to-face visits as appropriate

    3.   Base Audiences

        A.   Gauge their "temperament" through telephone contact
        B.   Arrange small group meetings where appropriate
        C.   Continue prodding information as needed

    4.   Undecideds

        A. Treat the same as "soft opposition"

    5.   General Communications Programs

        A.   Prepare to react to developments in the legal strategy
        B.   Link "supporters" with public policy process in the state Capitol

03/30   6.   Evaluation

        A.   Conduct follow-up survey research project to measure changes in attitudes
        B.   Full evaluation of the viability of the project based on our achieving a turn-around in public opinion

Figure 4-12. (*Continued*)

MODEL DOCUMENT #5:

SAMPLE FORMAT FOR A SUPERFUND COMMUNITY RELATIONS PLAN
(including sample tables of contents for a
Communications Control Book and a Contacts List Data Base)

Note: The format which follows reflects both good practice and parallels that are advocated by the U.S. Environmental Protection Agency. For other models we recommend you obtain the publication, Community Relations in Superfund: A Handbook, *published by the U. S. Environmental Protection Agency, Office of Emergency and Remedial Response, Document Number EPA/540/R/92/009, January 1992 (PB92-963341).*

I.   *Overview of Community Relations Plan*

Purpose: Provide a general introduction by briefly stating the purpose of the Community Relations Plan and the distinctive or central features of the community relations program planned for this specific site. Note any special circumstances that the plan has been designed to address. Do not repeat general program goals (e.g., "Keep the community informed.").

Length: One paragraph to several pages.

II.   *Capsule Site/Situation Description*

Purpose: Provide the historical, geographical, and technical details necessary to show why the site was put on the National Priority List (NPL).

Suggested topics:

- Site location and proximity to other landmarks;
- History of site use and ownership;
- Date and type of release;
- Nature of threat to public health and environment; and
- Responsibility for site (e.g., State- or Federal-lead).

Length: One page.

III.   *Community Background*

Purpose: Describe the community and its involvement with the site. Cover three topics:

A.   Community Profile: The economic and political structure of the community, and key community issues and interests.

B.   Chronology of Community Involvement: How the community has reacted to the site in the past, actions taken by citizens, and attitudes toward government roles and responsibilities. Discuss actions taken by any government agencies or government officials, such as public meetings or news releases.

C.   Key Community Concerns: How the community regards the risks posed by the site or the remedial process used to address those risks. One approach: break down the analysis by

Figure 4-12. *(Continued)*

community group or segment (e.g., public environmental interest groups; nearby residents; and elected officials).

In all three sections, but particularly in the last, focus on the community's perceptions of the events and problems at the site rather than the technical history of the site.

Length: From three to seven pages, depending on the history and level of community involvement in the site.

IV. *Highlights of Program*

Purpose: Provide concrete details on community relations approaches to be taken. This should follow directly and logically from the discussion in Section C of the community and its perceptions of the problems posed by the site. Do not restate the goals or objectives of conducting community relations at Superfund sites. Instead, develop a strategy for communicating with a specific community.

Suggested topics:

- Resources to be used in the community relations program (e.g., local organizations, meeting places);
- Key individuals or organizations that will play a role in community relations activities;
- Areas of sensitivity that must be considered in conducting community relations.

Length: One page.

V. *Techniques and Timing*

Purpose: State what community relations activities will be conducted at the site and specify when they will occur. Suggest additional techniques that might be used at the site as the response action proceeds, as well as when these techniques are likely to be most effective.

Length: Two to three pages. Matrix format may be suitable.

Attachments:

- List of Contacts and Interested Parties
- Locations for Information Repository and Meetings

(Names and addresses of individuals should not be included in the Community Relations Plan made available in the information repository for public review. Names and addresses should, however, be compiled for a mailing list as part of the Community Relations Coordinator's files.)

Figure 4-12. (*Continued*)

SAMPLE TABLE OF CONTENTS
COMMUNICATIONS CONTROL BOOK

TAB	DESCRIPTION	TAB	DESCRIPTION
1	Current Project Directory	12	Q&A: Siting Process
2	Project Goal/Philosophy	13	Q&A: Political Process
3	Project Components/Tasks	14	Q&A: Environmental Issues
4	Communication Timelines		• Public Health and Safety
5	Message/Theme Development		• Natural Environment
			• Social/Cultural Environment
6	Community Relations		• Technical Considerations
	• Supporters/Friendlies		• Economic Considerations
	• Opposition		• Reprint: Areas of Public
	• Options for Contacting		Interest and Inquiry
	− Door-to-door	15	Relevant Clips/Editorials/
	− Mailing		Commentary/Transcripts
	− Community Meetings		• Section Index
	− Phone Bank		
	− Special Events	16	Reports:
	• Community Environment Chart		• Index to Existing Reports (with
	• Influencing Public Attitudes Chart		brief summary)
			• Index to Upcoming Reports
7	Project Checklists		(with due dates)
8	Worst Case Scenarios	17	Public Opinion Research
	• "What-If" Scenarios		
	• Misconceptions to Correct/Guard	18	Response Statements
	Against		• Key General Messages
	• Vulnerabilities		• Health Study Response
	• Key Issues		• Landfill Issue Response
	Resource Information	19	Direct Contact Letters
	• Dealing with Confrontation		
	• Rules for Radicals	20	Public Official Contact Reports
	• Managing Bad News		
		21	Letters to Public Officials
9	Historic Event File		
	• Develop Key Event Chronology	22	Innovative Technologies
10	Q&A: Site	23	Media Relations Program
11	Q&A: Chemicals/Hazardous	24	Meeting Notes
	Substances		

Figure 4-12. *(Continued)*

PART 3

The Growth of EHS Auditing as a Discipline

5

Environmental, Health, and Safety Auditing Standards

Gilbert S. Hedstrom
Arthur D. Little, Inc.

As this book goes to press, perhaps no aspect of environmental, health, and safety auditing is more in flux than audit standards. Audit standards have been gradually evolving over the past decade, but the last several years have seen a rapid escalation of activity. Today, literally dozens of organizations around the world are working to influence the shape of audit standards. Looking back, environmental auditing professionals will likely regard the 1990s as the decade in which auditing standards took shape.

Audit Standards: A Context

To understand why standards have recently come to the top of the agenda in environmental auditing, we should return for a moment to the field's early days. When Arthur D. Little worked with Allied-Signal, formerly Allied Chemical Company, on that company's pioneer environmental audit program in 1977, one of the key initial questions was whether the audit program could be comparable to a financial audit, verifying, on an ongoing basis, that the company's environmental, health, and safety performance conformed with internal standards and applicable laws.

The answer was no. In the absence of clearly defined standards, an environmental, health, and safety audit could not be completely analogous to a financial audit. This drawback, however, did not discourage many companies from developing environmental, health, and safety audit programs. They saw that even without established standards of the kind that exist in financial auditing, an environmental audit could offer companies substantial benefits by providing vital information about the environmental aspects of their operations.

In the years since then, as environmental, health, and safety auditing has taken hold among companies worldwide, the issue of standards has grown in importance. But even as experience has accumulated about approaches to conducting audits, consensus on standards has been slow to emerge. While many organizations have established environmental, health, and safety audit programs, considerable diversity still exists in the precise manner in which audits are organized and conducted. Auditing professionals recognize that differences in organizational objectives, policy, structure, and culture can appropriately lead to differences in audit staffing, scope, coverage, and approach.

Despite these differences, a number of common auditing practices were defined during the 1980s. At the same time, auditing professionals increasingly discussed audit standards and auditor certification. Throughout these discussions, Arthur D. Little offered the point of view that standards and certification should be the last two stages of a four-part evolution of performance criteria for environmental auditing. This progression sees the first step as the development of a common set of auditing practices. This leads to a shared sense of basic principles, which in turn provides the foundation for audit standards, and, finally, auditor certification (Fig. 5-1). In our view, auditor certification is only appropriate after auditing practitioners have agreed on a set of standards that define the performance to be certified.

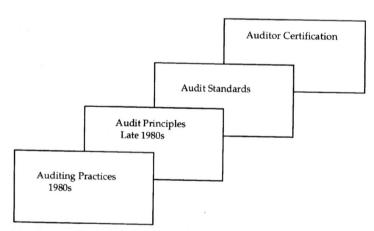

Figure 5-1. Evolution of professional codes in environmental auditing. (*Source: Arthur D. Little, Inc.*)

Standards Today

As audit practitioners look at where the profession is in the mid-1990s, this is what they see:

- A range of excellent documents about standards developed by governments, industry associations, and independent bodies
- A broadening base of support for underlying assumptions, as indicated by the many overlaps among existing standards
- The rapid development of more standards by standards-writing bodies
- Global initiatives that are driving the expansion of public disclosure of verified industry performance data

The development of standards is encouraged by the broad agreement that now exists on auditing practices and principles. An early effort in this area that has stood the test of time is an environmental audit policy developed by the U.S. Environmental Protection Agency in close collaboration with environmental audit professionals and published in 1986. A more recent set of guidelines that embodies much of the consensus that has developed around environmental auditing is the International Chamber of Commerce *Guide to Effective Environmental Auditing,* prepared in 1991 by auditing experts from Arthur D. Little and Allied-Signal.

Currently, there are many efforts under way worldwide to define auditing standards. Some of these efforts, like the Environmental Auditing Roundtable's "Standards for Performance of Environmental, Health, and Safety Audits" and the joint project of the Canadian Environmental Auditors' Association and the Canadian Standards Association to develop "Environmental Auditing Principles and Practices," have been carried forward by auditing practitioners and specifically address the auditing aspect of environmental management. Others, such as the British Standards Institute's BS 7750 and the European Union's Eco-Management and Audit Regulation, include auditing standards within the broader context of environmental management.

In fact, recent efforts by auditing practitioners to define standards are driven in part by the rapid emergence of these national and transnational environmental management initiatives. The Eco-Management and Audit Regulation has also prompted many European countries, including France, the Netherlands, and now Germany, to develop parallel standards. Moreover, the International Organization for Standardization, whose ISO 9000 quality management standard has won wide approval, is now working on its parallel environmental management standard, which, because of its applicability worldwide, is likely to be more influential than the Eco-Management and Audit Regulation. All of these standards include provisions about auditing.

Watching these developments, environmental, health, and safety auditors are concerned that unless careful attention is paid to defining the right standards, with the consensus of knowledgeable practitioners, standards could cause more problems than they solve or quickly become out of date. Some auditing professionals fear that defining standards too rigidly may reduce the incentive to advance the discipline of auditing. Real progress may be stifled. Companies may become content simply to meet the standards and reluctant to refine and further develop auditing in order to increase its potential value to the organization and its stakeholders. A slowdown in progress might, in turn, discourage skilled practitioners who would leave the field to seek more challenge elsewhere.

If leaders in the field of environmental auditing continue to participate in standards development, better solutions will be reached for many of these areas of concern. These areas include:

- Requirements for public disclosure of audit results
- Governmental testing and certification of auditors
- Required third-party audits

Another danger is that auditing standards writers in government or industry who are not professional auditors may focus too narrowly on specific auditing functions, overlooking the broader environmental management context. Particularly important in standards development of any kind is a focus on what the standards should accomplish.

Recent Initiatives: A Closer Look

As standards for environmental auditing emerge, they tend to focus consistently on a number of key issues. These issues closely parallel those covered in the environmental auditing principles first proposed by Arthur D. Little in 1988. Today we are seeing broad consensus about auditing principles and standards in the following areas:

- *Management support.* A key component of any successful audit program, it involves consistent, strong backing from senior management for the goals of the audit and appropriate follow-through when the audit is complete.

- *Clear objectives and scope.* Typical audit objectives include verifying compliance with applicable standards, assessing management systems, and identifying risks and liabilities. Audit scope is defined in terms of categories such as organization, geography, and location.

- *Auditor proficiency.* Includes knowledge of the issues and areas to be covered in the audit and sufficient audit training and proficiency to achieve the stated objectives.

- *Auditor independence and objectivity.* An essential requirement for any audit program, this is achieved through means such as adherence to professional standards of conduct and the development of an appropriate organizational structure.

- *Due professional care.* Involves conducting the audit with "reasonable care and diligence" and in accordance with accepted professional standards.

- *Formal plans and procedures.* Provide uniform guidance in audit preparation, fieldwork, and reporting.

- *Planned and supervised fieldwork.* Ensures that audit activities efficiently achieve consistent results.

- *Audit quality control/quality assurance.* Assures accuracy and encourages continuous improvement of audit management systems, procedures, and implementation.

- *Clear and appropriate reporting.* Characterizes the audit report, which should communicate information that is consistent with the audit scope and objectives, contains enough detail to facilitate corrective action, and is factual and objective in style.

- *Follow-up procedures.* Evaluate line management's provisions for responding to audit results and assess how exceptions from previous audits were managed.

- *Third-party verification.* This has been sought to date by only a handful of companies for environmental audits. It involves procedures and approaches for obtaining an independent verification of audit findings.

- *Public disclosure of findings.* This is not currently characteristic of environmental audits, although it is expected to develop in the future.

- *Auditor certification.* Does not currently exist for environmental auditing. It would require broad agreement on audit standards and the establishment of independent certifying bodies.

An examination of a range of major initiatives, conducted by Ralph Rhodes of Allied Signal in 1992, tracks which of these issues is addressed in each initiative (Fig. 5-2). The initiatives come from a range of sources, including:

A multinational organization—the European Union

A national government agency—the U.S. Environmental Protection Agency

An international business and professional organization—the International Chamber of Commerce

International and national standards-setting organizations:
- The International Organization for Standardization
- The Canadian Standards Association
- The American National Standards Institute

Environmental auditing associations
- The Environmental Auditing Roundtable
- The Institute of Internal Auditors

An environmental consulting firm—Arthur D. Little.

Criteria	Document	1 EU	2 ICC	3 ADL	4 IEA	5 CIA	6 EPA	7 IIA	8 USA	9 GAAS	10 CBI	11 CSA	12 EEI	13 ANSI	14 ISO	15 EAR
1.	Management Support	X	X	X	X	X	X	X	X		X	X	X	X		X
2.	Objectives & Scope Clearly Defined & Communicated	X	X	X	X	X	X	X	X	X	X	X	X	X	X	X
3.	Auditor Proficiency	X	X	X	X	X	X	X	X	X	X	X	X	X	X	X
4.	Auditor Independence & Objectivity	X	X	X	X	X	X	X	X	X	X	X	X	X	X	X
5.	Due Professional Care By Auditor			X	X		X	X	X	X		X	X	X	X	X
6.	Formal Plans & Procedures	X	X	X	X	X	X	X	X	X	X	X	X	X	X	X
7.	Planned and Supervised Fieldwork	X	X	X	X	X	X	X	X	X	X	X	X	X	X	X
8.	Audit Quality Control/Quality Assurance	X	X	X	X		X		X	X		X	X	X		X
9.	Clear and Appropriate Reporting	X	X	X	X	X	X	X	X	X	X	X	X	X	X	X
10.	Follow-up Procedures	X	X	X	X	X	X	X	X		X	X	X	X	X	
11.	Third-Party Verification	X			X			X	X							
12.	Public Disclosure of Findings	X														
13.	Auditor Certification	X														X*

*Optional

Figure 5-2. Current status of key provisions of criteria applicable to environmental auditing. (*Source: Ralph L. Rhodes*)

Key to Organizations

EU:	European Union
ICC:	International Chamber of Commerce
ADL:	Arthur D. Little
IEA:	Institute of Environmental Auditors
CIA:	Chemical Industries Association
EPA:	Environmental Protection Agency
IIA:	Institute of Internal Auditors
USA:	Comptroller of the United States
GAAS:	General Agreement on Auditing Standards
CBI:	Confederation of British Industry
CSA:	Canadian Standards Association
EEI:	Edison Electric Institute
ANSI:	American National Standards Institute
ISO:	International Organization for Standardization
EAR:	Environmental Auditing Roundtable

Figure 5-3. Key to organizations.

To get a better sense of how these initiatives address auditing issues, we can look more closely at examples from government, independent standard-setters, and the environmental auditing profession.

Eco-Management and Audit Regulation

In early drafts, the European Union's environmental management and audit scheme focused more exclusively on auditing, and, indeed, was called the "Eco-Audit Scheme." Published in June 1993, the voluntary regulation will take full effect in April 1995. It establishes an important precedent by describing both the form and the content of the information companies should disclose publicly. This information will be presented in an annual environmental statement to the public, covering pollutant emissions, waste generation, energy, water, raw materials consumption, noise, and other significant environmental aspects of operations.

The requirements surrounding the environmental statement are one critical factor in the global drive to establish standards for environmental auditing. The reason is that the Eco-Management and Audit Regulation requires that companies submit the statement for independent, third-party verification. That requirement, in turn, intensifies the need to develop agreed-on standards for accrediting the certifiers now.

Another key component of the Eco-Management and Audit Regulation is that it is structured to accommodate the concurrent development of "national, European, and international standards for environmental management systems." The regulation makes a provision for other standards to be recognized by the European Commission because "companies should not be required to duplicate the relevant procedures."

These developments ensure that in the near future, auditing will be defined to a much greater extent by widely accepted standards and accreditation procedures. In five years, discussions of the next stage for auditing standards will be shaped in part by the experience of companies, governments, and the public with the reporting provisions of the Eco-Management and Audit Regulation.

The Environmental Auditing Roundtable's "Standards for Performance of Environmental, Health, and Safety Audits"

In February 1993, the membership of the Environmental Auditing Roundtable approved a set of standards for the performance of environmental, health, and safety audits. Over 90 percent of the responding members voted in favor of the standards. With the organization's membership at more than 500 members, and growing rapidly, this is a strong vote of approval.

After a decade of discussion and debate on the subject, the Environmental Auditing Roundtable began developing a set of standards in 1990. A standards committee was charged with defining the minimum criteria for the performance of internal and external environmental, health, and safety audits.

The Environmental Auditing Roundtable's standards define auditing narrowly, as do the U.S. Environmental Protection Agency and the International Chamber of Commerce. The standards focus on audits against established criteria. They concentrate on the audit itself, from preparation through report writing. They do not include standards for the design of an audit program (the subject of a current Environmental Auditing Roundtable initiative). Nor do they offer standards for acquisition/divestiture (due-diligence) audits, risk assessments, management system assessments, or pollution prevention audits.

Through their standards, the Environmental Auditing Roundtable seeks to make sure that the voice of auditing practitioners is heard as national and international standards are considered. Several members of the Environmental Auditing Roundtable are working to help define both North American standards for auditing (in conjunction with the National Standards Foundation International and the Canadian Standards Association) and international standards through participation in work groups on the U.S. technical advisory group to the environmental management initiative of the International Organization for Standardization (ISO).

ISO's Environmental Management Standard

As this book goes to press, momentum is gathering for the completion of ISO's environmental management standard. The success of ISO 9000 has focused much attention on this standard.

ISO has set up six working groups, each responsible for a specific area within environmental management. The Netherlands holds the secretariat for the environmental auditing group. The first meeting of the subgroup on auditing, consisting of representatives from 25 nations, agreed on the following points in late 1993:

- A title and scope for the project
- A general plan of work with priorities
- The first two meeting dates for task groups
- An agreement to use the material drafted by the Strategic Advisory Group on the Environment on environmental auditing
- An overall timetable for drafting the standard by the spring of 1994 and publishing it in final form by January 1996

ISO's environmental standard is being written to have a high degree of compatibility with ISO 9000. It is also expected to parallel the British environmental standard, BS 7750, in many key respects. The purpose is to enable companies that wish to win ISO certification for their environmental management systems in the future to be able to achieve that within the context of management systems that are already in place. We can expect that the auditing provisions of the ISO effort will also parallel the auditing provision of other management standards.

A Look Ahead

It is clear that in the next several years, standards will become more fully established. We can already see many areas where broad agreement exists. But there are also several key areas that will require further thinking from the environmental auditing profession. To ensure that auditing standards continue to develop in ways that serve all of a company's stakeholders, practitioners will need to:

- Work with specific national standards where they are more prescriptive than international standards
- Determine what constitutes a certified auditor and establish when (or whether) certified auditors are needed
- Determine how much companies are willing to pay for audit insurance and refine the understanding of what actual costs are
- Determine what the public needs to know in order to be confident about the environmental audit process.

But perhaps the most important task of all will be ensuring that standards, as they emerge, do not signal an end to the process of discussion, questioning, and growth that has brought auditing so far in such a short span of time. The next decade will see rapid changes in the way industry reports its environmental performance internally and externally. New regulations, such as the U.S. Environmental Protection Agency's Enhanced Monitoring Proposal for the Clean Air Act, will continue to expand what the public knows about company environmental performance. As this new, more open climate takes hold, environmental auditing will need to be ready to evolve with it.

6

Applying Total Quality Management (TQM) to Health, Safety, and Environmental Compliance Auditing at Union Carbide

Paul D. Coulter
Director, Compliance Audit Program,
Union Carbide Corp.

As new laws were passed reflecting the increasing concern for the environment during the 1970s and early 1980s, it became apparent to Union Carbide that full compliance with all applicable laws was not sufficient to be among the leaders in health, safety, and environment (HSE). During 1985, with support and encouragement from Union Carbide's board of directors, we undertook the development of a new HSE program. The new program would not only ensure full compliance with applicable regulatory requirements but would also ensure that a consistent application of Union Carbide policies and procedures, which generally exceed the legal requirement, be implemented wherever Union Carbide and its affiliate companies might operate in the world. Our goal was to achieve and maintain a level of health, safety, and environmental performance for Union Carbide locations that is among the best in our industry.

We determined what we believed the characteristics of a leading HSE program should be:

- Program management should not be part of any business organization but report directly to senior corporate management.

- The program needed to be designed so that it would be strongly supported by both the senior corporate management and line management.

- Periodic reports should be made to senior management and to the board of directors.

- The establishment of management systems as well as compliance with internal and governmental requirements must be an integral part of the program to ensure continued compliance.

- These management systems must be regularly audited.

- There must be an effective mechanism to escalate HSE concerns to the highest management levels.

- Whenever practicable, state-of-the-art techniques should be used.

- The program must be able to withstand intensive external reviews.

The program Union Carbide developed meets all of the requirements we felt a leading program should include. A corporate vice president—who reports directly to the chief executive officer (CEO)/chairman of the board—was given sole responsibility for the Health, Safety, and Environmental Program. The vice president reports regularly to the board of directors on the status of various health, safety, and environmental programs and performance of all Union Carbide locations. Additionally, the board of directors contracted independent health, safety, and environmental consultants to oversee the implementation of Union Carbide's program. Today they continue to provide independent oversight and assurance to the board that Union Carbide's HSE programs are among the best and that the related reports provided to the board give an accurate reflection of Union Carbide's HSE performance.

A leading corporate HSE program requires an independent audit program to give an unbiased report on the status of the HSE program. The auditors should be full-time professionals and not have reporting responsibility to any business management organizations. The auditors not only must be skilled HSE professionals who keep their skills current by ongoing training, but they also must have excellent written and verbal communication and interpersonal skills.

Union Carbide's independent compliance audit program was established with its director reporting to the HSE vice president and giving quarterly reports to the board of directors on the audit program results. The audit program's goal is to help facilitate line management's implementation of its HSE responsibilities and to assure that the corporation's operations achieve HSE performance excellence and consistency. The program's independent assessment of HSE compliance performance has resulted in:

- Improved overall HSE performance
- Increased employee awareness and acceptance of HSE responsibilities
- Avoidance of HSE risks and liabilities
- An improved public and business image

The program assists senior management and the board of directors in fulfilling their fiduciary duties and exemplifies leadership and good corporate citizenship. Although Union Carbide's Health, Safety, and Environmental Program encompasses a great deal more than the audit function, this chapter will focus on auditing and cover other areas only when it is necessary for the understanding of our audit program.

Audit Overview

The Union Carbide HSE audit program is based on the financial audit model. It uses sampling techniques rather than a wall-to-wall-type audit. It does not replace business or location audits/assessments, and its principal purpose is to assure compliance.

Each of our major businesses has staff professionals who are responsible for a full range of traditional health, safety, environmental, and product safety functions. The individual businesses are responsible for carrying out in-depth and business-specific audits and developing and implementing the procedures and programs for meeting corporate HSE standards and full compliance with applicable laws and regulations. The major elements of Union Carbide's Audit Program are shown in Fig. 6-1.

Two types of audits are performed—manufacturing/distribution/laboratory site audits and product responsibility (safety) audits. The objective of both types of audits is to provide independent evaluation of line management and location performance with respect to:

- Compliance with applicable internal standards and governmental requirements
- Management systems that assure continued compliance
- The Responsible Care® requirements that have been incorporated into Union Carbide's internal standards

In addition, assessments are made of non-Union-Carbide-controlled businesses and contractors such as terminals, warehouses, and toll processors. The objective of the assessment is to reduce Union Carbide's exposure to risk and adverse

- **Information Systems**
  - Facility Profile
  - Hazard Ranking
  - Audit Document Life Cycle
  - Action Plan
- **Standards**
  - Standard Operating Procedures
  - Protocols
- **Audit Process**
  - Site Selection
  - Pre-Audit Activities
  - On-Site Audit Process
  - Post-Audit Activities
- **Follow-up**
  - Action Plan
  - Periodic Action Plan Update
  - Cause/Findings Analyses
- **Quality Control**

Figure 6-1. Main components of Union Carbide's Compliance Audit Program.

publicity by reviewing the supplier's compliance with legal requirements as well as Union Carbide contractional requirements and Responsible Care® standards in those parts of their business dealing with Union Carbide materials.

Information System

Union Carbide developed the Episodic Risk Management System (ERMS) as a management tool for making consistent and cost-effective decisions regarding prioritizations of risk reviews and methods for minimization of risk. Some elements of the system are:

- Hazard identification
- Operation ranking
- Risk estimation
- Risk mitigation

The audit program and the ERMS program share a database that supplies detailed information about each Union Carbide site and other locations that process, handle, or store Union Carbide materials. A facility profile is maintained for each location. Using the facility profile database and a hazard ranking model, a relative hazard ranking of all locations is developed. This ranking is used as one element of the audit site selection process. During an audit, the facility profile data are reviewed for accuracy.

The compliance audit program maintains a database on the various aspects of the audit process. The current year's schedule, with the auditor and team leader assignments, as well as requirements for contract or consultant auditors, is maintained and updated as changes are made. These data drive the notification letter process, which ensures that our locations and other appropriate people receive timely notice of the audit.

The database also includes the audit steps following the notification letter, such as the dates of the audit, dates of the responses when appropriate, the date the final report was issued, and the date when the action plan was received. The steps of the action plan process are similarly recorded, including the semiannual reviews, until all items of the action plan are completed. The audit findings/ analysis coding that is encoded in each audit report is extracted to a database for subsequent statistical analysis.

These databases are used to generate reminders and to issue warnings when an item is overdue. A report showing the status of the various steps of the process and which steps are past due is generated on a regular basis. We can use these data to develop a statistical analysis of corporate HSE performance and that of the individual audited site. The overall system helps ensure that nothing of importance is overlooked.

Audit Standards

A comprehensive set of HSE auditing standard operating procedures (SOPs) has been developed to assure uniform application and conduct of the audit process. These procedures give detailed instructions to the auditors regarding the planning, execution, and completion of the auditing process. There are general SOPs that cover the overall procedures for the department and the purpose, scope, and key elements of the audit program. The fundamental SOPs address such items as:

- Preaudit planning
- Audit preparation
- On-site compliance audit activities

On-site activities include:

- The opening meeting
- Understanding and assessing management systems
- Gathering audit evidence
- Sampling strategies (random number tables) and techniques

- Evaluating the audit evidence
- Management feedback meetings
- Preparing and issuing the audit report

The SOPs also describe:

- Interviewing techniques
- Working-paper requirements
- The postaudit dispute resolution process
- Record-keeping retention requirements
- Good travel practices
- Emergency procedures
- Security

The SOPs are regularly reviewed and updated.
  Audit protocols have been developed to cover the 10 HSE functional areas of:

- Occupational health and medicine
- Plant and employee safety
- Operational safety
- Product distribution
- Air emissions
- Waste management
- Water protection
- Environmental emergency preparedness
- Prior audit findings review
- A more general protocol used to direct the auditors' review of management systems and miscellaneous issues

For assessments there is a general protocol specifying which parts of the 10 functional protocols should be executed. Additionally, there is a protocol for conducting product responsibility (safety) audits that are non-site-specific, where such issues as the following are covered:

- Product risk assessments
- Product safety communications
- Product quality
- Toxic Substances Control Act (TSCA)
- Food and Drug Administration (FDA)

- Federal Insecticide, Fungicide, and Rodenticide Act (FIFRA)
- Product distribution
- Corresponding programs outside the United States

Each auditor has responsibility for one or more of the protocols, which he or she regularly reviews and updates as appropriate. Each protocol is maintained on a computer network disk. Before each audit, according to the functional areas they are to cover, the auditors print out the most recent version of the applicable protocols to take along with them.

Site Selection

To maximize the benefits and intent of the auditing process, sites must be selected and audited in a timely manner. Only then can management derive the maximum benefit from the program consistent with the resources employed. The principal components of Union Carbide's scheduling criteria are hazard potential, compliance potential, facility size, and past audit performance, to which is added a special site-specific component. In addition, a representative group of locations from all businesses and geographical areas is selected. Because of the large number of sites covered by our program, a mathematical model for prioritizing audit sites for scheduling purposes has been developed. The site data used for the model and to calculate the site's hazard ranking is contained in the shared facility profile database.

The audit scheduling algorithm is shown in Fig. 6-2. It is referred to as the *audit timeliness index* (ATI). The five components are each related to one of the scheduling criteria:

- The *hazard component* is based on the location score from the Hazard ranking model.
- The *prior-audit component* is based on performance classifications from the last two audits.
- The *time-since-last-audit component* is based on the number of years since the last audit.
- The *plant-population component* is proportional to the number of employees and on-site contractors at the location.
- The *special site-specific component* allows us to fine tune the model for site-specific factors, such as laboratories with special considerations, new businesses, or other situations that are not adequately covered by the other four components.

The scaling constants are chosen so that sites that should be audited during the current year receive an ATI numerical score of 10 or higher. The effect of prior-audit performance (explained later) and hazard ranking is illustrated in Fig. 6-3. In general, the better the prior-audit performance and lower relative

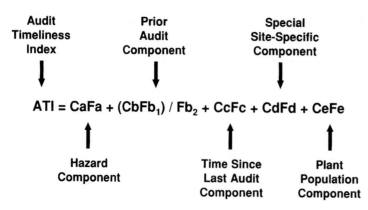

Figure 6-2. Audit scheduling algorithm.

Location A (GM, GM) = 11
         (SM,M)   = 9
         (M,M)   = 7

Location B (GM, GM) = 13
         (M,M)   = 9

Location C(x,RSI)   > 10
       x= don't care

Figure 6-3. Effect of prior audit performance and hazard ranking.

hazard, the longer the time interval between audits. Any location that receives our lowest audit classification (grade), i.e., *Requires Substantial Improvement* (RSI) on its audit receives an ATI score higher than 10 to ensure that it is audited after only 1 year.

Information about the site prior audits and ATI index are consolidated on a computer printout, an illustration of which is shown in Fig. 6-4. The information for each site includes:

- A description
- Business group code
- Internal audit unit number code
- Total number of sites contained within that audit unit
- The date of the most recent audit
- Year and performance classification of the two most recent audits

Description	Group	Audit Unit	Lin Cnt	Last Visit	Most Recent Year	Most Recent Perf	Prev Year	Prev Perf	Log SRS	Hazard	Prior Audit	Last Audit	Site	Population	ATI
Location A	CP	183	2	03/89	89	GM	87	N/A	4.1 -	6.7	0.0	1.5	0.0	2.4	10.7 *
Location B	CP	143	1	06/89	89	N/A	87	RSI	4.6 +	7.3	0.0	1.5	0.0	1.8	10.6
Location C	CP	255	1	05/91	91	GM	88	GM	4.6	7.6	0.0	0.5	0.0	2.5	10.6
Location D	CP	292	3						3.5	5.7	0.0	3.5	0.0	1.3	10.6 *
Location E	AT	162	5	10/91	91	SM	89	GM	4.9	7.8	0.0	0.5	0.0	2.3	10.6
Location F	AT	105	1	06/91	91	SM	90	GM	4.6 %	7.3	0.0	0.5	0.0	2.8	10.6
Location G	GA	625	1	10/88	88	M			3.5	5.7	-2.0	2.0	2.0	2.7	10.4

Figure 6-4. An example of Union Carbide's audit timeliness report.

The log of the hazard ranking score is in the numerical section of the printout plus flags, such as +, −, or %, which indicate whether any data are missing from either the facility profile or chemical thesaurus (used in the calculation of risk).

The next items are the components calculated from the hazard ranking score:

- Prior audit
- Last audit (proportional to the numbers of years since the last audit)
- Any site-specific data
- The population factor
- The ATI score itself. An asterisk after the ATI score indicates that an audit has been scheduled for the current year

Included in the ATI algorithm are all the considerations that were used in the previous scheduling process that was done manually. The use of the ATI index has greatly facilitated the scheduling process and led to real work simplification while increasing the quality of the product. The use of the ATI score as a tool by an experienced audit team has allowed Union Carbide to reach its goal of making site selection less subjective and more objective. This technique for scheduling is more understandable to the audited location, and they perceive it to be more fair. It also helps to motivate locations to work just that extra bit harder to get the highest possible classification so that the time between audits is longer.

Preaudit Activities

During the scheduling process, the audit team and team leader are chosen. Since the entire corporate full-time auditor staff is experienced and well-trained, the team leader responsibilities are rotated among the auditors. The team reviews past audits, if any, and determines which protocols should be used during this specific audit. Prior audit findings review, management systems, and operational safety protocols are administered on all audits. Three additional protocols are chosen from the following seven protocols:

- Plant and employee safety
- Occupational health and medicine
- Air emissions
- Waste management
- Water protection
- Environmental emergency planning
- Product distribution

For assessments of non-Union Carbide locations, the separate protocol that contains selected elements of the 10 protocols is used to give a broad-brush evaluation of HSE performance. These sites are generally assessed less often than Union Carbide locations, and breadth rather than depth is more appropriate for protecting Union Carbide's interest.

For product responsibility audits, a separate and special protocol is used. After the protocol selection is completed, the team leader assigns the responsibility for various parts of the protocols to each of the auditors on the team. The most recent versions of the protocols are downloaded from our computer system for use during the audit.

The team leader is responsible for sending the notification letter to the site and other appropriate managers, informing them of the intended audit and requesting preaudit information. The preaudit information is used by the auditors in carrying out some of the steps of the protocol and planning their on-site activity. For non-North American sites and for non-owned Union Carbide facilities, notification is given 30 to 45 days prior to the scheduled audit. All other locations receive a 10-day notice. By requesting preaudit information, the time spent on site is more effectively used, and more can be accomplished in the short time period of the audit. Each team member is responsible for reviewing the preaudit information, appropriate regulations, and corporate standards for the areas that he or she will be reviewing during the audit or assessment. The team leader defines and, where appropriate, arranges the necessary travel plans and any preaudit meetings with audit management or other Union Carbide health, safety, and environmental personnel that might be appropriate.

On-Site Audit Activities

Once the audit team is on site, the audit is conducted in a manner that is consistent with the five-step process described in the book *Environmental Auditing* by Greeno, Hedstrom, and DiBerto,[1] and in accordance with Union Carbide's SOPs. A non-Union Carbide consultant auditor participates in approximately 25 percent of our audits to provide an independent assessment of our performance.

The schedule for a normal 1-week audit starts on Monday morning and runs through midday Thursday. Thursday afternoon and evening are used to prepare the final report for a closing session held on Friday morning. For larger sites, two-week audits are scheduled. A few smaller locations require only 2 or 3 days, and the schedule is appropriately shortened.

The normal auditing staff consists of three auditors. On large locations we use four or more auditors, and on very small locations we may need only a single auditor. When we do not have sufficient full-time auditors of our own, we use either a consultant or independent contractor auditor whom we have trained and keep current through our periodic training courses. In many cases, the independent contractor is a retired Union Carbide auditor or HSE professional.

The on-site audit activities start with an opening meeting during which the audit team leader presents the objectives of the audit and gives some background on the audit process. The location gives the auditors an overview of the work being performed at the location and introduces the auditors to the key people they will be working with during the process of the audit. Following the opening meeting, an orientation tour allows the auditors to learn the physical layout of the plant. The audit team reviews reporting relationships with site management, which guides them to the appropriate persons for developing an understanding of the location's management systems. The location's management systems are evaluated to determine if they conform to the policies and standards of the corporation.

The audit proceeds by gathering compliance data from interviews with key staff members; reviewing appropriate documents; observing the work process and practice; and testing the system to ensure that the training, understanding, and knowledge of locations of the appropriate valves, controls, and related items are satisfactory. At each stage the auditor confirms that sufficient information has been collected to make an evaluation of the overall performance level.

Each day, the auditors have a short feedback session with the audited location to make them aware of the findings for that day. This helps ensure that the auditors' understanding is correct and that the location receives no surprises in the final report. In many cases, during the daily feedback session, the location is able to explain or provide additional information that avoids revision of the final report.

The audit findings are divided into three categories: exceptions, observations, and local attention items:

- *Exceptions* are clear deviations from internal or external requirements
- *Observations* document deviations from best practice or deviations from known or anticipated future requirements
- *Local attention items* are observations on housekeeping, items not in within the audit scope, very minor or anomalous deviations from requirements, and trends in the site programs which, if continued, could lead to an exception. Local attention items are not included as part of the report, but copies are left at the site and with the auditor's notes. Copies of local attention items are also

given to audit program management and Union Carbide's outside consultant for their review.

Each of our auditors has a laptop computer, which is used to write individual findings and, ultimately, to compile the audit report. Each audit team also carries a portable printer. One the afternoon and evening before the closeout meeting, the team meets, merges all their findings, prioritizes them within each functional area, and prioritizes the functional areas. The classification is determined, and the audit summary is written by the team leader and combined with the findings and a standard boilerplate to generate the report. If the performance classification is an RSI or the audited location requests time to respond to some of the findings, the report is issued in draft form. Otherwise, the report, with only minor editorial changes, becomes the final report and is left with the site and issued to business management.

In those unusual and special situations where an RSI, the lowest classification, is proposed as a performance classification, the report is reviewed by audit management, issue specialists in the corporate Health, Safety, and Environment Department, and the Law Department. When the performance classification is confirmed, the location is notified and the final report is issued to business group management. In all cases, the final report is entered into our computer database and sent to the responsible Union Carbide managers.

Audit Classification

During the preparation of the report, the current level of the location's health, safety, and environmental protection performance is assessed. A four-tier classification system, as shown in Fig. 6-5, is used to rank performance. Performance classification allows Union Carbide to set benchmarks for health, safety, and

M: MEETS Governmental and Internal Compliance Requirements

SM: SUBSTANTIALLY MEETS Governmental and Internal Compliance Requirements

GM: GENERALLY MEETS Governmental and Internal Compliance Requirements

RSI: REQUIRES SUBSTANTIAL IMPROVEMENT to Meet Governmental and Internal Compliance Requirements

Figure 6-5. Compliance performance classifications.

environmental objectives and to measure improvement in performance from year to year.

Several steps are taken to establish a performance classification. As described earlier, this is one of the last on-site audit process steps. It is a team activity and reflects the team's impression of location HSE performance. To improve the consistency and objectivity of the classification process, the process has been formalized in an SOP. The team develops a list of exceptions, which is used to integrate and summarize the deviations within a functional area. For each exception, we determine a severity rating:

- *Imminent concern* (demands immediate management attention)
- *Priority concern* (has the potential for serious adverse health, safety, environmental, or corporate impact)
- *Other* (all other situations)

For each functional area, we determine whether the degree of compliance is high, substantial, general, or limited. The functional area's severity rating is based on the highest rated exception. Using the severity and compliance factors and the table shown in Fig. 6-6, we determine the performance classification for each functional area. Performance classifications for each functional area are evaluated, and the overall location performance classification is determined.

For an audit rating of *meets,* at least 75 percent of the functional areas must be *meets* with no functional area *generally meets* or *requires substantial improvement,* and there is no other relevant information that suggests that the overall rating should be different than a *meets.* Similarly, *substantially meets* and *generally meets* are determined. The location receives a *requires substantial improvement* classification if more than 25 percent of the functional areas are classified as RSI.

By assigning a numerical value to the various classifications, an index for groups of audits can be developed. Visual representation of progress can be made by plotting the index for each year's audits versus time, as shown in Fig. 6-7. Generally, such plots indicate an increasing improvement in performance classification. However, if an analysis is made by grouping the audits for each year into three groups, based on the classification of a location on the first audit, second and third audits, and four or more audits, a slightly different story emerges, as is shown in Fig. 6-8.

Compliance	Priority Concern	Other Concern
Full	GM	M
Substantial	GM	SM
General	RSI	GM

Imminent Severity or Limited Compliance - RSI

Figure 6-6. Functional area performance classification.

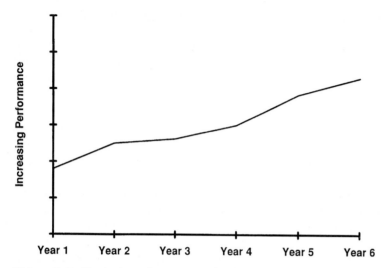

Figure 6-7. Illustration of average audit classification.

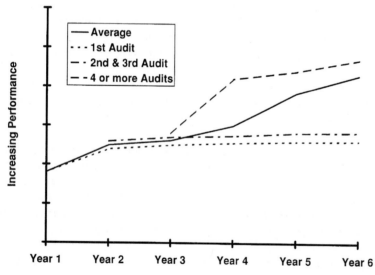

Figure 6-8. Illustration of audit classification, grouped by number of audits.

Over a period of time, the results from first audits for a location are nearly constant, while second and third audits and four or more audits of the same location show a higher degree of improvement. Because of the progressively greater number of four or more audits in later years as locations receive repeat audits, the overall yearly audit performance classification index climbs steadily. This illustrates the importance of visiting sites periodically to help them identify weaknesses. Since most of the improvement occurs in the early audits and tends to plateau after repeated audits, sites that have been audited repeatedly—and show a high degree of compliance—should have less frequent audits.

Follow-up

Within 60 days of the issuance of the final report, each audited location is required to develop a detailed action plan to address the findings noted by the audit. This action plan must reference each finding, the action being taken to correct the deficiency, the projected completion date, and the person responsible for each step of the action plan. The location's action plan must be reviewed by health, safety, and environment professionals in the business unit and forwarded to the accountable vice president for endorsement and submittal to the audit group. All action plans are reviewed by the audit team leader. If the team leader has questions about some of the proposed actions, the action steps in question are reviewed with the appropriate issue manager within the corporate HSE Department. If it is deemed that the proposed actions are not adequate or, less commonly, overstated, the action plan is sent back to the location for revision.

Approximately 10 percent of the submitted action plans are subjected to an in-depth review. This in-depth review includes a review of each item by the team leader, appropriate issue managers, and audit program management. This process is diagramed in Fig. 6-9.

Each location is required to submit a semiannual update on its action plan until all items are completed. On repeat audits, the auditors verify that physical or system revisions noted as complete in the action plans generated in response to previous audits are actually complete and that they correct the problem.

Recently, we have developed a method for the analysis of findings. The auditors, during the preparation of the final report, encode each finding as to its type, whether it is a deviation from country, state, local, business, or Union Carbide requirements, and possible causes. A comprehensive list of causal factors is used so that we can analyze recurring weaknesses within our overall health, safety, and environmental program. The analysis can be performed by business, type of plant, process, or geographical area. We believe this technique will allow Union Carbide to continue progressing toward excellence in its health, safety, and environmental programs.

Quality Control

In order to maintain a quality audit program, it is important for audit management to carefully monitor and maintain each aspect of the audit process. Oversight by an outside consulting firm is also very useful. We have developed our standard SOPs to prescribe for the auditors and related personnel the exact

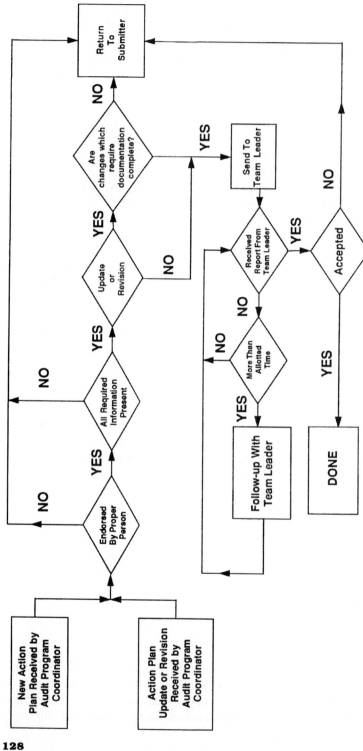

Figure 6-9. Action plan review process.

steps that are to be taken in the audit process. These SOPs are regularly reviewed to verify that they represent the latest thinking and incorporate improvements/changes in the audit process. In a similar manner, the protocols are reviewed by the individual auditors responsible for maintaining them and the appropriate issue managers in those areas where they have expertise. The protocols must accurately represent legal considerations as well as Union Carbide standards, policies, and procedures. They must also be user-friendly for the auditors in the field.

The best SOPs and protocols are not sufficient if the audit is not conducted in accordance with them. For this reason, it is important for audit management to spend time in the field with audit teams, observing both the team's and individual auditor's performance. The use of outside consultants on some audits to give an independent evaluation of the audit process is helpful. Feedback from the audited sites also must be considered in assessing the performance of the audit process. No one expects the audited sites to enjoy being audited, but they should feel that they get value for the time that they spend involved in the audit process.

There should be a regular review of the auditors' working papers. The audit team leader is responsible for reviewing the quality of the working papers after the audit is complete. We have developed a method for evaluating the completeness, accuracy, and consistency of the working papers. An outside consultant is used to evaluate our performance in this area by randomly choosing a number of sets of working papers and grading them. From this, we have been able to develop a program to improve the quality of working papers and, as a by-product, have developed a working paper guidance manual for the use of our auditors.

Management must regularly review the audit reports to ensure that they meet the high standards set for the compliance audit program. Audit management must plan regular training sessions for the auditors. Union Carbide holds two 3-day training sessions per year (spring and fall). Our contract auditors are invited to participate in these sessions. During these sessions, new/revised laws and Union Carbide requirements, as well as audit methods and techniques, are reviewed. The sessions are held off-site to ensure undivided attention and a relaxed atmosphere.

Summary

Union Carbide's HSE audit program has evolved through the method of continuous improvement, the use of quality action teams, and the help of outside consultants, to be in a leadership position. Management has given the program full support, requiring locations with weak performance to either improve or have personnel changes made at the location. Performance in the HSE area is an important factor in determining compensation. The board of directors continues to give oversight and to monitor the HSE program through periodic meetings with the corporate HSE vice president, the compliance audit director, other corporate HSE managers, and the board of directors' outside consultant. The concepts of Responsible Care® have been integrated into our internal standards and procedures.

Hunt and Auster[2] outlined the evolution of HSE management, which can be expressed as:

- Level 1, none
- Level 2, reaction (damage control)
- Level 3, compliance (laws and regulations)
- Level 4, prevention (compliance plus)
- Level 5, leadership (prevention plus)

We believe that we have reached level 4 and are well on our way to level 5, where we prevent adverse situations from occurring. Over all, our audit program has resulted in a more rapid and thorough meeting of our corporate expectations. It is a key element in Union Carbide's overall health, safety, and environmental plans.

References

1. Maryann DiBerto, J. Ladd Greeno, and Gilbert S. Hedstrom, *Environmental Auditing: Fundamentals and Techniques*, Cambridge, Mass.: Arthur D. Little, 1987.
2. Christopher B. Hunt and Ellen R. Auster, "Proactive Environmental Management: Avoiding the Toxic Trap," *Sloan Management Review*, Winter 1990.

7

Auditing for Sustainability: The Philosophy and Practice of The Body Shop International

David Wheeler

General Manager, Environment, Health, and Safety,
The Body Shop International

It may seem a rather commonplace proposition, but the most important question to address before commencing an environmental management or audit program is Why do it? Indeed, the justifications for auditing are listed so frequently by commentators and experts[1,2,3] that practitioners may be forgiven for completely ignoring them. Driving forces for environmental audits—i.e., stakeholder pressures, legal obligations, opportunities to save money (or avoid liability), and the need to identify market opportunities—are numerous and well-established in the literature.

But there is a danger here. Familiarity may in due course lead to complacency. Responding to stakeholders could become simply a convenient mechanism for ensuring a quiet life for corporate executives. Complying with legal obligations and avoiding liabilities are routine defensive functions that are never likely to excite anyone. Even saving money loses its attraction when the law of diminishing returns kicks in and managers realize that they have made all the investments they can in energy efficiency and waste minimization.

That just leaves market positioning—a mechanism for responding to consumer pressure and maximizing commercial opportunities.[4] This technique is notorious for its association with the exploitation of social and environmental concerns held by the ordinary citizen. One hopes that it will in due course lose its appeal—especially for companies committed to high ethical standards. So why should forward-thinking companies do environmental auditing?

The Body Shop is a manufacturer and retailer of naturally based skin and hair care preparations and cosmetics. Via subsidiaries and franchises, The Body Shop trades in more than 40 countries, in 19 languages and in more than 1000 retail outlets worldwide. The company is not a significant source of pollutants or a major user of energy and raw materials. Manufacturing at its principal site in West Sussex, England, produces no airborne emissions and less than 30 m^3 of wastewater per day. Energy consumption by our entire U.K. operation (including distribution and retail outlets) has been estimated at only 0.003 percent of total U.K. emissions of CO_2 (around 18,000 tonnes per annum). And the use of plastics in packaging represents only around 0.01 percent of total European Union (EU) demand.[5]

Nevertheless, environmental auditing has a very high profile in the organization. On the main site, it involves all staff and managers in continuous data collection, frequent reviews of priorities and targets (on a department-by-department basis), and an annual process of public reporting of results. The process extends to all retail outlets in the United Kingdom, and, during 1993–94, reviews were completed in all subsidiaries and overseas franchise operations.

The reason why all of this effort is expended has almost nothing to do with external pressure—legal or otherwise. Where they exist, such pressures are relatively undemanding. Environmental auditing at The Body Shop has little to do with cost saving—most efficiencies have already been gained. And the activity is certainly not driven by commercial considerations. The Body Shop completely rejects the notion of green marketing or any other form of ecological opportunism.[6,7]

The overriding factor for The Body Shop is the perception of a moral obligation to drive toward sustainability in business.[8] It is impossible to measure progress toward this ideal without a systematic process of data gathering and public reporting. Hence, auditing activities are considered absolutely essential to the company's long-term mission to become a truly sustainable operation— effectively to replace as many of the planet's resources as are utilized. The Body Shop wants to play a full part in handing on a safer and more equitable world to future generations.

This goal, sustainability in business, is how The Body Shop relates to the broader concept of sustainable development—a term first popularized by the Brundtland Commission[9] and which was subsequently adopted as the rallying cry of the Earth Summit in Rio de Janeiro in June 1992.[10] It is the strong belief of

The Body Shop that the moral burden of achieving sustainability in business will become the principle driving force behind environmental auditing and environmental management systems in the 1990s. The evidence for this is growing almost daily.

Available Methods

Three trends may be discerned in environmental auditing practice in recent years. They have been advanced by different constituencies, but there is considerable overlap in methodology. The first approach, and the one with the longest track record in industry, is that based on safety auditing.

The recognition that industrial processes that go wrong can lead to injury to both workers and the surrounding community has been understood since the advent of the chemical industry. Today managing and reducing the risk of such occurrences has become standard practice in responsible companies dealing with oil, petrochemicals, fine chemicals, mining, and power generation. So with the development of environmental concerns in the 1960s and 1970s, it was natural for companies in those industries to extend their concern for the local community to a broader concern for the local environment. Even in the late 1970s and 1980s disasters like Love Canal, Seveso, Bhopal, Sandoz, Chernobyl, and *Exxon Valdez* only served to confirm in the minds of the public that the environment was not necessarily safe in the hands of industry and that risks needed to be brought more fully under public control.[11]

In response, those industries concerned developed internal procedures that were aimed at measuring and reducing risks to the health and safety of workers as well as to neighboring communities and their local environments. Environment, health, and safety (EHS) auditing was the technique, and in the chemical industry it eventually became known as "Responsible Care®." It was all about measuring management effectiveness, ensuring legal compliance, and reducing risks and liabilities. It may be noted in this context that the cost of the Seveso disaster has been estimated to be no less than 15,000 times what it would have cost to prevent.[12] Meanwhile, the costs of Bhopal and Chernobyl are probably incalculable. Thus, environmental and safety audits were no more than common sense for high-risk businesses wishing to protect their bottom lines.

The best early practitioners of environment and safety auditing were in the United States and northern Europe. But via industry associations and consultancies, best practice gradually spread worldwide—though typically not as effectively to the subsidiaries of transnational companies operating in countries where regulations and standards were less strict. And of course, safety auditing in Eastern Europe and in the states of the former Soviet Union still leaves something to be desired.

A second and more recent trend in environmental auditing was that associated with the public sector (especially local government), service organizations (including retailers), transport companies, and light manufacturers.[13,14,15,16,17]

These organizations were not necessarily associated with high-risk working environments. In some cases, e.g., office-based organizations, they were not even associated with significant environmental impacts—either through polluting emissions or the consumption of raw materials.

Thus the second-wave environmental audits tended to focus on management systems: issues of policy, internal resources, purchasing, product or service design, communications, and education. Of course, questions like impacts on air and water, energy and waste management, and legal compliance were still examined. But in many cases greater emphasis was placed on the effectiveness of systems to manage and control a diverse range of issues of environmental consequence. Eventually, this management-systems approach led to proposals for the development of frameworks for comprehensive audits at national, European, and international levels.[18,19,20] A clear indication of this trend came in late 1992 when the European Union renamed their "Eco-Audit" Regulation as "Eco-Management and Audit."

The original mandatory directive had been aimed at major polluters and manufacturing industry. Sadly, following considerable lobbying by industry, the regulation became voluntary, and much of the onus on industry was dissipated.[21] This was further reflected by a proposal from the U.K. government that the scope of the regulation be formally extended beyond industry to the service sector.

Latterly, a third type of environmental audit system (and one that is still nascent) has emerged. It is advocated by ecologists, academic commentators, and a limited number of progressive commercial organizations. This is the process of auditing for sustainability. It is a holistic approach predicated on a clear world view and an understanding of the need for a paradigm shift in business culture.[22,23,24]

Organizations auditing for sustainability are especially keen to link environmental performance to wider issues of global ecology. Thus energy efficiency should be focused on the need to minimize NO_x, SO_x, and CO_2 emissions and to avoid nuclear waste. Waste minimization, reuse, and recycling should be driven by the need to conserve nonrenewable resources. Product design should prioritize the use of renewable resources. Processes should be designed to ensure zero toxic emissions for precautionary reasons. Sourcing of raw materials should have no negative impacts on global biodiversity, endangered habitats, animal welfare, or the rights of indigenous peoples.

In many cases ecological performance can be measured quantitatively. Indicators of sustainability may be developed that emphasize their relevance both to staff and to external audiences. In due course *sustainability indicators* may be sufficiently comparable within and between industries to allow establishment of best-practice benchmarks.

It is further proposed by some that environmental resource inputs and outputs to and from an organization may be amenable to full-cost accounting.[25] True costs of environmental impacts (frequently externalized by industry) could then be included in annual reports and financial statements of both private and public sector corporations. One Dutch company has already attempted to do this,[26] although the effort raised as many questions as answers.

At this stage the full-cost accounting option remains speculative and highly controversial. There are many ecologists and environmental managers who doubt both the wisdom and practicality of attempting to reconcile all ecological impacts with conventional financial indicators—whether in the accounts of national economies or commercial enterprises.[27] The Body Shop shares this concern. First, it is not possible to make realistic financial estimates of the intrinsic value of numerous important ecological assets, e.g., unique habitats, endangered species, the ozone layer, or the homelands of indigenous peoples. Secondly, it is not possible to predict what value will be placed on these assets by future generations: What price the last blue whale?

Thus it is impossible to envisage all key indices of sustainability emerging from cost accountancy. For this reason, The Body Shop's approach to environmental management and auditing and reporting is never likely to be more than passing attention to financial indicators. However, it will devote increasing efforts to the definition of ecological impacts with respect to true indices of sustainability.

The rest of this chapter is devoted to a description of The Body Shop's internal environmental management system, its methodology for auditing, and its approach to public disclosure of environmental information.

Structure, Responsibilities, and Motivation of the Environmental Management System

The Body Shop maintains a very decentralized system of environmental management. A corporate team of environment, health, and safety specialists acts as a central resource for networks of environmental advisers and coordinators in headquarters departments, subsidiaries, retail outlets, and international markets. Environmental advisers and coordinators are usually part-time, fulfilling their role in environmental communications and auditing alongside normal duties.

In headquarters departments, there is approximately one adviser for every 20 staff members. Every individual department has at least one adviser as does every retail outlet in the U.K. Subsidiaries have full-time environmental coordinators. Internationally, each country has one or more environmental coordinators. A number of these are full-time positions. In every case environmental coordinators and advisers have specific environmental training—usually via a combination of in-service training, annual educational events, and monthly or bimonthly updates, i.e., newsletters and training notes. In addition there are a wide range of general sources of environmental information circulating with The Body Shop. These include videos, slide sets, notes, leaflets, broadsheets, and booklets. Regular internal communications systems, e.g., newsletters and weekly news videos, often carry environmental issues of internal or external significance.

The corporate EHS team serves the networks of EHS coordinators and environmental advisers, ensures that policies and guidelines are disseminated effectively, coordinates specific terms of reference for auditing and environmental management, and receives and collates data on environmental indicators relevant to each part of the operation. The corporate EHS department also coordinates environmental policy development in direct contact with the main board of The Body Shop and liaises with senior managers on a regular basis to ensure constant support for the environmental management system. On an annual basis (during the formal audit process) liaison with the board and senior managers includes the setting of targets and objectives at corporate and departmental levels.

In line with the flexible and nonhierarchical management structure of The Body Shop, it is common for environmental advisers, managers, and representatives of the EHS department to convene ad hoc task groups to address issues of broad interest. Recently these task groups have included waste management, energy efficiency, product stewardship/life-cycle assessment, environmental statistics, environmental emergencies, and environmental communications.

The terms of reference for environmental advisers at headquarters level are agreed between the individuals concerned, their managers, and the EHS department. They include reference to all auditing duties as well as any responsibilities specific to their departments. Ad hoc task groups may also have terms of reference in order to help define the scope of their activities. However, written responsibilities are not seen as a dominant influence, and it is typical for new developments or responsibilities to be adopted on an informal basis. Because of this organic development of the environmental management system, it is usual for terms of reference to be updated annually.

The issues included in terms of reference reflect a comprehensive list of potential environmental concerns. This list is consistent with relevant regulatory requirements for environmental auditing (for example, the list of "Issues to be Covered" in Annex 1, Sec. B, of the European Union Eco-Management and Audit Regulation). Other environmental concerns may be added to an individual's or a task force's terms of reference if deemed appropriate from a general appraisal of the potential environmental impacts involved.

In summary, then, EHS at The Body Shop is decentralized, it is based on key networks of environmental advisers and representatives, and it is geared to a very active two-way flow of communications. Board-level and senior management support is delivered via the coordinating activities of the corporate EHS department. Informality and creativity is promoted via ad hoc groupings, regular meetings and training inputs, and, where appropriate, active participation in external campaigns.

As noted above, the primary purpose of environmental auditing at The Body Shop is to measure progress toward environmental sustainability. This is a strong motivating force for staff involved in the process. But another strong motivating force is the opportunity to contribute to more specific campaigns that involve The Body Shop in a general mobilization of staff effort.

Since 1985, The Body Shop has collaborated with environmental organizations and others on a wide variety of single-issue campaigns. These have included opposition to commercial whaling; demands for action on ozone depletion, global warming, and acid rain; petitions against rainforest destruction and nuclear waste; and promotion of the reuse and recycling of postconsumer waste. In each case considerable resources have been devoted to the campaigns. Shop windows have run posters, members have been recruited for environmental organizations, cash has been collected to help in cleanups from pollution incidents, thousands of letters have been written, and millions of signatures have been collected.

The knowledge that The Body Shop is prepared to campaign as a corporation at public and political levels has a profoundly energizing effect on the environmental management and auditing process. And constant contact between the corporate EHS department and environmental campaign groups ensures that The Body Shop remains at the cutting edge of best environmental practice in industry.

Audit Methods at The Body Shop

In recent years, The Body Shop has formalized its system of environmental auditing. A preliminary review was undertaken in 1989, when the organization had less than 1000 U.K. employees. This review was useful insofar as it recommended a number of qualitative improvements in policies and procedures. However, in 1991 the company decided that its increasing size and development required the inception of a more quantitative environmental management and audit system. The networks of environmental advisers at headquarters, retail, and international levels were formalized, and a member of the corporate EHS team was allocated to each. A full-time consultant was appointed to assist the materials purchasing function in the auditing and accreditation of raw materials suppliers. Meanwhile, a regular program of environmental reviews and audits was instituted in line with the requirements of the (then) draft E.U. Eco-Audit Regulation.

Because the Eco-Management and Audit Regulation requires a site-specific assessment of environmental impacts, the first environmental audit of The Body Shop's operations concentrated on the headquarters, manufacturing, warehousing, and distribution facilities located in Littlehampton, England. The site employs more than 600 people and is the central location for The Body Shop's international business. Audits were also carried out in two-thirds of the U.K. shops, and the results of these were summarized for inclusion in the main audit report.

The main audit was based on face-to-face interviews and data collection at every level of the organization. Departmental environmental advisers were interviewed individually to ensure that agreed terms of reference were being met. These interviews were followed by discussions with managers to confirm their familiarity with environmental policies relevant to their department or division. These discussions also resulted in the agreement of specific targets for the future. Finally, a series of internal discussions involving board members and senior managers provided the impetus for companywide targets ranging from a commitment to eventual energy self-sufficiency (using wind power) to the phase-out of polyvinyl chloride (PVC) in products and packaging.

During this process of interviews and target setting, a series of specialist consultancy inputs was commissioned. These covered energy efficiency/self-sufficiency, education and training, staff attitudes, and audit verification. The latter was (and remains) a requirement of the EU Eco-Management and Audit Regulation and is intended to provide independent corroboration that relevant issues have been addressed and that impacts have been properly recorded and expressed.

The final part of the process involved the board of directors endorsing the environmental statement and agreeing to its publication in full.[16] At no point did a board member or senior manager request the amendment of a conclusion or commitment except to strengthen it or request a shorter time scale for its implementation.

One of the most important targets adopted with the publication of The Body Shop's first comprehensive environmental statement was to ensure that environmental representatives' audits or reviews were conducted in every one of the company's subsidiaries and overseas franchises within 2 years. This commitment has resulted in further development and refinement of the audit system.

In order to establish uniformity in environmental audit procedures in more than 40 countries around the world, a working group of international environmental representatives was convened. With input from leading international environmental consultancies, a process of environmental audits and reviews was designed and adopted. It was clear that a system of self-assessment checklists was the most efficient and appropriate way to implement audit procedures worldwide (Fig. 7-1).

Audit checklists have to cover markets ranging from a single retail outlet (in countries like Antigua and Bahrain) to those with more than 100 (e.g., Canada, the United States, and the United Kingdom). In the larger markets, manufacturing facilities are also involved, so checklists need to reflect this. Similarly, nearly all markets have warehousing and distribution functions, hence the need to include assessment of these operations in most cases. The checklists are carefully designed to reflect global ecological concerns and the strong corporate commitment to environmental excellence.

The checklists all involve "closed" questions to maximize objectivity and allow the allocation of scores in each category of questions. There are four sections dealing with every environmental or management issue that could conceivably arise in The Body Shop's operations anywhere in the world. The audit process described above was repeated in 1992–93 and 1993–94. Annual indepen-

INTERNATIONAL ENVIRONMENTAL
AUDIT CHECKLISTS 1994/5

SECTION A
(All staff interviewed to be asked one checklist from this section, according to responsibility)

I   Individual (General Staff)
II   Managers (Senior/General Managers Non-Retail)
III   Directors
IV   Departmental Environmental Advisors (DEAs)
V   Environmental Coordinator
VI   BDMs/Area Managers
VII Retail Managers

SECTION B - Head Office/Warehousing Sites
(Managers/Site Service staff/ Environmental Co-ordinator)

I   Purchasing (Office Furniture, Stationary etc)
II   Sundries Purchasing
III   Waste Management (Offices)
IV   Waste Management (Warehouse/Site Services)
V   Energy
VI   Site Management
VII Transport: Company Cars
VIII Transport: Distribution Vehicles

SECTION C - Retail Support from Head Office
(Retail Managers/ Area Managers/Environmental Co-ordinator)

I   Purchasing (Shops)
II   Waste Management (Shops)
III   Energy (Shops)
IV   Site Management (Shops)

SECTION D - Manufacturing Sites

I   Manufacturing and Waste Water
II   Air, Noise and Odour
III   Waste Management
IV   Environmental Emergency Planning

Figure 7-1. List of self-assessment checklists. See Fig. 7-6 for all of the checklists.

dently verified public statements of The Body Shop International's environmental performance are now an established part of the company's activities.[28] It is envisaged that independent verification of the worldwide audit program will be feasible by 1994–95.

Product Stewardship

One area of environmental audit and verification that is largely confined to head office activities is the assessment of suppliers and their environmental performance. It is one thing to ensure good environmental performance of operations within the direct control of an organization. It is quite another to ensure that no unacceptable indirect impacts are occurring in the procurement and supply chain for raw materials and finished goods used by that organization.

In order to ensure due environmental diligence in The Body Shop's supply chain, the company has set up a program of product stewardship that takes a life-cycle, or "cradle-to-grave," approach to the sourcing, manufacture, and use of products. The program has four elements.

Risk Assessment

The first component of product stewardship is risk assessment. Risk assessment involves consideration of raw materials used in greatest quantity and those where there may be ecotoxicity or biodegradability issues to address. The Body Shop has always taken great care in the specification of raw materials in order to minimize any unwanted impacts of this nature. However, a formal review was undertaken in 1992 in order to confirm the validity of previous policies and assumptions. This process is now incorporated in the new-product development program.

Supplier Accreditation

The second element of The Body Shop's program for reducing environmental impacts associated with its products is an environmental accreditation scheme for suppliers. The scheme, run centrally by the materials purchasing function, involves a star-rating system (0 to 5 stars), under which suppliers are graded according to predetermined criteria. Positive responses to questions on a general first-stage questionnaire (Fig. 7-2) relating to environmental management and auditing enable suppliers to achieve one or two stars, depending on their overall score. This gains the supplier a certificate and encouragement to do more. Two-star suppliers are requested to complete an additional, more detailed, questionnaire (Fig. 7-3) relating to compliance, emissions, and waste management. Satisfactory progress with this questionnaire, and provision of other more detailed information, leads to the award of a three-star certificate or higher.

THE BODY SHOP ENVIRONMENTAL AWARD SCHEME

CHECK LIST I

◊ CONFIDENTIAL ◊

COMPANY NAME

NAME OF RESPONDENT

POSITION HELD

DESCRIPTION OF COMPANY BUSINESS

Please complete the following check-list, using additional sheets If neccessary.

Figure 7-2. First-stage questionnaire.

THE ENVIRONMENTAL AWARD SCHEME

	YES	NO

1 Would you be willing to attend a Body Shop
   Seminar on Environmental Management ?

2i Does your Company have a formal Environmental
    Management System (EMS) ?

2ii If no, would you be willing to adopt an Environmental
     Management System based on British, European, or
     International standards ?

3 Would you be willing to provide environmental
   information for the Body Shop Life Cycle
   Assessment (LCA) scheme ?

4 Do you have an Environmental Policy ?
   (If so please include a copy)

5i Have you carried out an Environmenal Audit ?
    (If so please include a copy of the audit statement)

5ii If yes, are the audit report findings available to the
     general public ?

Figure 7-2. (*Continued*)

THE ENVIRONMENTAL AWARD SCHEME

	YES	NO

6i Does your company have a Board Member / Senior Manager responsible for Environmental Affairs ?
(If yes, please provide name and position)

6ii If yes, approximately how much time would be allocated per week to oversee environmental affairs ?

	YES	NO

7 Does your Company have a formal system of quality assurance based on British, European or International (ISO 9000) standards ?

8 Please list 3 Pro-Environment activities or initiatives undertaken recently that you were most proud of (e.g, recycling, funding of environmental groups etc).

Please return your completed form to:

SCORING:

Figure 7-2. (*Continued*)

CHECK-LIST I SCORING

1) Would you be willing to attend a Body Shop Seminar on
Environmental Management?   Yes = 2 points

2a) Does your Company have a formal Environmental
Management System (EMS)?   Yes = 3 points

   b) If no, would you be willing to adopt an Environmental
Management System (EMS) eg British Standards Institution
scheme?   Yes = 1 points

3) Would you be willing to provide Environmental information
for The Body Shop Life Cycle Analysis (LCA) system?   Yes = 5 points

4) Do you have an Environmental Policy?   Yes = 2 points

5a) Have you carried out an Environmental Audit?   Yes = 4 points

   b) If yes, are the audit report findings available to the
general public?   Yes = 2 points

6a) Does your Company have a board member / senior   = 3 points between
Manager responsible for Environmental Affairs? (If yes   6a and 6b
please provide name and position).

   b) If yes, approximately how much time would be allocated
per week to oversee environmental affairs?

7) Does your Company have a formal system of quality
assurance, eg BS 5750 or European equivalent?   Yes = 1 point

8) Please list 3 Pro-Environmental activities / initiative
undertaken recently that you were most proud of (ie Recycling
funding etc)   = 3 points

Figure 7-2. (*Continued*)

DATE SENT: ....

DATE RECEIVED: ....

The BODY SHOP
Skin & Hair Care Preparations

THE ENVIRONMENTAL AWARD SCHEME

TWO STAR SUPPLIERS

CHECK LIST II

◊ CONFIDENTIAL ◊

COMPANY NAME

```
┌─────────────────────────────────────────┐
│                                         │
└─────────────────────────────────────────┘
```

NAME OF RESPONDENT

```
┌─────────────────────────────────────────┐
│                                         │
└─────────────────────────────────────────┘
```

POSITION

```
┌─────────────────────────────────────────┐
│                                         │
└─────────────────────────────────────────┘
```

Please complete the following checklist, using additional if necessary.

Figure 7-3. Checklist for two-star suppliers.

THE ENVIRONMENTAL AWARD SCHEME

◊ Please provide a summary of your emissions inventory
(to air, land, and water) with percentage compliance with
regulatory requirements noted for each emission:

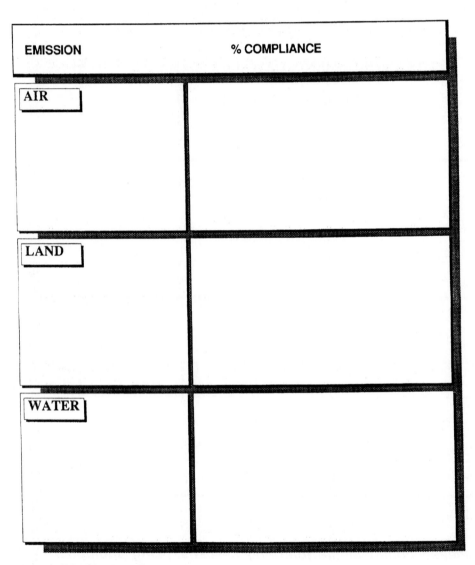

EMISSION	% COMPLIANCE
AIR	
LAND	
WATER	

Figure 7-3. (*Continued*)

THE ENVIRONMENTAL AWARD SCHEME

◊ Please provide a summary of your solid wastes inventory
with quantities diverted to re-use and recycling stated as
percentages of total arisings:

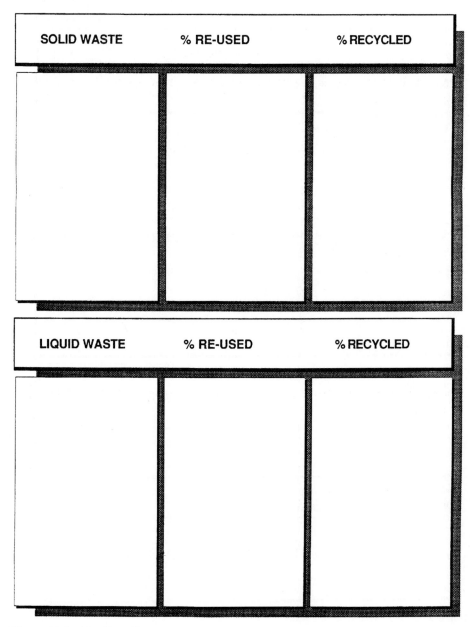

Figure 7-3. *(Continued)*

THE ENVIRONMENTAL AWARD SCHEME

◊ **Do you employ any of the following as components of raw materials to The Body Shop.**

* *(Please specify)*

	YES	NO
i Mineral - based components	☐	☐

☐

ii Petroleum - based components   ☐ ☐

☐

iii Animal derived materials   ☐ ☐

☐

iv Non-biodegradable materials (defined by OECD criteria)   ☐ ☐

☐

v Materials appearing on list of potentially toxic or ozone depleting chemicals (see attachment)   ☐ ☐

☐

Figure 7-3. (*Continued*)

THE ENVIRONMENTAL AWARD SCHEME

◊ **Please provide details of any prosecutions or legal actions taken, or pending, in respect of wastes or emissions.**

PLEASE RETURN YOUR COMPLETED FORM TO:

Figure 7-3. (*Continued*)

Life-Cycle Assessment

The third part of product stewardship is life-cycle assessment (LCA). This process is necessarily complex and detailed. It requires the active cooperation of suppliers, and satisfactory collaboration is a prerequisite for progression to higher levels in the accreditation scheme. LCA is a science that is in its infancy. It is the subject of technical and ethical debate, and it is doubtful whether a standardized methodology acceptable to all parties will emerge in the near future.[29]

SUPPLIER ACCREDITATION -

5 STAR SUPPLIER AWARD SCHEME

__1 STAR CRITERIA__ - To achieve 1 star status, suppliers will need to demonstrate clear evidence of a willingness to improve environmental performance beyond simple compliance with regulatory obligations. In addition, there should be no unacceptable or intractable environmental impacts associated with processes or raw materials in The Body Shop's supply chain.

- A minimum of 11 points needed on answers to checklist 1

__2 STAR CRITERIA__ - A minimum of 17 points needed on answers to checklist 1 (a supplementary checklist will be issued at this stage to cover emissions and raw materials)

__3 STAR CRITERIA__ As 2 Star, plus

- Provision of summary emissions and waste information (checklist 2)

- submission of satisfactory Life Cycle Analysis information

__4 STAR CRITERIA__ As 3 Star, plus

- Published Environmental Audits show that all identified impacts are being addressed to the satisfaction of an independent external authority

- Auditing is done periodically (in line with national or international regulations)

__5 STAR CRITERIA__ As 4 Star, plus

- All significant environmental problems overcome

- Systems to evaluate supplier performance in place, with evidence that BSI's supplier chain is environmentally secure

- Will be registered for EC Eco-audit regulation/or registered for equivalent environmental management systems standard (BS7750 or ISO equivalent)

Figure 7-3. (*Continued*)

The Body Shop is especially concerned that LCA may become a nonecological activity. It is clearly in the interest of suppliers of bulk commodities to draw the boundaries of LCA quite tightly in order to focus attention on those factors that are most easily controlled: wastes, polluting emissions, and energy consumption. However, for The Body Shop, a full ecological consideration of product life cycles also has to take into account the impact of raw material procurement on biodiversity, endangered habitats, human and animal welfare, and nonrenewable resources (Fig. 7-4). Ignoring these issues may be convenient (especially to

Product Life Cycle Assessment

Environmental Questionnaire covering
Manufacture, Use and Disposal of Products

Product: [   ]

*Product Category: [   ]

Contract Manufacturer: (If Applicable) [   ]

1. Product Concept

(i) Regarding the product formulation, what is the percentage composition for the following raw material categories:

       (a) naturally-occurring,# (i.e. undergoing no alteration to chemical structure) [   ]

       (b) naturally-derived materials? (i.e. based on a natural feedstock) [   ]

       (c) synthetic [   ]

	YES	NO
(ii) Are any ingredients currently restricted by the food and/or pharmaceutical industries in any country where The Body Shop trades?	☐	☐
(iii) (a) Does the formulation require a preservative in addition to those needed in the supply of raw ingredients?	☐	☐
(b) Can adequate preservation be achieved with naturally-occurring or naturally-derived materials?	☐	☐
(c) If synthetic preservatives are required, are they present at less than 1.5%?	☐	☐

*Enter one category from those listed in the Product Information Manual,
e.g. soap, facial bar, hair shampoo, hair conditioner, etc....
 # Exclude water

Figure 7-4. Life-cycle assessment questionnaire for product manufacture, use, and disposal.

YES   NO

*(iv) Is there information accompanying the product or normally available in the
shop which can educate the customer or raise their environmental awareness
about how:*

    *(a) the product reduces waste and resource consumption compared to others in the
same category*   ☐   ☐

    (b) she/he can reduce waste, eg correct dosage, return container or recycle locally, etc   ☐   ☐

    *(c) the product avoids the use of non-renewable resources, eg by using vegetable
feedstocks, savings in the consumption of fossil fuels, etc*   ☐   ☐

    *(d) material(s) have been procured through fair trading channels with sufficient
details about product sources and business practices to enable the public to assess
the social and financial effectiveness of the trade.*   ☐   ☐

    *(e) natural habitats and their living components are protected through
sustainable methods of extraction*   ☐   ☐

    *(f) she/he can act on their environmental and social responsibilities/concerns, eg by
leading a less wasteful and polluting lifestyle in ways that don't necessarily relate
to the use of the product, by lending support to an environmental/social cause or
campaign, etc*   ☐   ☐

YES   NO
or DON'T
KNOW

*(v) Has the product been designed around at least one ingredient for which fair trade
links have been, or could be, established where:*

    (a) such ingredients are renewable or inexhaustible.   ☐   ☐

    *(b) such ingredients are cultivated or extracted from an environment which can
sustain the traditional manner and level of exploitation.*   ☐   ☐

    *(c) projected levels of demand for such ingredients compare favourably with
methods and levels of production positively identified as being environmentally
sustainable*   ☐   ☐

    *(d) such ingredients are purchased from a community based organisation with an
established mechanism for equitable distribution of revenue and involvment of
members in decision making processes*   ☐   ☐

Figure 7-4. (Continued)

	YES	NO *or* DON'T KNOW
(e) no restrictions are posed on the social customs or mores of the community	☐	☐
(f) trade does not deny or endanger indigenous methods or knowledge	☐	☐
(g) such ingredients can be produced without risks to human health and safety	☐	☐
(h) the producer organisation has control over the levels of production, the quality and the price of the product	☐	☐
(i) increased control of the production has been or will be made possible by the provision of appropriate equipment and training at reasonable prices and without obligation	☐	☐
(j) bridging finance is available in the form of a repayment against confirmed purchase contracts for up to 60% of the contract value	☐	☐
(k) a fair price is paid, which reflects cost of production and the quality of the product plus a margin for investment and development	☐	☐
(l) a stable trading relationship on the basis of quality, continuity and mutual support are guaranteed	☐	☐
(m) the producer organisation has suffered because of poverty, discriminatory processes and restrictions on trade	☐	☐
(n) the producer organisation is committed to supporting the particpation and needs of working people and especially of women and of racial and social groups which suffer from discrimination and exploitation	☐	☐
(o) the producer organisation encourages equal rewards for women and men	☐	☐
(vi) Are any ingredients organically produced (with substantiation from the Soil Association or other competent body)?	☐	☐
(vii) Does the product offer environmental advantages through savings on packaging and the unnecessary shipping of water compared to earlier products in the same category?	☐	☐

Figure 7-4. (Continued)

2. *Manufacture & Filling*

YES   NO

(i) Is the product manufactured and filled in-house?   ☐ ☐

(ii) How many ingredients in the product?   ☐

(iii) Does the product require any of the following treatments on a routine basis to ensure satisfactory stability: ☐ ☐

☐ *(a) gamma-irradiation,*   ☐ *(b) ethylene oxide,*

☐ *(c) electron beaming,*   ☐ *(d) heat treatment?*

(iv) Is the product a cold mix formulation?   ☐ ☐

(v) Is filling done at ambient temperature?   ☐ ☐

(vi) Is any pre-treatment of an ingredient required before it can be mixed (eg melting of wax)?   ☐ ☐

(vii) What volume of trade effluent is generated (a) for each batch mixed   ☐

(b) for each batch filled?   ☐

(viii) What is the average chemical oxygen demand (COD) for liquid effluents from (a) manufacture   ☐ *mg/l*

(b) filling   ☐ *mg/l*

(ix) What is the average concentration of total synthetic detergents in the effluents from (a) manufacture   ☐ *mg/l*

(b) filling   ☐ *mg/l*

(x) Are anti-foaming agents, such as iso-propanol, ever required when washing out either mixing vats or filling lines?   ☐ ☐

Figure 7-4. (Continued)

3. *Packaging*

YES   *NO*

(i) Is the pack design refillable or re-usable? ☐ ☐

(ii) How many components are there in the pack design? ☐☐☐☐☐

(iii) Is any material used which does not fall into one of the categories presently sorted and recycled by BSI or by local authority waste recovery schemes? ☐ ☐

(iv) If packaging contains a plastic component, is it either polyethylene or polypropylene? ☐ ☐

(v) Are any components moulded in-house? ☐ ☐

(vi) Does any component contain post-consumer recycled material? ☐ ☐

(vii) Does any component contain waste obtained from an industrial process? ☐ ☐

(viii)   (a) Has there been any formal risk assessment of toxic substances being leached from the packaging into the product, ☐ ☐

(b) is there adequate routine monitoring for contaminants likely to enter the product in this way? ☐ ☐

(ix) Are materials of all components easily identified? ☐ ☐

(x)   (a) Are all components made out of the same material? ☐ ☐

(b) If different, can they be manually separated or otherwise remain compatible for material recycling? ☐ ☐

(xi) Are there any disposable components in the pack design? ☐ ☐

(xii) Is the container (not the lid) any colour other than 'natural' or white? (In the case of polypropylene single-walled jars, are they any colour other than 'natural', white, or black?) ☐ ☐

(xiii) When a batch of product is unfit for sale, is it easily removed from the container, allowing the container to be recycled without undue complication? ☐ ☐

Figure 7-4. (Continued)

	YES	NO

(xiv) Does the printing of containers or labels involve inks which are:

☐ *(a) solvent based,*   ☐ *(b) dried using u-v light,*   ☐ *(c) water based?*

(xv) Is a clear indication given on the packaging as to whether or not it is
(a) refillable   ☐ ☐

(b) recyclable?   ☐ ☐

(xvi) Does the product arrive at the shops with secondary packaging, ie packaging that only serves to 'dress' or present the products as part of merchandising?   ☐ ☐

4. Use of the Product

(i) Is it easy to totally empty the container of product?   ☐ ☐

(ii) What mechanism is used for dispensing the product, eg

(a) scooping out with finger or hand, a combination of gravity and squeezing,   ☐

(b) controlled application with a plug or ball,   ☐

(c) a winding mechanism or a finger-operated pump or   ☐

(d) any mechanism that uses or involves the release of a volatile organic compound (see Appendix 3)?   ☐

(iii) Does application or removal of the product require the use of a disposable accessory item, eg cotton wool, paper tissue, etc?   ☐ ☐

(iv) Does the product require a special routine involving additional products (other than soap and water) for removal?   ☐ ☐

(v) Is the product refillable?   ☐ ☐

Figure 7-4. (Continued)

5. Disposal and Fate in the Environment

	YES	NO

(i) Is there any suggestion that the product (as trade effluent), or its intermediate breakdown products, inhibits the bacterial digestion of sewage? ☐ ☐

(ii) In the event of product becoming unfit for sale, is it useful for another purpose? ☐ ☐

(iii) As part of household waste water, how long does it take for 95% of the product to be biodegraded (detoxified and mineralised)? [*days*]

(iv) As industrial waste, can the product be mixed with other solid or liquid wastes to be landfilled or incinerated without any subsequent risk to the environment or human safety and without any complication for the waste disposal company? ☐ ☐

Figure 7-4. (*Continued*)

the agrochemical, petrochemical, chemical, and mining industries), but it is not tolerable from an ecological perspective. For this reason, The Body Shop's approach to LCA attempts to cover the full range of potential ecological impacts associated with the sourcing of ingredients as well as the manufacture, distribution, and use of products.

Product and Commodity Guidance Notes

The final element of product stewardship is a very practical one. It is the provision of specific guidance notes on commodities and products used by The Body Shop in its operations worldwide (Fig. 7-5). These guidance notes are available for use by all buyers. They are simple *aides-memoires* that a buyer can refer to when renewing an order or placing new business with a supplier. The guidance notes include a blacklist of substances and processes that should be avoided in all cases.

Disclosures of Environmental Information in Audit Reports

The EC regulation on ecological management and auditing requires publication of all relevant data on environmental impacts. This places the regulation alongside measures such as the U.S. Toxic Releases Inventory—a procedure that has clearly had an enormous effect on U.S. industry and its relations with the wider community.[30] Although the EC regulation is not mandatory at the present time,[31] it is considered entirely likely that it will become binding on the most important sectors of industry in due course. This should lead to the systematic and periodic public disclosure of all relevant environmental impacts by a very large number of industries based in the EC by the late 1990s.

Product Life Cycle Assessment

Environmental Questionnaire for Suppliers of Raw Materials and/or Packaging Components.

1) Tracking Back

Please provide a flow chart to describe the production pathway for the ingredient or item. In order to cover the whole life cycle, production should be traced back to the point at which all material feedstocks are taken from the Earth's lithosphere, biosphere, oceans or atmosphere. Please include as many of the following details as possible:

   (i)   name of the company (optional) and the country in which the operation takes place making it clear which parts of the production pathway are under your company's immediate control.

   (ii)   every step or process involved in production

   (iii)   common, trade and scientific names or formulae of starting feedstocks and intermediates

2) Describing the Origin of the Base Materials

(These questions concern the site and means of extraction or cultivation of base materials (see "Explanatory Notes - Bulletin 1 for a working definition of the term base material). Please provide the following for each base material used in production:

   (i)   country or marine area

   (ii)   nearest town

   (iii)   geographical description (see "Explanatory Notes")

   (iv)   brief description of activity

Product Stewardship Programme 1.

Figure 7-5. Product life-cycle assessment questionnaire for suppliers.

	YES	NO	DON'T KNOW

(i)   *Has there been any conflict with local residents over traditional land use, access, occupancy, or ownership?* □ □ □

(ii) *Has the activity caused displacement of local people?* □ □ □

(iii) *Has the site ever been recognised (officially or by local residents) as having importance for the purposes of the conservation of biological diversity?* □ □ □

(iv) *Have there ever been complaints from local residents about the effects of the said activity on general living conditions (eg health, drinking water, dust, noise, etc), the productivity of their own food growing activities, or the neighbouring environment?* □ □ □

(v) *Were assessments of environmental and social impact made before the activity commenced?* □ □ □

(vi) *Is there a programme which routinely monitors environmental and social impact?* □ □ □

   (if the answer to either (v) or (vi) is 'yes', are there any summary reports which would be available to The Body Shop.) □ □ □

(vii) *Do these assessment exercises involve local people?* □ □ □

(viii) *Is the site of extraction or cultivation a well defined area?* □ □ □

(ix) *Does the operation involve the progressive encroachment on to new territory?* □ □ □

(x) *Does the source of this base material involve the use of others' wastes?* □ □ □

3) Methods for Extraction and/or Cultivation
Section A. Mineral-derived Base Materials

(i)   *What is the estimated life-span of this resource at present rates of global exploitation and present knowledge of reserves?* □

                                 SEA   LAND

(ii) *Does mining take place at sea or on land?* □ □

Product Stewardship Programme 2.

Figure 7-5. *(Continued)*

(iii) What method of mining is used. (If other, please specify).

☐ Open Cast   ☐ Dredging   ☐ Drilling

☐ Strip   ☐ Deep mining   ☐ Other: ....

		YES	NO
(iv)	*Is the target mineral hazardous in any way? (see Appendix 1)*	☐	☐
(v)	*Does extraction result in increased levels of exposure to ionising radiation?*	☐	☐
(vi)	*Are there other minerals found in association with the target mineral which are hazardous in any way?*	☐	☐
(vii)	*Are any hazardous materials used or produced in isolating or chemically processing the target mineral?*	☐	☐
(viii)	*Are any hazardous materials released into the aquatic environment?*	☐	☐
(ix)	*Are solid wastes left on site without any provisions for containment of hazardous materials?*	☐	☐
	(If the answer to any of the questions (iv) to (viii) inclusive is 'yes', please provide details.)	☐	☐
(x)	*Does the mining operation necessitate clearing natural vegetation or the sea bed?*	☐	☐
(xi)	*Is there any evidence to suggest that the living communities will regenerate naturally following cessation of extraction activities?*	☐	☐
(xii)	*Can spoil heaps support artificial colonies of plants?*	☐	☐

Figure 7-5. (Continued)

Section B. Plant- (or animal-) derived Base Materials   YES   NO

(i)   *Is the plant species being cultivated indigenous* to the source area? If not, is there documented evidence which concludes that introduction by humans in the last 10 years has resulted in neither the loss of biodiversity nor interfered with the ability of local people to produce food?*   ☐   ☐

(ii)   *Is production dependent on sources of seed beyond the means or access of local residents (eg patented high-yielding varieties developed through the use of genetic engineering)?*   ☐   ☐

(iii)   *Are synthetic pesticides used?*   ☐   ☐

(iv)   *Are artificial fertilisers used?*   ☐   ☐

(v)   *Does the method of cultivation involve maintaining a monoculture?*   ☐   ☐

(vi)   *Is there crop rotation?*   ☐   ☐

(vii)   *Is irrigation required that goes beyond the level traditionally practiced by local residents for food production?*   ☐   ☐

(viii)   *Does cultivation rely on an irrigation scheme made possible by capital-intensive engineering projects that involved the construction of large dams or water abstraction from subterranean reservoirs?*   ☐   ☐

(ix)   *Is cultivation taking place in a region vulnerable to soil erosion?*   ☐   ☐

* *the term 'indigenous' is defined in this scheme as not having been introduced by humans in the last 500 years*

4. Processing

(i)   *Does processing rely on material inputs from the petro-chemical industry?*   ☐   ☐

(ii)   *Does processing use any hazardous materials (see Appendix 1)?*   ☐   ☐

Product Stewardship Programme 4.

Figure 7-5. (*Continued*)

(iii)   *Are any ozone depleting chemicals used (see Appendix 2)?*   ☐ ☐

iv)   *Does processing involve the use of volatile organic compounds (see Appendix 3)?*   ☐ ☐

(v)   *If the answer to (iii) and/or (iv) is 'yes', are there systems in place to prevent and monitor leaks.*   ☐ ☐

(vi)   *How many individual processes are required to produce the raw ingredient? (see Explanatory Notes)* | No. |

(vii)   *Does the manufacture of the raw ingredient involve the use of gamma-radiation?*   ☐ ☐

(viii)   *Does the raw ingredient in its finished form need to be stabilised by:*

☐ *gamma-irradiation*   ☐ *electron beaming,*   ☐ *heat treatment*

☐ *treatment with ethylene oxide,*   ☐ *the addition of a preservative made known to The Body Shop?*

(ix)   *Does the ingredient need a specialised form of packaging (beyond simple physical containment) to prevent spoilage, or to reduce risks to human safety during transit?*   ☐ ☐

(x)   *Has a formal risk assessment exercise been carried out for toxic or pathogenic contaminants which could be present in the ingredient?*   ☐ ☐

(xi)   *Is there adequate routine monitoring for any toxic or pathogenic contaminants likely to be present in the ingredient?*   ☐ ☐

THE BODY SHOP

Product Stewardship Programme 5.

Figure 7-5. *(Continued)*

5. Resource Consumption

Please list the resources consumed per kilogramme of ingredient produced using a table in the manner shown below. If complete quantitative information for a given process is simply not available, please provide as detailed an inventory of inputs as possible for the individual process(es) identified in section 1 (iv).

Process	(1) Energy (MJ per Kg)	(2) Efficiency of power generation %	Nuclear Contribution %	Renewable Contribution %	(3) Fossil carbon Feedstock (kg/kg)	(4) Fresh Water (kg/kg)

(1) Energy
Please distinguish between the energy which is externally applied to "power" a process, such as heat or electricity, and the calorific energy value of any of the material components taking part in the process. Where the energy required for the process is derived in part from the heat of reaction, please indicate what percentage of the total energy requirement is derived in this way. Where steam is used to apply heat, please ensure figures entered are for units of energy per kilogramme, not for units of mass of steam.

(2) Efficiency of power generation
Efficiency of power generation is largely a function of the country and the mix of power generation methods. Where electrical power is tapped from the national grid, a national average should be used on the relative contributions from coal, natural gas, oil, nuclear, wind, hydro-electric, biomass and other power sources. Where power is generated on site, please enter the relative details. Please do not confuse the efficiency of electrical power transformation with that of power generation. Transformation efficiency can be ignored unless it is much below 95%. Generation efficiency for a power plant on site is unlikely to exceed 80-85%; the national average for elecrical power generation is likely to be between 40 and 50%, depending on the mix of energy sources.

(3) Fossil Carbon feedstock
This should include anything derived from coal, oil, natural gas, bitumen, orimulsion, etc. The production of ethylene oxide, for example, required in the manufacture of some detergents, should be entered as so many kg of the particular fossil carbon per kg of the final raw material product, in this case a synthetic detergent.

(4) Fresh Water
Water consumption figures should include feedstock and water needed for cleaning/washing, deodorising (particularly for oils and butters), cooling and steam generation. It is appreciated that steam heating systems do not essentially "consume" water, but there can nevertheless be leaks and pressure release losses. Please include any fresh water used to dilute trade effluents in order to satisfy discharge consents.

Product Stewardship Programme 6.

Figure 7-5. (*Continued*)

6. *Waste Generation*

YES   NO

(i)   Does production and/or processing yield any hazardous materials
(see Appendix 1), ozone depleting chemicals (see Appendix 2), or any
volatile organic compounds (see Appendix 3)?

☐   ☐

(ii)   For hazardous wastes that require treatment and/or disposal, is this carried out

☐ on company premises   ☐ by another company licensed to handle
and dispose of hazardous wastes

☐ in the country
where generated   ☐ exported to another country

iii)   What is the preferred method of disposing of hazardous waste:

☐ chemical treatment,   ☐ landfill,   ☐ incineration,

☐ discharge to
environment   ☐ other (please specify)?

(iv)   Is there a set of environmental criteria used to select and monitor waste
collection/disposal contractors which address the 'Duty of Care' principle
in the Environmental Protection Act 1990, or equivalent?

☐   ☐

(v)   For organic wastes discharged to the aquatic environment, how long does it
take to achieve 95% biodegradation? (see Explanatory Notes for working
definition of the term "biodegradation").

☐

(vi)   Is there any evidence to suggest that wastes, or intermediate products of
biodegradation, are toxic to a municipal sewage treatment works?

☐   ☐

(vii)   To satisfy consent limits on liquid effluents, does the manufacturer need
any waste water treatment facilities?

☐   ☐

(viii) (a) Do gaseous emissions occuring along the productionpathway contain any acidic
components (eg oxides or other compounds of S, N, P, F, Cl, Br, etc]?

☐   ☐

(b) Is there any equipment fitted to remove them ?

☐   ☐

(ix)   Have the environmental risks of dust been assessed and has the fitting of equipment
to control such releases been considered?

☐   ☐

Product Stewardship Programme 7.

Figure 7-5. (Continued)

YES   NO

☐ ☐

(x)   Does the re-use or recycling of wastes feature anywhere in the production of the raw ingredient?

(xi) For each step in its production, please list the quantities (in kg) of each waste material generated per kilogramme of the final raw ingredient produced, indicating the percentage discharged to the environment. Please present the information in the following manner.

Process	Waste material* or heat	Quantity (kg per kg) or (MJ per kg)	Percentage discharged to environment %

* Any waste materials generated at a rate of less than 0.0005 kg per kilogramme of final raw ingredient produced will be excluded from this exercise.

7. *Distribution*

(i)   How many different sites are involved in the manufacture of the ingredient?

☐

(ii) What is the total distance covered by the base material(s) in reaching your premises?

km

(iii) What container systems are used for supplying the raw ingredient to The Body Shop or contract manufacturer?

☐ *Open*   ☐ *Bulk Tanker*   ☐ *plastic/metal drum*

☐ *Other , Please specify*

(iv) Are there any systems for the recovery, re-use and/or recycling of containers in operation?

☐ ☐

THE BODY SHOP

Product Stewardship Programme 8.

Figure 7-5. (*Continued*)

Appendix 1.   Hazardous Materials

Metals and their compounds

| Antimony | Arsenic | Cadmium | Chromium (VI) | Copper |
| Lead | Mercury | Nickel | Silver | Zinc |

Miscellaneous metallic compounds

Metal carbonyls
Tributyl and triphenyl tin

Organic compounds

Aldrin	1,2-Dichloroethane	Hexachlorobutadiene
2-Amino-4-chlorophenol	Dieldrin	Hexachloroethan
Anthracene	2,4-D	Linuron
Atrazine	Demeton-O	Malathion
Azinphos-ethyl	1,4-Dichlorobenzene	Mevinphos
Azinphos-methyl	1,1-Dichloroethylene	Organophosphorus pesticides
Biphenyl	1,3-Dichloropropan-2-ol	Polyaromatic hydrocarbons
Benzene	1,3-Dichloropropene	Polychlorinated biphenyls,
Carbamate pesticides	Dimethoate	terphenyls and naphthalenes
Carbon tetrachloride	Dioxins (including 2,3,7,8-	Pyrazon (Chloridazon)
Chloroacetic acid	TCDD and congeners thereof)	Selenium and compounds
2-Chloroethanol	Ethylbenzene	Simazine 2,4,5-T
Chloroform	Endosulfan	Trichlorobenzene
Chlorophenols (such as	Endrin	1,1,1-Trichloroethane
trichlorophenol & pentachlorophenol, PCP)	Fenitrothion	Trifluralin
Chloroprene	Fenthion	Vinyl chloride
3-Chlorotoluene	Furans (and congeners	
4-Chloro-2-nitrotoluene	thereof)	
Cyanuric chloride	Hexachlorocyclohexane	
DDT (all isomers)	(all isomers)	
Dichlorvos	Hexachlorobenzene (HCB)	

Miscellaneous materials
Asbestos and other mineral fibres

Appendix 2.   Ozone depleting chemicals

Carbon tetrachloride or tetrachloromethane; Methyl chloroform or 1,1,1 -Trichloroethane;
Chlorofluorocarbons
Halons
Any man-made halogenated alkanes, eg CFCl3, CF2Cl2, CFCl2CF2Cl, and CF3Br

Appendix 3.   Volatile organic compounds

Basic definition for this exercise: Any organic compound which is present as a vapour at or above room temperature (15 degrees Celsius). The main chemical groups can be petrochemicals or biologically synthesised compounds and include:

> alkanes (eg methane), alkenes (eg ethylene), alkynes (eg acetylene or ethyne)
> alcohols (methanol, ethanol, propanol, etc)
> aldehydes (eg formaldehyde, ethanal, etc)
> chlorinated organic compounds (eg tetrachloroethylene, trichloroethane)
> cyclic organic compounds (eg dioxan, benzene, toluene, xylene, naphthalene, etc)
> dimethoxymethane
> ethers (general formula R-O-R' such as diethyl ether)
> ketones (general formula RR'C:O such as methyl ethyl ketone)
> petrol and other fuel oils
> white spirit and turpentine.

Product Stewardship Programme 9.

Figure 7-5. (Continued)

The importance of public disclosure of environmental information cannot be overstated. A number of surveys of industrial auditing practice have been undertaken in recent years. In 1991, it was estimated that only 8 percent of U.K. industries outside the chemical sector engaged in environmental audits.[32] In contrast, a survey of 1420 major companies in the U.K. in 1992 revealed that 35 to 56 percent claimed some involvement in environmental auditing. The precise figure depended on the size of company concerned.[33] But whatever the proportion of companies engaged in environmental auditing in the United Kingdom, only a very small proportion publish the type of independently verified statements of their environmental impacts required by the EU regulation.

It is salutary to note that important industrial sectors still view the idea of mandatory disclosure of environmental information with alarm.[33] In a survey conducted by management consultants Touche Ross and others,[34] two-thirds of industrial respondents in North America, Europe, and Japan admitted that companies will "hardly ever" release disadvantageous environmental information voluntarily. More than 60 percent of respondents accepted that voluntary reporting will "never be adequate" and that stricter legislation will be necessary.

It is for precisely these reasons that there is so much public interest and legislative pressure for more honesty and integrity on the part of industry. The momentum for this in Europe is probably now unstoppable.[28]

The European Union auditing regulation is only one part of a much wider set of demands for transparency in environmental reporting. For example the CERES (Valdez) Principles require such disclosure.[35] Agenda 21 encourages business and industry to "report annually on their environmental records."[36] The 5th Environmental Action Programme of the European Union requires that the public "must have access to environmentally relevant data to enable them to monitor the performance of industry and regulators alike."[37] In the United Kingdom, even the government-appointed Advisory Committee on Business and the Environment has accepted the need for mandatory environmental reporting.[38]

Given the forces now assembled in favor of mandatory environmental disclosure, it can only be a matter of time before the practice becomes commonplace. As noted above, The Body Shop has published annual comprehensive statements in line with the EU Eco-Management and Audit Regulation since 1992.

The Body Shop's position on environmental reporting is clear. Considered by external commentators to be wholehearted and comprehensive in its own approach to environmental reporting,[39,40] The Body Shop has petitioned the U.K. government to accept the inevitability of mandatory disclosure in the context of the EU auditing regulation.[21] It remains to be seen how long those transnational corporations that now resist such appeals for comprehensive environmental reporting can maintain their current unsustainable position.

Appendix

The complete set of self-assessment checklists (referred to in Fig. 7-1) is presented in Fig. 7-6.

SECTION A - I

INDIVIDUALS (General Staff)

	QUESTIONS		SCORE
1	In your Letter of Recruitment/Appointment, was any mention made of the Company's attitude to the environment?	Yes.............................3 No..............................0	
2	Was the environmental policy and responsibilities of all members of staff/departments mentioned in your induction?	Formal Induction..........3 Info Pack to Read.........2 Brief/Poss Mention.......1 NotMentioned 0	
3	If you had an idea to improve the Company's environmental performance, do you feel that Management would be receptive?	Yes.............................3 No..............................0	
4	Do environmental responsibilities feature in your job description/objectives?	Written Definition ......3 Verbal Description ......1 Nothing 0	
5	Do environmental responsibilities form part of an individual job appraisal system?	Yes.............................3 No..............................1	
6	Are you advised of Departmental environmental objectives and targets?	Yes.............................3 No..............................0	
7	Have you received any environmental training whilst with the company? Specify training received: (Notes: Time; Hours: eg Training/Awareness Raising Packs)	Yes.............................3 No..............................0	
8	In your opinion, how many people in your department take environmental issues seriously?	All................................3 Most (>50%)................2 Some(<50%)................1 None 0	
9	In your opinion, do the Managers in your department take environmental issues seriously?	All................................3 Most (>50%)................2 Some (<50%) ....1 None 0	
10	How often do you take steps to reduce the impact of your activities on the environment at work? List examples: (Notes: energy, waste, water, transport, using paper on both sides etc)	Daily.........................3 Weekly......................2 Monthly....................1 Never........................0	

Figure 7-6. Self-assessment checklists.

SECTION A - I

INDIVIDUALS (General Staff) - cont'd

	QUESTIONS		SCORE
11	How do you pick up environmental information within the (x) Company? HOW OFTEN Communication Meetings .... News letters .... Environmental Co-ordinator/Adviser .... Noticeboards .... BSTV .... Training Sessions	6 Areas......................3.0 5 Areas .... 2.5 4 Areas......................2.0 3 Areas......................1.5 2 Areas......................1.0 1 Area 0.5	
12	Do you know about the following publicly available documents? Yes No Have you read it? Green Book 1 Green Book 2 Environmental Policy	6 Areas......................3.0 5 Areas......................2.5 4 Areas......................2.0 3 Areas......................1.5 2 Areas......................1.0 1 Area 0.5	
		TOTAL SCORE _____ X 100 = ...% 36	

Figure 7-6. (*Continued*)

SECTION A - II

MANAGER'S (Senior/General Managers- Non Retail)

	QUESTIONS		SCORE
1	Have you read The Body Shop International's Corporate Environmental Policy? Comments:	Yes...............................3 No................................0	
2	Does the (x) Company have a written environmental policy signed by the Board of Directors? Comments: (Notes: Hard copy to be provided if in existence)	Yes...............................3 No................................0	
3	Are you familiar with the policy and what it means to you as a manager? Please specify those areas of the Policy that are of particular relevance to your work: (Notes: Have copy of Body Shop International's policy and discuss if the (x) company does not have own policy.	Yes...............................3 No................................0	
4	Do environmental responsibilities feature in your own job description/objectives?	Yes...............................3 No................................0	
5	Is there a main Board member responsible for the environmental management of the (x) Company? Name:	Yes...............................3 No................................0	
6	Is there a Coordinator with specific responsibility for the environmental performance within the (x) Company? Name and Job Title:	Yes...............................3 No................................0	
7	In your opinion, are the environmental responsibilities of this Coordinator clearly defined? Comments:	Formal written.............3 Informal verbal............1 No................................0	
8	Is there a network of environmental advisers at head-office? Name: (Notes: If answer to this question is no, go to question 11)	Yes...............................3 No................................0	
9	If so, please specify how often environmental matters are discussed with the environmental adviser in your department:	Weekly........................3 Fortnightly ....2 Monthly.......................1 Less than Monthly/ Not applicable 0	

Figure 7-6. (*Continued*)

SECTION A - II

MANAGER'S (Senior/General Managers Non-Retail) - cont'd

	QUESTIONS		SCORE
10	Are the environmental responsibilities of the DEA clearly defined in their job description/objectives? Comments:	Written definition........3 Verbal definition..........1 None/NA.....................0	
11	Do environmental responsibilities form part of the job appraisal system for your staff?	All staff......................3 Some staff ....2 DEA only....................1 None 0	
12	Do environmental responsibilities form part of your own job appraisal? Comments:	Yes.............................3 No...............................0	
13	How many staff (in your department,) do you think are aware of, and responsible for the environmental implications of their actions?	All...............................3 Most (>50%).................2 Some(<50%).................1 None 0	
14	Are staff (in your department) provided with training to improve their environmental awareness?	Quarterly.....................3 Annually......................2 Less than Annually.......1 Never...........................0	
15	Have you received any environmental training whilst with the company? Specify Training received (Notes: when and how long)	Yes.............................3 No...............................0	
16	In recruiting new staff, do you mention the Company's commitment to the environment?	Yes.............................3 Sometimes....................1 No................................0	
17	Is the environment covered during the induction of new staff?	Yes.............................3 Sometimes ....1 No................................0	
18	Does the (x) Company monitor environmental performance? Specify type of monitoring: (Notes: Waste production, water usage, energy usage)	Yes.............................3 No...............................0	
19	Does the (x) Company have formal environmental targets with deadlines? Comments:	Yes.............................3 No...............................0	

Figure 7-6. (*Continued*)

SECTION A - II

MANAGER'S (Senior/General Managers Non-Retail) - cont'd

	QUESTIONS		SCORE
20	State targets, if in existence, that have been set for your department and those that have been achieved. List:	No Score	
21	Is progress on departmental and company environmental performance communicated on a regular basis to all staff? Specify how : (Notes: Communication Meetings, Newsletter)	Monthly......................3 6 Monthly...................2 Annually......................1 Never..........................0	
22	Are you aware of the following environmental reports and policies which have been produced by Body Shop International? Green Books 1 & 2 Energy Policy Purchasing Policy Transport Policy Packaging Policy Waste Policy (Notes: Specify which policies and strategies are of relevance to your area. Have copies at interview)	6 Areas.......................3.0 5 Areas.......................2.5 4 Areas.......................2.0 3 Areas.......................1.5 2 Areas.......................1.0 1 Area 0.5	
23	Are you able to identify specific environmental costs and/or benefits in your area? Clarify with examples: (Notes: training, pollution, legislation, waste, energy)	Yes...............................3 No................................0	
24	Do you have a budget assigned to you specifically for environmental projects within the (x) Company?	Yes...............................3 No................................0	

TOTAL SCORE

$$\frac{\underline{}}{69} \times 100 = \text{............}\%$$

Figure 7-6. (*Continued*)

SECTION A - III

DIRECTORS

	QUESTIONS		SCORE
1	Have you read The Body Shop International's Corporate Environmental Policy? Comments:	Yes...............................3 No..............................0	
2	Does the (x) Company have a written environmental policy signed by the Board of Directors? Comments: (Notes: Hard copy to be provided, if in existence)	Yes...............................3 No..............................0	
3	Are you familiar with the policy and what it means to you as a Director? Please specify those areas of the Policy that are of particular relevance to your work: (Notes: Have copy of Body Shop International Policy and discuss if the (x) company does not have own policy.)	Yes...............................3 No..............................0	
4	Do environmental responsibilities feature in your own job description/objectives?	Yes...............................3 No..............................0	
5	Is there a Coordinator with specific responsibility for the environmental performance within the (x) Company? Name and Job Title:	Yes...............................3 No..............................0	
6	In your opinion, are the environmental responsibilities of this Coordinator clearly defined? Comments:	Yes...............................3 No..............................0	
7	Is there a system of non-management environmental advisers in the (x) Company?	Yes...............................3 No..............................0	
8	Are the environmental responsibilities of the environmental advisers clearly defined in their job description/objectives? Comments:	Written Definition.......3 Verbal Definition..........1 None/NA....................0	
9	Do environmental responsibilities form part of the job appraisal system for your staff?	All Staff.....................3 Some Staff...................2 DEA only....................1 None...........................0	

Figure 7-6. (*Continued*)

SECTION A - III

DIRECTORS - cont'd

	QUESTIONS		SCORE
10	Are staff provided with training to improve their environmental awareness? Comments:	Quarterly......................3 Annually......................2 Less than Annually.......1 Never..........................0	
11	Does the (x) Company monitor environmental performance? Specify type of monitoring: (Notes: Waste production, water usage, energy usage)	Yes..............................3 No................................0	
12	Does the (x) Company have formal environmental targets with deadlines? Comments:	Yes..............................3 No................................0	
13	Is progress on departmental and company environmental performance communicated on a regular basis to all staff? Specify when and how:	Monthly......................3 6 Monthly....................2 Annually......................1 Never..........................0	
14	OPTIONAL Are you aware of the following environmental reports and policies which have been produced by Body Shop International: Green Books 1 & 2 Energy Policy Purchasing Policy Transport Policy Packaging Policy Waste Policy Please specify which policies & strategies are of particular relevance to your area: (Have copies at interview)	6 Areas......................3.0 5 Areas......................2.5 4 Areas......................2.0 3 Areas......................1.5 2 Areas......................1.0 1 Area 0.5	
		TOTAL SCORE ____ X 100 = ...% 42	

Figure 7-6. (Continued)

SECTION A - IV

ROLE OF DEPARTMENTAL ENVIRONMENTAL ADVISER

	QUESTIONS		SCORE
1	Are your environmental responsibilities clearly described in your job description/objectives? Comments:	Formal written .... 3 Informal verbal .... 1 None 0	
2	Was the environmental policy and responsibilities of all members of staff/departments mentioned in your induction?	Formal Induction............ 3 Info Pack to read............ 2 Brief/Poss Mention .... 1 Not mentioned.............. 0	
3	Have you received any internal/external training whilst with the company, to carry out your environmental responsibilities? Specify the amount of training received:	Yes............................... 3 No................................ 0	
4	How frequently are environmental matters discussed with your manager? Specify:	Weekly........................ 3 Fortnightly .... 2 Monthly....................... 1 Less than monthly 0	
5	How much time do you spend on your environmental responsibilities? .... % of time In your opinion, is this sufficient?	Yes............................... 3 No................................ 0	
6	Are you/specified individual(s) responsible for monitoring the environmental performance of your department? Specify Name:	Yes............................... 3 No................................ 0	
7	Do you play any part in the environmental auditing of the (x) company? Specify Involvement:	Yes............................... 3 No................................ 0	
8	Are you involved in giving training to raise environmental awareness of management and staff? Comments:	Quarterly..................... 3 Annually...................... 2 Less than annually........ 1 No................................ 0	
9	Are your environmental responsibilities and performance assessed during your job appraisal? Comments:	Formally...................... 3 Verbally .... 1 Nothing 0	

Figure 7-6. (Continued)

SECTION A - IV
ROLE OF DEA - cont'd

	QUESTIONS		SCORE
10	In your opinion, do the Managers in your department take environmental issues seriously?	All..............................3 Most (>50%)..................2 Some (<50%) ....1 None 0	
11	In your opinion, how many people in the (x) Company take environmental issues seriously?	All..............................3 Most(>50%)...................2 Some (<50%) ....1 None 0	
12	How frequently are environmental matters discussed with the Environmental Coordinator of the (x) Company?	Fortnightly ....3 Monthly......................2 Quarterly.....................1 Less than quarterly.......0	
13	In your opinion, how many people in your department are knowledgeable about environmental issues?	All..............................3 Most(>50%)...................2 Some (<50%) ....1 None 0	
14	How often do you inform staff in your department about environmental issues at communication meetings?	Weekly........................3 Fortnightly ....2 Monthly......................1 Less than monthly 0	
15	How do you circulate general environmental information within the (x) company? HOW OFTEN Communication Meetings .... News letters .... Word of Mouth .... Noticeboards .... BSTV .... Training Sessions	6 Areas.......................3.0 5 Areas ....2.5 4 Areas ....2.0 3 Areas.......................1.5 2 Areas.......................1.0 1 Area 0.5	
16	When you have ideas to improve the (x) Company's environmental performance, do you feel your Manager is receptive?	Yes...............................3 No.................................0	
17	What environmental issues do you feel need to be covered in further training sessions? List: (Eg: Photocopier procedures, recycling procedures, use of packing materials etc)		

Figure 7-6. (*Continued*)

SECTION A - V

ENVIRONMENTAL COORDINATOR/MANAGER

	QUESTIONS		SCORE
1	Does the (x) Company have a written environmental policy signed by the Board of Directors? (Notes: Hard copy to be provided, if in existence)	Yes..............................3 No................................0	
2	Do you fully understand how to implement the (x) Company's environmental policy?	Yes..............................3 No................................0	
3	Is there a main Board member responsible for the environmental management of the (x) Company? Name:	Yes..............................3 No................................0	
4	Are your environmental responsibilities clearly defined in your job description/objectives: Comments:	Formal Written ....3 Verbal Description ....1 Not Included................0	
5	Have you any experience or qualifications in the following areas? Please specify what level of qualifications and/or experience you have in the following areas? Environmental ☐ Communication ☐ Managerial ☐	3 Areas........................3 2 Areas........................1 1 Area 0	
6	Do you have adequate time to address your environmental responsibilities? How much time do you spend on your environmental responsibilities? .... % of time. In your opinion, is this enough?	Yes..............................3 No................................0	
7	Do you have responsibilities other than environmental ones? Specify responsibilities and time spent on them.	Yes..............................3 No................................0	
8	Do you take responsibility for overseeing environmental auditing within the (x) company?	Yes..............................3 No................................0	

Figure 7-6. (*Continued*)

SECTION A - V

ENVIRONMENTAL COORDINATOR/MANAGER - cont'd

	QUESTIONS		SCORE
9	Have you received any internal/external training whilst with the company, to carry out your environmental responsibilities? Specify the amount of training received:	Yes.................................3 No...................................0	
10	How frequently are environmental matters discussed with your manager?	Weekly.........................3 Fortnightly ....2 Monthly.......................1 Less than Monthly........0	
11	Do you ensure management and staff receive training to improve their environmental awareness? Comments:	Quarterly.....................3 Annually.......................2 Less than Annually.......1 None 0	
12	Are your environmental responsibilities and performance assessed during your job appraisal? Comments:	Formally.......................3 Verbally ....1 Nothing 0	
13	Are you responsible for a network of Environmental Advisers? Specify: DEAs SEAs	Yes.................................3 No...................................0	
14	Do you communicate with this network on a regular basis? Specify method(s) of communication and how frequently:	Yes.................................3 No/Not Applicable......0	
15	Are there Environmental Advisers in every department of the (x) Company? (Notes: List departments without EAs)	Yes.................................3 No...................................0	
16	Are the environmental responsibilities of the Environmental Advisers clearly defined in their job descriptions/objectives?	Yes.................................3 No...................................0	
17	Does the (x) Company have formal environmental targets with deadlines?	Yes.................................3 No...................................0	
18	Are you the person responsible for reporting the environmental performance of the (x) company?	Yes.................................3 No...................................0	

Figure 7-6. (*Continued*)

SECTION A - V

ENVIRONMENTAL COORDINATOR/MANAGER - cont'd

	QUESTIONS		SCORE
19	In your opinion, to what extent is the Environmental Management System of the (x) Company developed? Please specify priorities for action:	Fully...........................3 Partially.....................2 Non-existent...............0	
20	How do you circulate general environmental information within the (x) company? HOW OFTEN Communication Meetings .... News letters .... DEA .... Noticeboards .... BSTV .... Training Sessions	6 Areas.....................3.0 5 Areas ....2.5 4 Areas ....2.0 3 Areas.....................1.5 2 Areas.....................1.0 1 Area 0.5	
21	How often do you contact The Body Shop Corporate Environment, Health and Safety Department? Specify frequency:	Frequently/Routinely...3 Occasionally/When Required......................2 Never...........................0	
22	How often do you have contact with external bodies for advice or guidelines? Specify frequency: (Notes: Local/National Environmental Groups, Local/Municipal Authorities, Businesses)	Frequently/Routinely...3 Occasionally/When Required......................2 Never...........................0	
23	Do you ensure that environmental literature is included in the recruitment and induction programme for the (x) company?	Yes...............................3 No.................................0	
24	Do you give external presentations to schools, businesses etc?	Monthly.......................3 Quarterly.....................2 Annually.......................1 Never...........................0	
25	Are you able to identify specific environmental costs and/or benefits in your area? Clarify with examples: (Notes: training, pollution, legislation, waste, energy)	Yes...............................3 No.................................0	
26	Do you have a budget assigned to you specifically for environmental projects within the (x)Company?	Yes...............................3 No.................................0	

Figure 7-6. (*Continued*)

SECTION A - V

ENVIRONMENTAL COORDINATOR/MANAGER - cont'd

	QUESTIONS		SCORE
		TOTAL SCORE $$\frac{}{78} \times 100 = \text{........ .....}\%$$	

Additional Questions

1.   What has been achieved in the (x) company over the last year?

Figure 7-6. (*Continued*)

SECTION A - VI

BDM'S/AREA MANAGERS

	QUESTIONS		SCORE
1	Are environmental responsibilities clearly described in your job description/objectives?	Formal Written ....3 Informal Written..........1 Nothing 0	
2	Do environmental responsibilities feature in the job description of anyone in your department/company? Name and Job Title of person with responsibility: Comments:	Yes............................3 No 0	
3	Does the (x) company have a written environmental policy signed by the Board of Directors? Comments: (Notes: Hard copy to be provided, if in existence)	Yes............................3 No.............................0	
4	Are you familiar with the policy and what it means to you as a manager? Please specify those areas of the policy that are of particular relevance to your work: (Notes: Have copy of Body Shop International policy at interview and discuss if the (x) company does not have its own policy.)	Yes............................3 No.............................0	
5	Are you encouraged to play an active role in improving the environmental performance of the shops in your area? Clarify:	Yes............................3 No.............................0	
6	Are you appraised on your environmental performance and that of the shops in your area? Comments:	Shops & Individual.....3 Shops Only.................2 Individual Only ....1 No.............................0	
7	Have you received any internal/external environmental training to enable you to carry out your environmental responsibilities? Specify Amount of Time:	Yes............................3 No.............................0	
8	Do the shops in your area have SEAs? Comments:	All.............................3 Most (> 50%) ....2 Some (< 50%)..............1 None 0	

Figure 7-6. (*Continued*)

SECTION A - VI

BDM'S/AREA MANAGERS - cont'd

	QUESTIONS		SCORE
9	During a shop visit, how much time do you spend on average checking the environmental performance of the shop and environmental awareness of the staff? Clarify:	Over 30 mins...............3 15-30 mins..................2 Less than 15 mins ....1 None 0	
10	Is there an adequate procedure for reporting the environmental performance of your shops so that improvements can be made? Comment on reporting system:	Formal system ....3 Informal system...........1 No system...................0	
11	In your opinion, how many shops in your area take environmental issues seriously?	All.............................3 Most (>50%)...............2 Less (<50%)................1 None 0	
12	How often do you or anyone else within the (x) Company review any of the following data to improve the environmental performance of the shops? Energy Consumption ☐ Waste Disposal Method ☐ No. of Refills ☐ Bottles Returned for Recycling ☐ Amount & Type of Cleaning Agents Used ☐ Other ☐ Specify "Other " Category/Comments: (Notes: Possibilities - water consumption, campaigns, etc)	6 Areas....................3.0 5 Areas....................2.5 4 Areas....................2.0 3 Areas....................1.5 2 Areas....................1.0 1 Area 0.5	
13	Do you or anyone else within the (x) Company regularly set targets to improve the environmental performance of the shops? Energy Consumption ☐ Waste Disposal Method ☐ No. of Refills ☐ Bottles Returned for Recycling ☐ Amount & Type of Cleaning Agents Used ☐ Other ☐ Specify "Other " Category/Comments:	6 Areas....................3.0 5 Areas....................2.5 4 Areas....................2.0 3 Areas....................1.5 2 Areas....................1.0 1 Area 0.5	
14	How often are environmental issues discussed with the Environmental Coordinator/Manager?	Weekly......................3 Fortnightly ....2 Monthly.....................1 Less than monthly 0	
		TOTAL SCORE $\dfrac{\quad}{42}$ X 100 = ...%	

Figure 7-6. (*Continued*)

SECTION A - VII

RETAIL MANAGERS

	QUESTIONS		SCORE
1	Are you aware of The Body Shop's Corporate Environmental Policy? Comments:	Yes..............................3 No..................................0	
2	Does the (x) Company have a written environmental policy signed by the Board of Directors? Comments: (Notes: Have copy of policy at interview)	Yes..............................3 No..................................0	
3	Are you familiar with the policy and what it means to you as a manager? Please specify those areas of the policy that are of particular relevance to your work: (Notes: Have copy of Body Shop International policy at interview and discuss if the (x) company does not have its own policy)	Yes..............................3 No..................................0	
4	Do environmental responsibilities feature in your own job description/objectives?	Yes..............................3 No..................................0	
5	Is there a co-ordinator with specific responsibility for the environmental management and performance of the Retail Outlets? Name and Job Title:	Yes..............................3 No..................................0	
6	In your opinion, are the environmental responsibilities of this co-ordinator clearly defined?	Formal Written ....3 Informal Verbal............1 No..................................0	
7	How often do you discuss environmental issues with the BDMs?	Fortnightly ....3 Monthly........................2 Quarterly.....................1 Six Monthly 0	
8	How often do you discuss environmental issues with the Environmental Coordinator?	Fortnightly ....3 Monthly........................2 Quarterly.....................1 Six Monthly 0	
9	Is there a system of non-management environmental advisers in the shops? Comments:	Yes..............................3 No..................................0	

Figure 7-6. (*Continued*)

SECTION A - VII

RETAIL MANAGERS - cont'd

	QUESTIONS		SCORE
10	Are the environmental responsibilities of the Shop Environmental Advisers clearly defined in their job description/objectives? Comments:	Yes...............................3 No................................0	
11	Have the Shop Environmental Advisers received formal environmental training? Specify training received: (Notes: when and how long)	Yes...............................3 No................................0	
12	Has a Shop Environmental Guide been issued giving comprehensive guidelines for the shops?	Yes...............................3 No................................0	
13	Does the (x) Company have formal environmental targets or objectives for the retail outlets? Comments:	Formal Written............3 Informal Verbal............1 No................................0	
14	Is progress on the shops' environmental performance communicated on a regular basis to all staff in the shops and at Head Office?	Monthly.....................3 6 Monthly.....................2 Annually.....................1 Never..........................0	
15	Is there a formal system of environmental auditing for the retail outlets?	Yes...............................3 No................................0	
16	Are environmental issues included in all training courses/sessions for the shops?	Routinely ....3 Mostly (>50%)..............2 Sometimes (<50%)........1 Never..........................0	
17	Have you received any environmental training whilst with the Company? Specify training received: (Notes: when and how long)	Yes...............................3 No................................0	
18	When recruiting new staff, do you mention the Company's commitment to the environment?	Yes...............................3 Sometimes.....................2 No................................0	
19	Do you ensure all new staff to your department attend the formal induction programme, which has an environmental section?	Yes...............................3 Sometimes.....................1 No................................0	
		TOTAL SCORE $\dfrac{}{36}$ X 100 = ...%	

Figure 7-6. (*Continued*)

SECTION B - I

PURCHASING (Office furniture, stationary, computers)
Any staff making purchases

	QUESTIONS		SCORE
1	Please list the items you have responsibility for purchasing at head office:		
2	Does the (x) Company have an environmental purchasing policy? Comments: (Notes: Hard copy to be provided, if in existence)	Yes.............................3 No...............................0	
3	Have you seen the Body Shop International Environmental Purchasing Policy & Appendices? Comments: (Notes: Have a copy of Body Shop International Policy and appendices at interview)	Yes ....3 No...............................0	
4	Are you familiar with the Policy and what it means to you ? Specify those areas of the Policy that are of particular relevance to your work: (Notes: Have a copy of the Policy to go through at interview)	Yes.............................3 No...............................0	
5	Do you have an active strategy to purchase renewable, reusable, recyclable or recycled materials wherever possible? Comments: (Notes: Specify any purchases through Trade-Not-Aid)	Yes.............................3 No...............................0	
6	Have you collected environmental information, with the use of the 'Environmental Checklists for Buyers', about suppliers, their products, materials, process and policies? Comments: (Eg: Information on the use of hazardous, environmentally damaging or scarce resources)	All Suppliers...............3 Most Suppliers(>50%)..2 Some Suppliers(<50%).1 None 0	
7	Have you established environmental criteria for assessing your suppliers? (Please supply a copy of criteria, if in existence.)	Formal written Questionnaire ....3 Informal verbal..........1 Nothing 0	
8	Are criteria on purchasing mandatory or voluntary?	Mandatory ....3 Voluntary....................1 Non-Existent 0	

Figure 7-6. (*Continued*)

SECTION B - I

PURCHASING (Office furniture, stationary, computers)
Any staff making purchases - cont'd

	QUESTIONS		SCORE
9	Have you required or encouraged suppliers to carry out environmental reviews of their own operations? Comments:	Required....................3 Encouraged ....2 No Action...................0	
10	Do you have inventories of hazardous products that are used by your business? Comments: (Notes: Hazardous symbol: eg bleaches, solvents, cleaning agents etc)	Yes............................3 No..............................0	

TOTAL SCORE

$$\frac{}{30} \ \times \ 100 \ = \ ....\%$$

Figure 7-6. (*Continued*)

SECTION B - II

SUNDRIES PURCHASING

	QUESTIONS		SCORE
1	Have you seen the Body Shop International Purchasing Policy? Comments:	Yes ....3 No 0	
2	Are you familiar with the main constituents of the policy and how they are being implemented? Clarify:	Yes ....3 No 0	
3	Have you seen the Body Shop International Packaging Policy? Comments:	Yes ....3 No 0	
4	Are you familiar with the main constituents of the policy and how they are being implemented within your department? Clarify:		
5	Have you collected environmental information, with the use of the 'Environmental Checklists for Buyers', about suppliers, their products, materials, processes and policies? Length of time this has been done: (Notes: Should include hazardous, environmentally damaging or scarce resources)	All Suppliers ....3 Most Suppliers (>50%) 2 Some Suppliers (<50%)1 None 0	
6	Have you established environmental criteria for assessing your suppliers? (Please supply a copy of the criteria, if in existence.)	Formal written Questionnaire..............3 Informal verbal ....1 Nothing......................0	
7	Are the criteria on purchasing mandatory or voluntary?	Mandatory ....3 Voluntary ....1 Non-Existent 0	
8	Have you required or encouraged suppliers to carry out environmental reviews of their own operations? Comments:	Required ....3 Encouraged ....2 No Action 0	
9	Do you have an active strategy to purchase renewable, reusable, recyclable or recycled materials wherever possible? Comments:	Yes ....3 No 0	

Figure 7-6. (*Continued*)

SECTION B - II

SUNDRIES PURCHASING- cont'd

	QUESTIONS		SCORE
10	Do you feel that there are any restrictions which prevent the purchase of environmentally acceptable sundry items? Comments: (Notes: Financial costs, lack of time to source)	Yes ....3 No 0	
11	What percentage of sundry items come from Trade-not-Aid projects?	50-100% ....3 10-50% ....2 <10% ....1 None 0	
12	Do you have a strategy in the Sundries range to use recycled and/or recyclable packaging? Clarify:	Yes ....3 No 0	
13	What input does your department have on the environmental design of the Christmas gift baskets? Comments:	Full Input ....3 Some Input....................2 Minimum Input.............1 No Input.......................0	

TOTAL SCORE

$$\frac{\quad\quad}{39} \ \text{X} \ 100 = \text{.............}\%$$

Figure 7-6. (*Continued*)

SECTION B - III

WASTE MANAGEMENT (Offices)
Managers/Site Services

	QUESTIONS		SCORE
1	Do you have a policy on Waste Management? (Notes: Hard copy to be provided, if in existence)	Formal Written ....3 Informal Verbal...........1 Nothing 0	
2	Are you aware of procedures in dealing with waste that you generate? Comments: (Notes: Attach copy)	Formal Written ....3 Informal Verbal...........1 Nothing 0	
3	Does the site have the correct waste handling licences? (Notes: Carriage, storing, handling of waste)	Yes............................3 No.............................0	
4	Have waste targets been set to reduce the amount of waste?	Formal Written ....3 Informal Verbal...........1 Nothing 0	
5	Are waste targets reviewed and revised?	Yes............................3 No.............................0	
6	Are on-site collection facilities for recycling sufficient? Specify categories of waste are collected:	Yes............................3 No.............................0	
7	Does the (x) Company re-use office materials/stationary? Examples:	Always ....3 Mostly(>50%) ....2 Sometimes(<50%)........1 Never..........................0	
8	Are staff encouraged to reduce, re-use and recycle waste through training or any other methods? Specify:	Yes............................3 No.............................0	
		TOTAL SCORE _____ X 100 = ...% 24	

Figure 7-6. (*Continued*)

SECTION B - IV

WASTE MANAGEMENT (Warehouse/Site Services)
Manager/Site Services

	QUESTIONS		SCORE
1	Do you have a policy on Waste Management? Comments: (Hard Copy to be provided, if in existence)	Formal Written ....3 Informal Verbal............1 Nothing 0	
2	Are you familiar with the policy and what it means to you as a manager. (Notes: Specify those areas of the policy that are of particular relevance to your work. Have copy of policy at interview)	Yes..............................3 No...............................0	
3	Does the site have the correct waste handling licences? (Notes: Carrying, storage, handling of waste)	Yes..............................3 No...............................0	
4	Are there documented procedures for managing waste on site? Comments: (Attach a copy of the procedures, if in existence)	Formal Written ....3 Informal Verbal............1 Nothing 0	
5	Are wastes properly segregated and stored in a tidy fashion to facilitate recycling? Comments:	Always ....3 Mostly (>50%)..............2 Sometimes (<50%)........1 Never...........................0	
6	Are you aware of legislation and company policies relating to the handling, storage, transport and disposal of all wastes generated? Comments:	Full Knowledge ....3 Partial Knowledge ....1 No Knowledge..............0	
7	Do you document quantities of waste generated and transferred from your control, within the (x) Company and externally? Comments:	Formal Written ....3 Informal Verbal............1 Nothing 0	
8	Does the (x) Company have a waste minimisation strategy?	Formal Written ....3 Informal Verbal............1 Nothing 0	
9	Does the (x) Company have targets for waste reduction, recycling and reuse of materials? Comments:	Published Targets.........3 Written, Unpublished...2 Verbal.........................1 None 0	

Figure 7-6. (*Continued*)

SECTION B - IV

WASTE MANAGEMENT (Warehouse/Site Services)- cont'd
Manager/Site Services

	QUESTIONS		SCORE
10	Are waste targets reviewed and revised? Comments:	Annually.....................3 Less than Annually.......1 Never N/A...................0	
11	Have you explored local opportunities to reuse waste? List examples: (Notes: paints, solvents, furniture, palasets)	Yes..............................3 No................................0	
12	Has a competent member of staff visited and reviewed the activities of all waste disposal contractors and disposal sites? Comments:	Formal Checklist Review..........................3 Observational Visit......2 Verbal Communication..1 Nothing 0	
13	Are building contractors made aware of any waste management policies of the company?	Yes..............................3 No /NA........................0	
14	Are building contractors requested to supply details on how they will dispose of any old materials during their time on site? Comments:	Formal Written Detail.3 Verbal..........................2 None 0	
15	Are staff encouraged to reduce, reuse and recycle waste through training or any other methods? Specify:	Yes..............................3 No................................0	
16	Are all staff involved in handling wastes, properly trained?	Formal Training............3 Informal Training ....1 None 0	
		TOTAL SCORE ____ X 100 = ...% 48	

Figure 7-6. (*Continued*)

SECTION B - V

ENERGY
Managers/Maintenance Staff.

	QUESTIONS		SCORE
1	Has an energy review/audit been done on the site in the last 2 years? Comments:	Yes..............................3 No................................0	
2	Is the total quantity of all sources of energy recorded on a regular basis? Comments: (Notes: Specify type of energy used - gas, electricity, wind etc)	Weekly.......................3 Monthly.....................2 Annually....................1	
3	Is the total quantity of all sources of energy reported to all users on a regular basis? Specify details and how often: (Notes: Quantity, Cost, CO_2)	Yes.............................3 No...............................0	
4	Do you have a list of electrical equipment, both portable and fixed, which includes the power ratings? Comments:	Formal list.................3 Informal list'...............2 None..........................0	
5	Is energy efficiency taken into account when upgrading/ordering new equipment? Specify:	Always.....................3 Sometimes..................2 Never........................0	
6	How often is equipment serviced?	According to Specifications.............3 Infrequently................1 Never.......................0	
7	Is the design and operation of the lighting known to be of maximum efficiency? Specify:	Yes.............................3 No................................0	
8	Is use of natural light maximised? Comments:	Yes.............................3 No................................0	
9	Do you use alternative or renewable sources of energy for electric power and/or heating requirements? State type and proportion of demand:	Using Above 50%.........3 Using Less than 50%.....2 Being Considered..........1 None...........................0	

Figure 7-6. (*Continued*)

SECTION B - V

ENERGY
Managers/Maintenance Staff cont.

	QUESTIONS		SCORE
10	Do you have energy efficiency targets to achieve? Comments:	Formal and Implementing ....3 Informal......................1 None 0	
11	Are the targets reviewed regularly? Comments:	Annually.....................3 Less than annually.......2 Never..........................0	
12	Are building contractors requested to incorporate and design energy efficiency features a part of their brief? Comments:	Yes.............................3 No...............................0	
13	Are energy efficiency features incorporated into the design/refurbishment of the buildings? Comments:	Yes.............................3 No...............................0	
14	Are staff encouraged to improve energy efficiency through training or any other methods? Specify:	Yes.............................3 No...............................0	
		TOTAL SCORE $\dfrac{\quad}{42}$ X 100 = ...%	

Figure 7-6. (*Continued*)

SECTION B - VI

SITE MANAGEMENT
Managers/Site Services

	QUESTIONS		SCORE
1	Is there a list of previous uses of this site? (Note: Please attach details)	Yes..................................3 No....................................0	
2	Is the site being contaminated in any way through the operation of the Company? (Notes: If yes, describe actions being taken to rectify)	Yes...................................0 No....................................3 Don't Know..................0	
3	Have actions been taken to develop and improve the wildlife/landscape value of the site, in particular waste, derelict or neglected land within the last 12 months? Clarify:	Yes-Wasteland..............3 Yes-Neglected Land......3 No....................................0	
4	Have tropical hardwoods been used in the construction of the premises since your occupation?	No....................................3 Yes...................................0	
5	Are building utilities properly operated and systematically checked/maintained (eg. heating, extraction units and air conditioning units)?	Formalised schedule.....3 Informal schedule.........1 Never................................0	
6	What is the frequency of complaints of noise, odour or visual intrusion from local residents or businesses? Actions if Complaints Received:	Never............................3 Monthly......................2 Weekly.......................1 Daily............................0 N/A.............................0	
7	Were energy efficiency measures taken into account in the construction/refurbishment of the building in the following areas? Lighting ☐ Heating/Air Conditioning ☐ Insulation ☐ eg. Double glazing wall cavity filling	3 Areas.........................3 2 Areas.........................2 1 Area ....1 None 0	
8	Has any action been taken to improve the internal working environment of the buildings? Specify:	Implemented ....3 Considered ....1 Nothing 0	
9	When contractors are refurbishing buildings are they advised against the use of hazardous materials/scarce resources? Comments:	Formal Written ....3 Verbal...........................2 No...................................0	

Figure 7-6. (*Continued*)

SECTION B - VI

SITE MANAGEMENT
Managers/Site Services - cont'd

	QUESTIONS		SCORE
10	Is there any monitoring done to prevent Sick Building Syndrome? Comments: (Notes: Specify any cases in the past 12 months)	Yes..............................3 No................................0	
11	Are building contractors made aware of any waste management policies of the company?	Yes..............................3 No................................0	
12	Are contractors instructed during the course of their work to ensure it is carried out with a minimum of waste materials and any waste is disposed of correctly? Comments:	Formal Written.............3 Verbal..........................2 No................................0	
13	Are contractors given guidelines to minimise/avoid the use of ozone depleting chemicals (refrigerants) or VOCs (paints/adhesives)?	Formal.........................3 Informal Verbal............2 No................................0	
14	Has the use of natural ventilation been maximised in the design of the building?	Yes..............................3 No................................0	
		TOTAL SCORE $\dfrac{}{42}$ X 100 = ...%	

Figure 7-6. (*Continued*)

SECTION B - VII

TRANSPORT : COMPANY CARS
(Managers/Transport Manager)

	QUESTIONS		SCORE
1	Is there a company car policy? (Hard Copy to be provided if in existence.)	Yes...........................3 No............................0	
2	Is the policy reviewed regularly?	Annual ....3 Bi-Annual ....2 5 Yearly ....1 Never........................0	
3	On what basis are company cars provided? Comments:	Essential & Car Pool....3 Essential Use..............2 Essential & Benefit......1 Benefit Only...............0	
4	Is a car pool system operated? Comments:	Formal ....3 Informal.....................1 Nothing 0	
5	Does the company have any environmental criteria/guidelines for the purchase of company cars? Comments: (Notes: Hard copy to be provided, if in existence)	Formal Written ....3 Informal Verbal..........1 Nothing 0	
6	Does the company support/provide alternative transport to get to work? Comments: (Notes: Subsidised bike scheme, coach service, season ticket loans, etc)	More than one scheme ....3 1 scheme.....................2 Nothing 1	
7	Are company bicycles made available for local use by staff?	Yes...........................3 No............................0	
8	Are company owned cars serviced regularly to maintain maximum fuel efficiency? Comments:	As specified...............3 Occasionally ....1 Never........................0	
9	Are distances travelled by company cars and their fuel consumption documented? Comments:	Per journey.................3 Monthly.....................2 Weekly......................1 Never........................0	

Figure 7-6. (*Continued*)

SECTION B - VII

TRANSPORT : COMPANY CARS
(Managers/Transport Manager) - cont'd

	QUESTIONS		SCORE
10	Are the quantities of CO_2 produced by the company cars quantified? Comments/Specify: (Notes: Reached by converting fuel efficiencies)	Yes..............................3 No...............................0	
11	Are staff with essential use cars given advanced driving training to reduce fuel consumption and vehicle wear and tear?	Yes..............................3 No...............................0	

TOTAL SCORE

$$\frac{\quad}{33} \quad X \quad 100 = \quad.............\%$$

Figure 7-6. (*Continued*)

SECTION B - VIII

TRANSPORT: DISTRIB. VEHICLES
(Managers/Transport Manager)

	QUESTIONS		SCORE
1	Is there an environmental strategy on transport and distribution? Describe? (Notes: Conservation of fuel/vehicle, use of alternative transport - rail ships, use of environmentally sensitive distribution routes and methods. Attach copy of strategy)	Formal written ....3 Informal verbal ....1 None 0	
2	Are distances travelled and the fuel consumption of distribution vehicles documented centrally by a designated member of staff? Comments: Name and Title of Member of Staff:	Per journey ....3 Monthly......................2 Annually.....................1 Never..........................0	
3	Does the company have fuel efficiency targets for the distribution vehicles? What are they?	Formal written ....3 Informal verbal ....1 None 0	
4	Are fuel efficiency targets reviewed by senior staff? Comments:	Monthly......................3 6 Monthly...................2 Yearly........................1 Never..........................0	
5	Have the distribution vehicles been improved to lessen their impact on the environment? Pollution Abatement Systems ☐ Air Management Systems ☐ Fuel Consumption Meters ☐ (Notes: Pollution Abatement: catalytic converters, filters)	3 Systems....................3 2 Systems....................2 1 System......................1 None/Don't Know........0	
6	Are the distribution vehicles used to collect raw materials or materials destined for recycling when delivering products to the same area? Comments:	Yes.............................3 No...............................0	
7	Has a full environmental impact assessment of your vehicle distribution system been conducted? Comments:	Yes.............................3 No...............................0	
8	Are opportunities to use alternati ve methods of transport (eg rail, ship) explored on a regular basis? Comments:	Yes.............................3 No...............................0	

Figure 7-6. (*Continued*)

SECTION B - VIII

TRANSPORT: DISTRIB. VEHICLES
(Managers/Transport Manager) - cont'd

	QUESTIONS		SCORE
9	Is the CO_2 produced by the distribution vehicles quantified? Comments: (Notes: reached by converting fuel efficiencies)	Yes...............................3 No...............................0	
10	Are drivers given training to enable them to reduce fuel consumption, and vehicle wear and tear?	Yes...............................3 No...............................0	
		TOTAL SCORE $\dfrac{\qquad}{30}$ X 100 = ...%	

Figure 7-6. (*Continued*)

SECTION C - I

PURCHASING (Shops)
Any staff making purchases for shops.

	QUESTIONS		SCORE
1	What items do you purchase for the shops? Please list: (Eg: Stationary, cleaning agents etc.)		
2	Does the (x) Company have an environmental purchasing policy that is relevant to the shops? Comments: (Notes: Hard copy to be provided, if in existence)	Yes..........................3 No..............................0	
3	Have you seen the Body Shop International Environmental Purchasing Policy & Appendices? Comments: (Notes: Have a copy of Body Shop International policy and appendices at interview)	Yes ....3 No..............................0	
4	Are you familiar with the Policy and what it means to you as someone who purchases items for shops? Specify those areas of the Policy that are of particular relevance to your work: (Notes: Have a copy of the Body Shop International Policy and discuss if the (x) company does not have a policy in existence.	Yes..........................3 No..............................0	
5	Do you have an active strategy to purchase renewable, reusable, recyclable or recycled materials wherever possible? Comments:	Yes..........................3 No..............................0	
6	Have you collected environmental information about your suppliers, their products, materials, process and policies? Comments: (Notes: Should include hazardous, environmentally damaging or scarce resources)	All Suppliers...........3 Most Suppliers(>50%) ....2 Some Suppliers(<50%) ....1 None.......................0	

Figure 7-6. (*Continued*)

SECTION C - I

PURCHASING (Shops) - cont'd
Any Staff Making Purchases For Shops

	QUESTIONS		SCORE
7	Have you established environmental criteria for assessing your suppliers? Comments: (Please supply a copy of criteria)	Formal Written Questionnaire..........3 Informal Verbal....1 Nothing...................0	
8	Are criteria on purchasing mandatory or voluntary?	Mandatory...............3 Voluntary................1 Non-Existent............0	
9	Have you required or encouraged suppliers to carry out environmental reviews of their own operations? Comments:	Required..................3 Encouraged..............2 No Action.................0	
		TOTAL SCORE $\frac{}{27}$ X 100 = ...%	

Figure 7-6. (*Continued*)

SECTION C - II

WASTE MANAGEMENT (Shops)
Retail Managers, Area Managers & Environmental Coordinators

	QUESTIONS		SCORE
1	Do you have a policy on Waste Management which is relevent to the shops? (Notes: Hard copy to be provided, if in existence)	Formal Written ....3 Informal Verbal............1 Nothing 0	
2	Are shops aware of procedures in dealing with waste? Comments: (Notes: Attach copy)	Formal Written ....3 Informal Verbal...........1 Nothing 0	
3	Do all shops have the correct waste handling licences? (Notes: Carriage, storing, handling of waste)	Yes.............................3 No..............................0 N/A...........................0	
4	Does anyone n the shop regularly monitor waste generaton? Please clarify details and how often monitoring occurs:	Yes.............................3 No..............................0	
5	Have waste targets been set to reduce the amount of waste?	Formal written...........3 Informal verbal...........1 Nothing.....................0	
6	Are shops aware of the targets?	Yes.............................3 No..............................0 N/A...........................0	
7	Are waste targets reviewed and revised on a regular basis?	Yes.............................3 No..............................0 N/A...........................0	
8	Are on-site collection facilities for recycling sufficient? Specify categories of waste collected:	Yes.............................3 No..............................0	
9	Are opportunities to reduce and reuse waste in the shops maximised? Please specify how:	Yes.............................3 No..............................0	
		TOTAL SCORE $\dfrac{}{33}$ X 100 = ...%	

Figure 7-6. (*Continued*)

SECTION C - III

ENERGY (Shops)
Retail Managers, Area Managers & Environmental Coordinators

	QUESTIONS		SCORE
1	Has an energy review/audit been done of the shops in the last 2 years? Comments:	Yes...............................3 No................................0	
2	Is the total quantity of all sources of energy used in the shops, recorded on a regular basis? Comments: (Notes: Specify type of energy used - gas, electricity, wind etc.)	Weekly........................3 Monthly.......................2 Annually......................1	
3	Is the total quantity of all sources of energy reported to all shops on a regular basis? (Notes: Quantity, Cost, CO_2)	Yes...............................3 No................................0	
4	Do you have a list of electrical equipment, both portable and fixed, which includes the power ratings for the average shop? Comments: (Notes: Provide List)	Formal list.................3 Informal list................2 None 0	
5	Is energy efficiency taken into account when upgrading/ordering new equipment for the shops? Specify:	Always ....3 Sometimes...................2 Never..........................0	
6	How often is equipment serviced? (Notes: Air conditioning units, portable appliances, electrical system, heating system)	According to Specifications..............3 Infrequently.................1 Never..........................0	
7	Is the design and operation of the lighting known to be of maximum efficiency in the shops? Specify:	Yes...............................3 No................................0	
8	Is use of natural light maximised in the shop and back-up areas? Comments:	Yes...............................3 No................................0	
9	Do you use alternative or renewable sources of energy for electric power and/or heating requirements? State type and proportion of demand:	Using Above 50%........3 Using Less than 50%.....2 Being Considered.........1 None 0	

Figure 7-6. (*Continued*)

SECTION C - III

ENERGY (Shops)
Retial Managers, Area Managers and Environmental Coordinators - cont'd

	QUESTIONS		SCORE
10	Does each shop have energy efficiency targets to achieve? Comments:	Formal and Implementing ....3 Informal......................1 None 0	
11	Are the targets reviewed regularly? Comments:	Annually....................3 Less than annually.......2 Never.........................0	
12	Are shop fitters requested to incorporate and design energy efficiency features a part of their brief? Comments:	Yes.............................3 No..............................0	
13	Are energy efficiency features incorporated into the design/refurbishment of the shops? Comments:	Yes.............................3 No..............................0	
14	Are staff encouraged to improve energy efficiency through training or any other methods? Specify:	Yes.............................3 No..............................0	
		TOTAL SCORE ____ X 100 = ...% 42	

Figure 7-6. (*Continued*)

SECTION C - IV

SITE MANAGEMENT (Shops)
Retail Managers, Environmental Coordinators & Shop Fitters

	QUESTIONS		SCORE
1	Is there a list of previous uses of each shop's site? (Note: Please attach details)	Yes.................................3 No..................................0	
2	Is each shop site being contaminated in any way through the operation of the Company? (Notes: If yes, describe actions being taken to rectify)	Yes..................................0 No...................................3 Don't Know...................0	
3	Have tropical hardwoods been used in the construction of the premises since your occupation?	No..................................3 Yes.................................0	
4	Are building utilities properly operated and systematically checked/maintained (eg. heating, extraction units and air conditioning units)?	Formalised schedule.....3 Informal schedule.........1 Never...........................0	
5	What is the frequency of complaints of noise, odour or visual intrusion from local residents or businesses? Actions if Complaints Received:	Never............................3 Monthly.......................2 Weekly.........................1 Daily............................0 N/A..............................0	
6	Were energy efficiency measures taken into account in the construction/refurbishment of the building in the following areas? Lighting ☐ Heating/Air Conditioning ☐ Insulation ☐ eg. Double glazing wall cavity filling	3 Areas.........................3 2 Areas.........................2 1 Area ....1 None............................ 0	
7	Has any action been taken to improve the internal working environment of the shops? Specify:	Implemented ....3 Considered ....1 Nothing 0	
8	Are shop fitters made aware of any waste management policies of the company?	Yes.................................3 No..................................0	
9	Are shop fitters instructed during the course of their work to ensure it is carried out with a minimum of waste materials and any waste is disposed of correctly? Comments:	Formal Written ....3 Verbal..........................2 No..................................0	
10	Are shop fitters given guidelines to minimise/avoid the use of ozone depleting chemicals (refrigerants) or VOCs paints/adhesives) ?	Formal ....3 Informal Verbal.............2 No..................................0	

Figure 7-6. (*Continued*)

SECTION C - IV

SITE MANAGEMENT (Shops)
Retail Managers, Environmental Coordinators & Shop Fitters - cont'd

	QUESTIONS		SCORE
11	Has the use of natural ventilation been maximised in the design of the building?	Yes................................3 No................................0	
		TOTAL SCORE ____ X 100 = ...% 39	

Figure 7-6. (*Continued*)

SECTION D - I

MANUFACTURING & WASTE WATER

	QUESTIONS		SCORE
1	Are sanitisers used on a regular basis to reduce risk of contamination of product? Comments: (Notes: Main mixing vessels/filling lines)	Always ....3 Mostly (>50%).............2 Sometimes (<50%).......1 Never.........................0	
2	Do you follow guidelines to reduce the risk of contamination of bulk product? Provide copy of guidelines:	Yes.............................3 No...............................0	
3	Do you follow guidelines to minimise the use of sanitisers? Comments: Provide copy of guidelines: (Notes: alternating/changing type of sanitiser regularly)	Yes.............................3 No...............................0	
4	Is the amount of out of specification bulk product which needs to be reworked quantified and recorded? Comments:	Formal Written ....3 Informal Mental ....1 No Record...................0	
5	Is the amount of out of specification bulk product which needs to be disposed, recorded and monitored? Comments:	Formal Written ....3 Informal Mental ....1 Nothing 0	
6	Are bulk raw materials and hazardous raw materials stored in adequate sized bunded areas? Comments:	Yes.............................3 No...............................0	
7	Are measures being taken to reduce the quantities of bulk product being disposed of? Comments:	Implemented ....3 Investigated................1 Nothing 0	
8	Is the effluent being discharged in the most efficient and effective manner? Comments:	Treated on Site-Full Compliance ....3 Treated off Site-Full Compliance ....2 Discharged to Foul Sewer under permit................1 Not Considered...........0	

Figure 7-6. (*Continued*)

SECTION D - I

MANUFACTURING & WASTE WATER - cont'd

	QUESTIONS		SCORE
9	Are volumes of water and waste water quantified regularly? Comments:	Weekly........................3 Monthly.....................2 Annually....................1 Never.........................0	
10	Is waste water quality monitored for the following? COD ☐ BOD ☐ pH ☐ Organics ☐ Metals ☐ Cations ☐ eg. sulphates Name and Job Title of member of staff keeping records: .... (Notes: Chemical O_2 Demand, Biological O_2 Demand)	6 Areas......................3.0 5 Areas......................2.5 4 Areas......................2.0 3 Areas......................1.5 2 Areas......................1.0 1 Area ....0.5 No Areas....................0	
11	Have your effluents complied with these regulations at all times within the last 12 months?	Yes..............................3 No................................0	
12	Do you have all the appropriate licences and permits relating to waste water discharge and disposal? Specify: (Observe permits) (Notes: Disposal: via tankered; Discharge: via drainage system)	Yes..............................3 No................................0	
13	Has the Company adopted a Water Conservation Strategy? (Notes: to include local availability of water, flow meters could include targets)	Formal Written ....3 Informal Verbal...........1 Nothing 0	
14	Are there targets for water reduction that are regularly reviewed? Comments:	Quarterly...................3 Six Monthly ....2 Annually....................1 Never /N/A................0	
15	Is the wastewater recycled? Specify methods:	Yes..............................3 No................................0	
16	Are water sources, pipe work and processes checked for leakage/wastage on a regular basis?	Monthly.....................3 6 Monthly...................2 Annually....................1 Never.........................0	

Figure 7-6. (*Continued*)

SECTION D - I

MANUFACTURING & WASTE WATER - cont'd

	QUESTIONS		SCORE
		TOTAL SCORE $\dfrac{\quad}{48}$ X 100 = ...%	

Figure 7-6. (*Continued*)

SECTION D - II

AIR, NOISE & ODOUR

	QUESTIONS		SCORE
1	Is there a list of all air emissions? Specify: (Notes: Air extraction unit filters)	Yes..............................3 No.................................0	
2	Is there a member of staff who is responsible for keeping and reporting records of these air emissions or a regular basis? Name and Job Title:	Weekly......................3 Monthly......................2 Annually.....................1 Never...........................0	
3	Is there a member of staff who is responsible for keeping and reporting records of noise emissions on a regular basis? Name and Job Title: (Notes: Machinery, Sirens, Radio, Vehicles)	Weekly......................3 Monthly......................2 Annually.....................1 Never...........................0	
4	Is all appropriate machinery systematically checked/maintained so they meet manufacturers specifications on noise and air? Specify:	Formalised Schedule ...3 Informal Schedule ....1 Never...........................0	
5	Does the company comply with any relevant air emission regulations/standards? (Notes: Specify relevant legislation)	Yes..............................3 No.................................0 Not sure......................0	
6	Does the company comply with any relevant noise regulations/standards? (Notes: Specify relevant legislation)	Yes..............................3 No ....0 Not sure......................0	
7	Is there a strategy to reduce or eliminate the escape of volatile organic compounds (VOCs), ozone depleting chemicals or greenhouse gases to the environment? Ozone ☐ VOCs ☐ Greenhouse Gases ☐ Specify:	3 Areas.......................3 2 Areas.......................2 1 Area ....1 Nothing 0	
8	Are any operations which cause noise and odour, located away from likely sources of complaint?	Yes..............................3 No.................................0	
9	Have there been any complaints of noise/odour from local residents or businesses? (Notes: Actions of complaints received)	Yes..............................3 No.................................0	

Figure 7-6. (*Continued*)

SECTION D - II

AIR, NOISE & ODOUR - cont'd

	QUESTIONS		SCORE
		TOTAL SCORE $\dfrac{\quad}{27} \times 100 = \ldots\ldots\%$	

Figure 7-6. (*Continued*)

SECTION D - III

MANUFACTURING WASTE MANAGEMENT

	QUESTIONS		SCORE
1	Do you have a policy on Waste Management? Comments: (Notes: Provide copy of policy)	Formal written ....3 Informal verbal ....1 Nothing 0	
2	Are you familiar with the policy and what it means to you as a manager? (Notes: Specify those areas of the policy that are of particular relevance to your work. Have copy of policy at interview)	Yes...............................3 No................................0	
3	Are there documented procedures for managing waste on site? Comments:	Formal Written ....3 Informal Verbal...........1 Nothing 0	
4	Do your wastes have to be labelled? Comments: (Notes: For reasons of safety/environmental protection)	All Automatically.......3 Most.........................2 Some ....1 None 0	
5	Are wastes segregated and stored in a tidy fashion? Comments: (Notes: Observe)	Always ....3 Mostly (>50%).............2 Sometimes (<50%).......1 Never.........................0	
6	Are liquid wastes stored in adequate sized, bunded areas to prevent spillage? Comments:	Always ....3 Mostly (>50%).............2 Sometimes (<50%).......1 Never.........................0	
7	Are there dedicated bunded and lockable areas for storage of hazardous/semi-hazardous waste materials? Comments:	Yes...............................3 No................................0	
8	Are there defined procedures to cater for spillages of water?	Formal Written ....3 Informal Verbal...........1 Nothing 0	
9	Are staff trained to implement these procedures?	Formal Written ....3 Informal Verbal...........1 Nothing 0	

Figure 7-6. *(Continued)*

SECTION D - III

MANUFACTURING WASTE MANAGEMENT - cont'd

	QUESTIONS		SCORE
10	Are wastes segregated and stored in a tidy fashion to facilitate recycling? (Notes: Observe)	Always ....3 Mostly (>50%)..............2 Sometimes (<50%).......1 Never..........................0	
11	Are you aware of legislation and Company policies relating to the handling, storage, transport and disposal of all wastes generated?	Full Knowledge ....3 Part Knowledge...........1 No Knowledge.............0	
12	Does the site have the correct waste handling licences? (Notes: Carrying, Storage, Handling of Waste)	Yes..............................3 No................................0	
13	Are all staff involved in handling wastes properly trained?	Formal Training...........3 Informal Training ....1 None 0	
14	Do you document quantities of each type of waste generated and transferred from your control, within the Company and externally?	Formal Written ....3 Informal Verbal...........1 Nothing 0	
15	Does the (x) Company have a waste minimisation plan?	Formal Written ....3 Informal Verbal...........1 None 0	
16	Does the (x) Company have targets for waste reduction, reuse and recycling of materials?	Published Targets........3 Written, Unpublished..2 Verbal..........................1 None 0	
17	Are waste targets reviewed and revised?	Annually.....................3 Bi-Annually................2 Never/NA 0	
18	Have you explored local opportunities to reuse waste? List examples: (Notes: Product barrels)	Yes..............................3 No................................0	
19	Do all contractors have the appropriate licences to handle and dispose of wastes? Comments:	Formal Documents........3 Informal Communication............1 No Evidence 0	

Figure 7-6. *(Continued)*

SECTION D - III

MANUFACTURING WASTE MANAGEMENT - cont'd

	QUESTIONS		SCORE
20	Has a competent member of staff visited and reviewed the activities of all waste disposal contractors and disposal sites? Comments:	Formal Checklist Review........................3 Observational Visit.....2 Verbal Communication....:........1 Nothing 0	
21	Do you have an approved list of waste contractors, as defined by formal criteria?	Yes..............................3 No...............................0	
22	Have possibilities for energy recovery in manufacturing processes been maximised? (Notes: Heat exchangers, re-circulating cooling or process waters)	Doing everything feasible.......................3 Taken some steps...........1 Has not been considered 0	

TOTAL SCORE
___ X 100 = ...% 66

Figure 7-6. (*Continued*)

SECTION D - IV

ENVIRONMENTAL EMERGENCY PLANNING

	QUESTIONS		SCORE
1	Have you identified and documented within your business, those operations which pose the greatest environmental risk? Comments:	Formal Written ....3 Informal Verbal...........1 No..............................0	
2	Have you identified possible environmental effects within your business and developed a plan or strategy to deal with emergencies eg. spillages, floods, fire etc. Comments:	Formal Written ....3 Informal Verbal...........1 None 0	
3	How many staff are trained and aware of their responsibilities in an emergency? Comments:	All Staff.....................3 Most Staff (>50%) ....2 Some Staff (<50%).......1 None 0	
4	Do you have written procedures for communicating with employees, the public and the press in the event of an emergency? Comments:	Yes..............................3 No..............................0	
5	Are the appropriate services aware of your emergency plans and procedures? Specify:	Yes..............................3 No..............................0	
6	Are emergency plans and procedures updated where necessary? Comments: (Notes: Telephones, addresses, etc)	Quarterly...................3 6 Monthly...................2 Annually....................1 Never.........................0	
		TOTAL SCORE ____ X 100 = ...% 18	

Figure 7-6. (*Continued*)

Acknowledgment

The author wishes to thank Sally Power of The Body Shop International Environment, Health, and Safety Department for preparing this manuscript.

References

1. Harrison, L. L. (ed.), 1984. *The McGraw-Hill Environmental Auditing Handbook: A Guide to Corporate and Environmental Risk Management,* (1st ed.). New York: McGraw-Hill.

2. Greeno, J. L., G. S. Hedstrom, and M. DiBerto, 1986. *Environmental Auditing: Fundamentals and Techniques* (2d ed.). Cambridge, Mass.: Arthur D. Little.

3. Business in the Environment, 1991. *Your Business and the Environment: A DIY Review for Companies.* London: Coopers and Lybrand.

4. Blumerfield, K., R. Earl III, and J. Shopley, 1991. "Identifying strategic environmental opportunities: a life cycle approach." *Prism,* 3d quarter 1991. Cambridge, Mass.: Arthur D. Little.

5. Wheeler, D., 1991. "Retail Packaging: In Search of More Sustainable Options." *Proceedings of PWMI Conference,* Brussels.

6. Wheeler, D., 1992. "Comment creer des produits ecologiques?" *Proceedings of IIR Conference: Gagnez la Bataille des Ecoproduits.* Paris: IIR.

7. Wheeler, D., 1992. "Environmental management as an opportunity for sustainability in business—economic forces as a constraint." *Business Strategy and the Environment,* 1(4), 37–40.

8. Roddick, A., 1992. "In Search of the Sustainable Business." *Ecodecision 7.*

9. World Commission on Environment and Development, 1987. *Our Common Future.* Oxford: Oxford University Press.

10. United Nations, 1992. *Rio Declaration on Environment and Development.* Principle 1. Geneva: United Nations.

11. Taylor, A., 1992. *Choosing our Future. A Practical Politics of the Environment.* London: Routledge.

12. Cowell, J., "Assessing Environmental Risk," 1991. In *Advances in Environmental auditing. Proceedings of the IBC/IEA Conference.* London: IBC.

13. Local Government Management Board, 1991. *Environmental Auditing in Local Government.* Luton, U.K.: LGMB.

14. Barwise, J. (ed.), 1991. *Local Authority Environmental Policy—A Framework for Action.* Guildford, U.K.: University of Surrey.

15. British Telecom, 1992. *BT and the Environment.* London.

16. The Body Shop International, 1992. *The Green Book: The Body Shop Environmental Statement 1991/92.* Littlehampton, U.K.: BSI.

17. British Airways, 1992. *Annual Environmental Report.* Hounslow, U.K.: British Airways plc.

18. British Standards Institution, 1992. BS7750, "Specification for Environmental Management Systems." Milton Keynes, U.K.: BSI.

19. Council of the European Communities, 1992. "Proposal for a Council Regulation (EEC) allowing voluntary participation by companies in the industrial sector in a Community Eco-audit scheme." Com (91) 459, *Official Journal of the European Communities*, C76 1-13.

20. International Chamber of Commerce, 1989. *Environmental Auditing.* Paris: ICC.

21. Wheeler, D., 1992. Memorandum by The Body Shop International. In: *A Community Eco-audit Scheme. 12th Report of the Select Committee on the European Communities*, House of Lords Paper 42. London: HMSO, 58–59.

22. Callenbach, E., F. Capra, and S. Marburg, 1991. "The Elmwood Guide to Eco-Auditing and Ecologically Conscious Management." Global File, Report No. 5. Berkeley, Calif.: The Elmwood Institute.

23. Smith, D., 1992. "Strategic management and the business environment: what lies beyond the rhetoric of greening?" *Business Strategy and the Environment*, 1(1), 1–9.

24. Commoner, B., 1990. "Can capitalists be environmentalists?" *Business and Society Review*, 75, 31–35.

25. Gray, R., 1993. *Accounting for the Environment.* London: Paul Chapman Publishing.

26. ENDS (1992). "BSO/Origin: Putting a price tag on environmental damage." ENDS Report 210, 19–21.

27. Victory, K. (ed.), 1993. "Companies begin using green accounting to pin down environmental costs." *Business and the Environment*, 4(1), 2–5.

28. Wheeler, D., 1993. "Two years of environmental reporting at The Body Shop." *Integrated Environmental Management.* October 1993, 13–16.

29. Wheeler, D., 1993. "The future for product life cycle assessment." *Integrated Environmental Management*, 20, 15–19.

30. Coles, T., 1992. Supplementary memorandum by the Institute of Environmental Assessment and Verbal Evidence. In: *A Community Eco-audit Scheme. 12th Report of the Select Committee on the European Communities.* House of Lords Paper 42, 39–50. London: HMSO.

31. Hillary, R., 1993. "Cleaner by choice." *Environment Risk,* December/January 1993, 30–31.

32. Coles, T., 1991. "Standards and contents for EC Eco-audits." In *Advances in Environmental Auditing. Proceedings of an IBC/IEA Conference.* London: IBC.

33. Schot, J., 1992. "Credibility and markets as greening forces for the chemical industry." *Business Strategy and the Environment*, 1(1), 35–44.

34. DTTI, Sustainability and IISD, 1993. *Business in the Goldfish Bowl.* "Corporate environmental disclosure: why sustainable development depends on it." London: Touche Ross.

35. Coalition for Environmentally Responsible Economics, 1990. *The 1990 CERES Guide to the Valdez Principles.* Cambridge, Mass.: CERES.

36. United Nations, 1992. Agenda 21. Geneva: United Nations.

37. European Community, 1992. *5th Environmental Action Programme. Towards Sustainability.* Brussels: Commission of the European Communities, 93–97.

38. Advisory Committee on Business and the Environment, 1993. *Report of the Financial Sector Working Group.* London: Department of Trade and Industry and Department of the Environment.

39. ENDS, 1992. "Body Shop sets precedent with pledge on renewable energy." ENDS Report 208, 5–6.

40. Jones, M., 1992. "Environmental reporting: will accountants lose out?" *Certified Accountant,* August 1992, 26–28.

PART 4

Establishing Your Own EHS Audit Program

8

Setting up Your Own Audit Program

Peter C. Chatel

Senior Environmental Auditor,
The Coca-Cola Company

So, you have been asked to set up your own audit program! Well, as the saying goes, There's good news and there's bad news. The bad news is that you and your organization are *way* behind the curve. Over the last 15 to 20 years, environmental, health, and safety auditing (I'll use "EHS auditing" or just "environmental auditing") has grown significantly and become an expected practice—an integral part of *everyone's* definition of environmental management systems. Now for the good news: you can learn from all that experience. The lessons learned over these years will help you avoid many of the pitfalls encountered by others.

This chapter presents a strategy for designing and implementing an EHS audit program—from establishing the program's mission, goals, and objectives to matching resources so that these goals and objectives are achieved, from selecting the tools for your auditors to avoiding problems already solved, and from communicating the audit results to ensuring that appropriate corrective actions are taken.

Establishing the Audit Program's Mission, Goals, and Objectives

Aligning the Program with the Company's Mission

Over the past 5 to 10 years, most companies have been through a process to identify the mission of the organization. As Stephen Covey has written, the mission statement can be thought of as the constitution of the organization—the expression of vision and values. As such, the mission becomes the fundamental yardstick which everything else gets measured against. If you aren't sure whether your organization has a written mission—find out! If it does not have one, encourage developing one.

If your organization has a mission statement, ensure that you have a copy and review it periodically. Before getting very far, take time to discuss the organization's mission with your management, peers, and direct reports. Internalizing this mission serves as the compass in designing the audit program.

Mission statements vary from organization to organization—always unique to the customers, products, and people. The mission may be any of the following:

- To be and stay first in market share
- To produce the best at the lowest cost
- To exceed all the customers' needs and expectations
- To protect and enhance the trademark and share owner value
- To have the products of choice and the workplace of choice

Just as any organization should have a mission statement, so too, should the audit program. The audit program's mission statement should be directly linked to the mission statement of the organization. In fact, if the audit program's mission statement cannot be linked to the mission statement of the organization, then the audit program probably will not survive—much less thrive. In these times of furious change (e.g., customer expectations, regulatory requirements, competitive advantages), organizations are frequently scrambling to assess the effectiveness, contribution, significance, and value of almost everything. If any program isn't explicitly contributing to fulfill the organization's mission then its time is up!

Obtaining Consensus on Program Goals

If the mission statement is described as the constitution, then the audit program's goals can be described as the law of the land. One thing to remember, particularly for those of us living in the United States, the more laws there are, the harder it becomes to enforce them! Stated another way, keep the audit program's goals achievable. If the audit program is pressed to be all things to all people, it will more likely be very little to anyone.

The audit program could have many potential goals:

- To reduce environmental impacts and protect the environment
- To improve legal performance
- To enhance profit
- To train/develop staff
- To protect management

While all of these may be desirable, the challenge is to identify the most significant. Pick the goals necessary to meet the mission. Recognize that the greater the number of goals and objectives, the greater resources, both financial and human, required.

In addition, consider whether the audit programs should have both short-term and long-term goals. In the short term, the audit program may focus on eliminating the problems operations have in meeting compliance requirements. In the long term, the audit program may identify management systems' weaknesses in a way that not only improves performance but also enhances efficiency and competitiveness.

While it may seem apparent that everyone in the organization knows what the goals of the audit program should be, frequently each individual has his or her own perspective. And, more times than not, these perspectives are quite different. Real value is achieved by having the appropriate decision makers meet to discuss the goals and objectives for the audit program. A well-facilitated meeting results in insightful discussions. These discussions illuminate the necessary goals and objectives and allow reaching consensus. Every other aspect of the program's design and implementation is driven by the need to fulfill the program's goals and objectives. Do not take a shortcut here—these discussions are critical.

Clarifying Individual Audit Objectives

The objectives for individual audits need to be clear. Audit team members need to know whether individual audits will:

- Verify compliance with applicable laws and regulations
- Assess environmental management systems
- Evaluate environmental impacts
- Train facility management
- Identify EHS hazards and propose control strategies
- Recommend improvements

Time and time again, questions and challenges arise during implementation of the audit program. When the mission, goals, and objectives are clearly and commonly understood, the answers are apparent.

Manager's Summary

When setting up your audit program begin by:

- Aligning the program with the company's mission
- Obtaining consensus on program goals
- Clarifying individual audit objectives

Matching Resources to Meet the Audit Program's Needs

Managing the Program

After clearly identifying the audit program's mission, goals, and objectives, the next task involves finding the right people to make it all happen. The program will need a manager, audit team leaders, and audit team members. The mission, goals, and objectives of the program, as well as the complexity of the organization and operations, will largely determine the number of people required in each of these roles. The responsibilities of each of these roles will also affect the type of person required. Figure 8-1 lists typical responsibilities for program managers, audit team leaders, and audit team members.

Some of the more important issues include whether to:

- Staff the audit program internally or externally
- Have dedicated audit staff or some other option
- Have a full-time or part-time program manager
- Make further decisions on how to involve the legal department in the audit program

To get the most out of your audit program, *find a way* to staff it from within the organization! The people who learn the most on any audit are the auditors. And, if you hire someone else to do the work, many of the audit lessons will leave when they do. By doing it yourself, the initial audits may be a little less thorough, comprehensive, and consistent, but there are ways to deal with these potential shortcomings. Hiring consultants in the short term may help. In particular, organizations having little internal EHS auditing expertise should consider hiring consultants to assist with program start-up. By supplying experienced team leaders and/or team members, consultants can help to ensure that the initial audits are performed properly.

Consultants serve three roles quite effectively:

- Providing expertise not currently within your organization
- Providing resources necessary for a quick start-up when they are not available internally
- Providing independence and objectivity

The choice between using full-time dedicated audit staff or part-time, collateral-duty staff may dictate the need for some form of ongoing outside assistance. Full-time audit staff offer some significant advantages, most notably, greater auditing proficiency and consistency. Part-time staff, on the other hand, allow the audit learning, referred to above, to be more quickly and broadly dissemi-

Audit Program Manager Responsibilities
Program Implementation

- Select and schedule audits
- Identify audit team leaders and members for individual audits
- Negotiate wording of audit results with legal and operational staffs
- Participate in selected audits
- Review audit team working papers prior to report issuance
- Review audit draft reports
- Provide feedback and development opportunities to audit program staff
- Conduct performance valuations of audit program staff
- Report periodically to senior management regarding the program results
- Coordinate the tracking of audit results and ensure follow-up occurs within the organization

Program Review

- Coordinate periodic quality assurance assessments of the audit program by either internal audit program managers or external specialists
- Analyze audit results to identify trends for management
- Evaluate audit mission, goals, and objectives to periodically update and modify

Audit Team Leader Responsibilities
Preaudit Activities

- Select team members, assign audit responsibilities, gather and distribute background information, and schedule team meetings
- Review federal, provincial, and local regulations and company policies and procedures
- Modify protocol to reflect facility-specific requirements and information gained during review of background information
- Determine and confirm travel/hotel arrangements with the team members and facility
- Prepare items for audit (necessary forms, supplies, protocols)

Figure 8-1. Typical program manager, team leader, and team member responsibilities.

On-Site Activities

- Lead opening meeting presentation
- Serve as liaison between team and facility personnel to ensure all team members are appropriately scheduled to meet with facility personnel
- Perform audit duties as determined by the audit plan and solicit feedback from each team member on the status of work accomplished
- Review assigned protocol steps with each auditor to ensure that all steps are covered appropriately and to document the rationale for changing the scope of the audit (if necessary)
- Understand the context for and meaning of each finding reported by the team
- Provide periodic feedback to facility personnel on the status of the audit and findings
- Prepare the exit meeting discussion sheets listing all findings, and ensure that all findings are reviewed by each team member
- Review all findings with the key environmental contact persons prior to the exit meeting to ensure the accuracy of all findings
- Lead exit meeting; summarize reporting schedule and format

Postaudit Activities

- Review all working papers to ensure that all topics were covered and that all findings are corroborated by working paper notes
- Prepare and distribute draft report; incorporate comments where appropriate into the final report
- Organize legal review of report
- Ensure compliance with any records retention policy in place

Team Member Responsibilities
Preaudit activities

- Make travel arrangements (if required)
- Attend preaudit team meeting (if required)
- Prepare for the audit by reviewing appropriate federal, provincial, and local regulations, company policies and procedures, and available background information

Figure 8-1. (*Continued*)

On-Site Activities

- Perform duties assigned by the team leader during the audit
- Serve as a resource for other audit team members during the audit
- Report on your progress to the team leader throughout the audit, including any problems encountered
- Share observations/concerns with other team members during the audit to ensure each is addressed appropriately
- Keep facility personnel apprised of findings as they are noted
- Summarize all your findings and report them to the team leader before the exit meeting
- Assist with preparing the exit meeting discussion sheets
- Ensure that all findings noted in your working papers are presented on the exit meeting discussion sheets and accurately reflect the facts as you understand them
- Contribute during the exit meeting when questions are raised about findings which you had discovered

Postaudit Activities

- Review draft audit reports for:

  Wording changes
  Suggested changes in placement of findings with the report
- Provide input as necessary when findings in the draft report are challenged.

Figure 8-1. (*Continued*)

nated. This is particularly true when audit team members are pulled from a pool of individuals located throughout the organization.

Frequently, the staffing available to the audit program is a given, with little flexibility. In this case, it is important to recognize the strengths, as well as the limitations of the given resources. Take time and steps to design the program to address such recognized limitations. For instance, develop audit protocols or questionnaires with greater detail to build in consistency when part-time auditors are being used.

The coverage (frequency and number) of operations will, in large part, prescribe the need for a full-time versus part-time program manager. As a rule of thumb, greater than 30 audits per year warrants a full-time program manager.

Identifying and Selecting the Audit Team

Regardless of whether full-time or part-time staff is used, the specific individuals selected to assist with the program will be critical. When considering individuals for program managers ensure that they have the experience and organizational stature to relate to plant managers and board members. In addition, when considering individuals for team leaders and members, look for the following experience:

- Plant operations
- EHS laws and regulations
- EHS technologies
- EHS auditing
- EHS training

Further, team members need to have strong interpersonal skills. Auditors face many difficult situations. The ability to keep cool under fire is crucial. Someone who is technically an expert should also have common sense, openness, flexibility, a sense of humor, compassion, and honesty—skills and qualities that will help to diffuse potential powder kegs.

There's a saying, You are only as good as your latest effort. Remember, identify the best possible candidates for each audit! One bad audit can undermine your past successes.

Obtaining Legal Assistance— It's Not All Bad

As noted above, experience with EHS laws and regulations is important. What better place to find that expertise than the legal department. This, of course, depends on the breadth and depth of your organization's legal department. But regardless of the level of EHS expertise you find, the legal department will serve an extremely important function to the audit program. For some, working with the legal department is undesirable—frustrations result for a number of reasons. However, working with the legal department is a necessary and important component of the audit program.

The legal department serves a vital role in any organization. Fundamentally, it protects the organization. The legal department can also help protect the audit program results through the attorney-client privilege. Further, in select cases where litigation is suspected, the attorney work-product doctrine may provide protection. Legal personnel can help protect the audit program internally, as well. The legal department can provide significant assistance, including:

- Identifying and obtaining applicable laws and regulations for the audit teams to use
- Reviewing and editing audit reports to ensure accuracy and clarity

- Reviewing and commenting on proposed corrective actions for sufficient due diligence
- Serving as audit team resources
- Providing legal interpretations, particularly in the event of a difference of opinion
- Helping to evaluate and analyze audit results
- Supporting the audit program in discussions and presentations to senior management

Manager's Summary

Selecting the people to carry out the audits is another critical step in setting up the program. Remember:

- Pick the best people you can.
- Design the audit tools to address potential resource shortcomings.
- Engage the legal department as an audit partner.

Designing and Implementing Audit Tools and Techniques

With resources known, the next process involves developing the tools and techniques necessary for consistent and effective audits. This section will focus on three tools and techniques that are fundamental to EHS auditing:

- Using audit questionnaires, checklists, and protocols that work
- Writing working papers worth reading
- Conducting meaningful interviews

Audit Questionnaires, Checklists and Protocols

In addition to the applicable laws and regulations, the principal tools used by the auditors in the field are the EHS audit questionnaires, checklists, and/or protocols (see Fig. 8-2 for examples of each). The specific design of each of these tools can vary. However, the content of what to consider is frequently quite similar and is driven by the audit scope.

Staffing of the audit program should force certain considerations when it comes to the tools to use. Using part-time auditors necessitates greater detail in questionnaires and checklists or, to facilitate greater consistency, audit protocols may be the preferred option. Audit protocols provide detailed instructions to the auditors, indicating not only what to look at but also how to look. The audit pro-

Environmental Management Checklist

INSTRUCTIONS: Use interviews (I), plant tours (T), and/or document review (D) to confirm that the statements apply to the facility being reviewed. Place an I, T, and/or D in the space to the left of each statement depending upon the method(s) of confirmation used.

A. <u>Overall Environmental Management</u>

___1. Environmental programs and regulatory requirements get management attention and monitoring in a similar way as production, quality, cost, safety, and other core performance requirements.

___2. Policy statements, job descriptions, and performance standards (e.g., Individual Management Agreements) reinforce line responsibility for environmental compliance.

___3. Total Quality Management, or similar techniques, are employed to search for ways to reduce environmental risks and enhance environmental performance on an ongoing basis.

___4. The Plant Manager has communicated and widely distributed the Company's environmental policy statement and ensured thorough employee understanding of the document.

___5. Facility procedures were modified/developed to ensure incorporation of Company policies into day-to-day operations.

___6. There is a formal incident reporting process, which defines internally and externally reportable events, including a process for investigating certain incidents.

___7. The facility has a formal system to assess employee training needs. The facility subsequently offers and monitors the required environmental training, especially for those employees whose activities may have an impact on environmental performance (line managers, engineering and maintenance staff, operators, craftsmen, etc.).

___8. The facility has a system to ensure required environmental documentation is properly maintained and controlled.

___9. Environmental requirements are built into the standard operating procedures and defined in specific procedures, including preventive maintenance of pollution control equipment, sampling/monitoring reports, hazardous materials and waste management, pollution prevention, etc.

___10. There is routine reporting to the Plant Manager and his direct reports, as well as to corporate environmental staff, as required, on the environmental status of operations (including issues and exceptions).

Figure 8-2. Sample audit tools.

Procedures and Controls Questionnaire
Environmental Management

QUESTION	YES	NO	NOT APPLIC- ABLE

A.   Water and Wastewater Management

1.   Is drinking water quality monitored at some
     frequency? (Step 12)   —   —   —

2.   Has the facility identified the drinking water
     sampling points using some rationale? (Step 9)   —   —   —

3.   Is monitoring equipment calibrated routinely?
     (Step 12c)   —   —   —

4.   Are samples taken and analyzed by trained
     personnel? (Step 12b)   —   —   —

5.   Does the facility have procedures to implement
     when monitoring results exceed acceptable levels
     that include notifying affected populations?
     (Step 13d)   —   —   —

6.   Does the facility retain drinking water monitoring
     records? (Step 14)   —   —   —

7.   Has the facility taken any steps to specifically
     identify all of its wastewater discharge points
     (e.g., reviewing piping diagrams, dye studies,
     tours of the plant, etc.)? (Step 18)   —   —   —

8.   Were all wastewater discharges characterized
     prior to submitting most recent permit/license
     application? (Step 19)   —   —   —

Figure 8-2. (*Continued*)

Environmental Audit Protocol

Facility Name:_____

Dates of Audit:_____

Team Members:_____

Review Period:_____

This protocol is intended as a guide for planning and conducting environmental audits at Company's locations. It may require additions, revisions, or other modifications in order to meet the needs of facility-specific audit objectives or other special circumstances. This document is a working document and will undergo changes over time in order to increase its usefulness.

This protocol consists of the following sections:

Figure 8-2. (*Continued*)

	Auditor(s)	Working Paper Reference

D. Aboveground and Underground Storage Tanks

Chemical Storage Tanks

1. Evaluate the facility's practices regarding the storage of hazardous chemicals in bulk tanks by performing the following:

    a. Tour the facility, review documents, and interview personnel to identify the location of bulk storage tanks.

    b. Confirm that adequate secondary containment is provided for all aboveground storage tanks (e.g., drainage dikes, or other containment systems are in place around storage tanks). *Note: Coordinate with auditor assigned to spill control and emergency planning protocol.*

    c. Confirm that all secondary containment is adequately sized (i.e., capable of holding the volume of the largest single tank as well as allowance for precipitation; and of suitable materials of construction).

    d. Confirm that all nearby ancillary equipment (e.g., pumps, piping, fill connections) are contained within the secondary containment.

    e. If the tanks are located inside buildings, verify the following:

        1) Each connection through which liquid normally flows is equipped with a valve as close as possible to the tank.
        2) The inlet of the fill pipe is located outside the building away from any building opening and free from any source of ignition.

    f. Confirm that area monitoring is conducted, remote shut-off valves are installed, and grounding/bonding devices are in place.

2. Confirm that during loading and unloading operations the following good management practices are followed:

    a. Prior to transfer adequate volume in the receiving tank has been confirmed.

Figure 8-2. (*Continued*)

Topic and Protocol Step:

Date:_____
(month/day/year)

Initials:_____

	Action Notes	Protocol Step	Working Paper Notes
1.			
2.			
3.			
4.			
5.			
6.			
7.			
8.			
9.			
10.			
11.			
12.			
13.			
14.			
15.			
16.			
17.			
18.			
19.			
20.			
21.			
22.			
22.			
23.			
24.			
25.			
26.			

Summary:_____

Figure 8-2. (*Continued*)

tocols can also provide detailed testing plans for auditors where audit proficiency is an issue. Full-time auditors require less detail. In fact, experienced professionals frequently prefer short checklists that highlight only the issues to consider.

Remember the customers of the questionnaires, checklists, and/or protocols are the auditors. Whatever is designed and implemented must be usable and effective for these people. By soliciting feedback from the auditors, you will make sure audit questionnaires, checklists, and protocols will continuously evolve and improve.

Once developed, the questionnaires, checklists, and/or protocols will need to be maintained and periodically updated. A temptation when developing these tools is to include regulatory, legal, or policy references for quick and easy use by auditors. This necessitates more frequent updating, which is time-consuming and tedious work.

Some consulting firms have developed computer-based applications that include detailed audit questionnaires, checklists, and/or protocols. These are worth investigating. Even if you decide against this approach now, the advancements in this technology over the next 5 years will lead more and more practitioners into this area. For this reason, it is worth finding out what is available and where the different vendors are heading with their products.

Working Papers

The auditor's field notes or working papers are a critical component of every audit. At the completion of the site visit, the working papers remain the *only* records of the audit team's efforts. These working papers become the basis of the audit report and serve to document the team's success in achieving the audit's objectives. Simply put, the working papers should fully document what the team accomplished while on site:

- What was looked at
- Who was spoken to
- What findings were identified and shared with plant management

In addition, for the audit program manager, the working papers provide a significant tool to measure the quality of an audit. During review, the program manager can evaluate completeness, clarity, and consistency. Encouraging team members to review each other's working papers can also help audit quality.

There are 10 basic habits to instill in auditors for developing effective working papers:

- Write while working; do not wait until later or procrastinate about developing working papers.
- Note who was interviewed, what documents were reviewed, and what equipment was inspected.

- Write so others can read the notes taken.
- Review the notes frequently to check that they make sense.
- Include exhibits (photocopies of key documents).
- State facts and avoid opinions.
- Avoid extreme language (e.g., This facility's hazardous materials storage is terrible).
- Ask others to read the notes taken to confirm completeness.
- Summarize the conclusions for each issue evaluated.
- Keep the notes organized.

In order to achieve these goals, develop working papers in a dedicated tablet for audit note taking. Sometimes it is helpful to have these tablets specially bound or distinctively colored so that everyone in the organization recognizes the audit working papers. Such recognition facilitates their proper management.

The working papers represent a significant asset to the audit program but also present a potentially significant liability to the organization. Therefore, store working papers in a controlled and secured manner. Work with your legal department and records management staff to devise the specific record retention policy for working papers. Keep working papers at least until issuing the final audit report. Beyond this point, each organization has its own variation.

Interviewing Technique

During any audit, a tremendous amount of information gathering occurs through interviews. As a result, interviewing skills dramatically affect the effectiveness of any auditor and audit team. While many people have received training on how to conduct effective interviews, the techniques are not always practiced during the heat of an audit. Emphasizing the following seven habits of highly effective interviewers, can significantly improve the effectiveness of each audit.

1. *Plan the interview—Begin with the end in mind.* By taking time to identify the desired outcome of the interview, the effective auditor is more likely to achieve it. Being clear about what information is needed will help to identify the best person to interview as well. You can save important time by avoiding the people who know the least about the issues of interest.

When planning the interview:
- Create an outline of the issues to cover with some initial questions
- Clearly identify the most appropriate person to interview
- Establish a time, place, and desired length for the interview

2. *Introduce the interview—Set the context for why the questions are being asked.* Before beginning any interview, provide a short but clear introduction. People are typically much less defensive when they have a greater understanding of

the purpose and scope of an interview. A common mistake made by auditors includes assuming that the interviewee knows why the audit is taking place and why the auditor wants the interview. Skipping this introduction may cause barriers to the free flow of information auditors would like. Offering this introduction eliminates any preoccupation the interviewee may have with issues unrelated to the topics covered—like How is this information going to be used?

3. *Ask open-end questions—Be a reporter.* Effective auditors use open-end questions. Frequently referred to as reporter's questions, these questions begin with who, what, when, where, why, and how. The intent is to get the interviewee to talk and tell the interviewer what is going on. Effective interviewers are guided by the Pareto principle of interviewing; that is: The interviewer should be doing about 20 percent of the talking while the interviewee is doing about 80 percent of the talking. Another way to look at it: if the interviewer is hearing his or her voice a lot, then he or she is asking the wrong questions.

Another common pitfall of auditors includes using (closed-end) questions, where the interviewer provides the answer—So, you do wash out the tank between each batch?, or That area is inspected weekly, right? Many auditors find such questions a little less threatening to ask. Others find that this is an excellent way to demonstrate their knowledge. Still others like to build rapport through such questions. While rapport is desirable, closed-end questions lead to inaccurate information being transferred, premature closure of the interview, and confusion of the interviewee. *Avoid closed-end questions, avoid closed-end questions, avoid closed-end questions!*

4. *Summarize the information received—Use active listening skills.* Periodically, each effective auditor will pause and summarize the information received to ensure an accurate and complete understanding. In social sciences, this is called "active listening." By repeating the information back to the interviewee, the interviewer can check to verify the information was accurately understood. While doing this, cautiously note the reaction of the interviewee. When the interviewee is adding onto the summary or restating it in some other way, some important distinction is being made. Take care to understand the distinction and the implications on the issue being discussed.

5. *Make sure the interviewee is comfortable—Be aware of the environment.* Always consider the comfort of the interviewee. The location and setup of the interview is extremely important in enabling the free flow of information between the auditor and interviewee. Choosing a location where the interviewee feels at home frequently induces more effective interviews. Rather than calling the interviewee to the conference room assigned to the audit team, go to the interviewee's office or machine area. Now, if the work area of the interviewee is too loud or otherwise distracting (e.g., constant phone calls, other workers interrupting) find some alternative. Also, avoiding the typical setting of talking across a table will improve the comfort level.

6. *Close the interview positively—Say thanks for the help!* Effective auditors always find the way to close an interview positively—even when the interviewee has

been less than cooperative. In almost every case, sincere appreciation can be expressed for the information and cooperation. Even when the interviewee has been challenging, the effective auditor will say, Thanks for the time. Just plain common courtesy and saying thanks will serve you well—particularly if you need to return to conduct a second interview with the same person.

7. *Document the results of the interview—Write during and after the interview.* Because we all have imperfect memories, the experienced auditor develops and cultivates the habit of documenting information gathered during interviews. The techniques used to document this information may vary. However, all effective auditors ensure that at the close of every interview, they set aside some time to document the important information. If you write during an interview, make sure you are not distracting the interviewee. Also, avoid scheduling interviews consecutively without allowing at least 5 minutes to focus and write down what you heard before it's forgotten.

The ability of each auditor individually, and the audit team collectively, to practice these seven habits will improve the quality and quantity of information gathered.

Manager's Summary

Design audit protocols, questionnaires, and checklists with the audit staff in mind—make them user-friendly.

Take time to review and comment on each team member's working papers. Provide meaningful feedback to each team member on how to improve. In addition, encourage the audit staff to review and critique each other's working papers.

Frequently review and practice interviewing skills and techniques.

Communicating the Audit Program Results

The general purpose of any audit is to review and evaluate an organization's performance so that the appropriate levels of management can be informed. Therefore, to achieve this purpose, the process of communicating the audit results becomes vital. The manner and form for communicating audit results has tremendous impact and, therefore, needs careful consideration. This section focuses on the steps you can take:

- Communicate on site
- Prepare audit reports
- Share audit information broadly within an organization

Communicating on Site

Taking time to communicate the audit findings while on site is essential. In fact, sharing the audit team's observations and concerns during, and not just at the completion of, the on-site activities will help to ensure a smooth exit meeting with management and an accurate final report. In addition, by confirming the accuracy of the team's understanding along the way, the team can avoid misunderstandings leading to wasted time. Further, keeping the facility management well-informed throughout the audit significantly reduces the anxiety level of those being audited. With reduced anxiety, those being audited are typically more open and candid.

Team meetings, held at either the beginning or the end of each day, provide the most effective forum for the on-site communication. These meetings should be open to all facility management but should definitely include the key site contact (e.g., environmental coordinator). Team members should be encouraged to share their important accomplishments, key concerns or findings, and the additional information and interviews required. Facility representatives participating in these team meetings should participate fully and help clarify any misunderstandings. The goals of these meetings include:

- Ensuring the audit team accurately understands the facility's management and performance
- Ensuring the facility is aware of the team's concerns
- Allowing the team leader an opportunity to determine the progress made by the team
- Facilitating the planning of further actions necessary to complete the on-site activities

Conducting the Exit Meeting

Before leaving the site, the team should conduct a formal exit meeting with facility management. During this meeting, the team should share the audit results in as detailed a fashion as possible—portable computers make having a typed draft report ready for this meeting quite feasible. The written list of audit findings should be completely reviewed with plant management. Every effort should be made to discuss each finding that will be presented in the final report. When issues require additional research before reaching a final conclusion, apprise management of these open items.

The goal of the exit meeting is to provide an opportunity for plant management to correct any misinformation obtained during the audit. Take the time necessary to clarify any questions or concerns before leaving the facility; the time required to resolve questions after leaving the facility is almost always an order of magnitude greater.

Preparing Individual Audit Reports

Audit reports should clearly and concisely present the audit results. The exact format and content of the final audit report varies between companies. The needs and expectations of the principal recipients should drive the format and content. Perhaps you will be best guided by checking with these individuals prior to preparing the first report.

Audit reporting has changed over time. Indeed, the nature of environmental audit reporting has evolved significantly over time. When this practice was just getting started, the reports were typically nothing more than a memo which summarized the trip to the plant along with some recommendations. Over time these reports took on greater formality and began to focus more on capturing the exceptions to required practices. Recently, several organizations have added an audit opinion to the report. The opinion is worded in a way that summarizes the audit team's judgment of the overall level of performance at the audited facility.

Some suggestions to improve the effectiveness of your audit reports follow:

- Keep the report brief—avoid describing everything that was done during the audit.
- Focus on the exceptions to required practices, but highlight noteworthy (positive) practices as well.
- Summarize and reference the applicable standard for each audit exception before describing the conditions found.
- Clarify whether recommendations are expected as part of the report before starting.
- Use simple language and avoid extreme wording.
- Have the legal staff review the report before issuing it.

An example table of contents for an environmental audit report is provided in Fig. 8-3.

Sharing Audit Information

Who should receive the audit report? The audit report's distribution provides assurance that the appropriate levels of management are aware of the audit results. This distribution should balance the need for the appropriate levels of management to receive the information with the need to provide some protection from potentially damaging audit results. Typically, the recipients of the audit reports include the individuals in the management chain who authorize the capital expenditures to address the audit results. These receipts may include:

- Plant manager
- Division manager

TABLE OF CONTENTS

Objectives and Scope of Examination ....

Approach...

Report Format..

Priority Ranking...

Noteworthy Programs ....

Audit Opinion ....

Findings and Recommendations...

      A. Water and Wastewater Management ....

      B. Solid and Hazardous Waste Management ....

      C. Storage Tank Management ....

      D. Spill Control and Emergency Response Planning ....

      E. Hazardous Materials Management ....

      F. Air Pollution Control...

      G. Energy Management ....

      H. Compliance Program Weaknesses...

      I. Environmental Improvement Opportunities ....

Figure 8-3. Sample table of contents for environmental audit report.

- Group manager, vice-president, and/or president
- Legal department
- Environmental affairs

Beyond the formal report, taking time to broadly communicate the trends of the audit findings can help the overall organization make swifter performance improvements. Each year, some organizations take time to share the top ten common audit exceptions throughout the organization with the hopes that when the audit team shows up, these issues will have already been addressed. In addition, this periodic analysis of the types of exceptions and the underlying root causes can help identify policy, standard, guidance, or training needs. And, when the same findings continue to reappear year after year, the organization will see the results of ineffective communication, or the failure to identify real root causes. Suggesting more significant system redesign may be necessary.

Manager's Summary

Make sure to communicate the audit results openly with facility management while on site through daily team meetings and with a formal presentation just prior to leaving the site.

Prepare a clear and concise final report that is targeted to meet the needs and expectations of the primary recipients.

Periodically evaluate the audit program's results and share common trends throughout the organization to facilitate swifter and broader performance improvement.

Corrective Action Planning and Audit Follow-up

Without the commitment to address every issue found during environmental audits—do not audit! In this field, anyone knowledgeable and experienced will provide you with this advice. When problems are identified, organizations only increase their liability by failing to responsibly address and correct these problems. When legal compliance related issues go uncorrected, organizations are subject to not only greater civil penalties but harsher criminal sentences. Obviously, the corrective action planning and audit follow-up phases of the audit program are vital.

Closing the Loop to Avoid a "Smoking Gun"

Corrective action plans must be developed for every audit. These plans should clearly identify:

- What actions will be taken to correct each issue identified during the audit
- Who will be responsible for managing the implementation of each action
- What resources will be required to complete each action
- When each action will be completed

An example corrective action plan form is provided in Fig. 8-4.

Corrective action plans should focus on root causes. Often, several findings result from a single root cause and can be effectively remedied with one corrective action addressing the root cause. Identifying root causes is frequently more

Environmental Audit Program
Proposed Corrective Actions

Facility: _____

Final Report Date: _____

Report Number: _____

Acknowledged: _____

The Corrective Actions set forth below will be undertaken by the Responsible Party within the dates specified.

Signature _____ Date _____

Finding #	Functional Area	Type	Title	Corrective Action	Proposed Completion Date	Responsible Party	Date Completed	Comments

Type:
R - Regulatory
P - Company Policy
W - Compliance Program Weakness
O - Environmental Opportunity

Distribution:
(*same as report distribution*)

Figure 8-4. Sample corrective action plan form.

difficult than it sounds. Several techniques have been developed to assist with the process. Perhaps the simplest but most effective approach can be called "five whys?" The premise of this approach is that the root causes will emerge after asking Why? at least five times to each audit exception. Encourage the audited facilities to address root causes in preparing the corrective action plan.

Corrective action plans must be responsive, timely, and cost-effective. Given its significance the plan should be reviewed for completeness. The corrective action plan review should include the audit team or, at least, the team leader. In addition, for legal-compliance-related issues the legal department should confirm appropriate timing of corrective actions. For budget reasons, appropriate line management should also review the corrective action plan. Review of the corrective action plan will help to ensure that all audit exceptions are addressed effectively and efficiently.

Ensuring That Due Diligence Is Demonstrated

After the corrective action plans have been reviewed and approved, follow up on the actual implementation of the actions to confirm that actions take place. The confirmation of corrective action helps to demonstrate the organization's due diligence in its efforts to correct known deficiencies. The confirmation of corrective actions should take place with periodic status updates. In addition, the confirmation should include some physical review of the facility's operations. This detailed confirmation step will help the organization protect itself against charges of negligence.

Manager's Summary

Develop a corrective action plan for every audit and then ensure the plan is followed.

Conclusion

You are embarking on a very important journey—a journey that others have taken. The lessons learned by those who have preceded you can serve you well. Take time to plan and prepare for this journey—be as clear as possible about your destination before you begin. The sights and insights you gather along the way will be invaluable—take time to share them with others. Good luck!

9

Computer Software for Environmental Auditing

Elizabeth M. Donley

Publisher, Environmental Software Report
and Environmental Software Directory

As a tool for identifying hazards and monitoring compliance with environmental regulations, environmental auditing is an essential part of an effective environmental management program. And computer programs, or software, for environmental auditing are the tools that can help the auditor conduct comprehensive audits of a facility and its equipment and processes in the most effective and efficient manner possible. As such, environmental audit software helps to reduce environmental risk, while demonstrating that the company, or facility, is serious about maintaining compliance. In addition, according to developers of environmental auditing software, using these systems will reduce the cost of an audit by 40 to 50 percent and save 65 to 70 percent of the time required to conduct an audit. These software systems can also serve as legal references or training tools for environmental managers, facility engineers, consultants, and corporate lawyers.

Although there are only a few software packages designed specifically for environmental audits, many systems provide legislative and regulatory information to help facility personnel stay up to date with environmental regulatory requirements. In addition, there are more than 1000 environmental software systems that can provide information or manage data for environmental compliance.

Software packages specifically designed for environmental audits are compliance-driven. That is, the software design begins with and focuses on the regulations and whether the facility meets those regulations. Reports generated are primarily for internal use. Regulatory compliance is also the objective of other software packages for environmental management; however, they focus on hazardous substances and use a chemical database as the core of the system. The basis of reports generated for the regulators is information in the database.

For many years, the only commercially available environmental auditing software was Audit Master by Utilicom, Inc. (Pittsford, N.Y.). Reviewing the history of this software package provides a brief history of environmental auditing software and how the environmental software industry has progressed.

Audit Master

Audit Master is interactive software that uses screening questions (Fig. 9-1) to walk the user through the auditing process. It guides the user through the process of determining data requirements, collecting input data, accessing appropriate regulations, developing reports, and providing a program for corrective action. The software creates an audit protocol, selecting areas of regulations applicable to the user's facility. Audit Master automatically generates a report of noncompliance issues. Features include: built-in word processing capabilities for creating working papers and additional reports, database management for tracking corrective actions, access to a digest of federal regulations, and "alert" screens for past-due corrective actions. The user can conduct a facilitywide or focused audit and add questions or references that reflect state and local regulations, company policy, or industry standards. Available modules include: air quality, water quality, solid and hazardous waste, spill prevention, storage tanks, polychlorinated biphenyl (PCB) management, Toxic Substances Control Act (TSCA), Emergency Planning and Community Right-to-Know Act (EPCRA), hazardous materials transportation, industrial hygiene, personnel safety, materials and equipment safety, and occupational medicine. The modular design adapts to individual facilities and operating procedures (e.g., industrial, medical, or educational). Audit Master runs on IBM-compatible microcomputers and is available in a network version.

Utilicom, Inc., originally a subsidiary of Rochester Gas and Electric Corporation, assisted by attorneys from Nixon, Hargrave, Devans, and Doyle (Rochester, N.Y.) developed Audit Master in 1988. The original version helped users comply with federal and New York state regulations. The 1988 version had only eight modules: air pollution, emergency planning and community right-to-know, PCBs, pretreatment, solid and hazardous waste, spill prevention and control, above- and belowground storage tanks, and water pollution and quality.

Here is a typical report that an auditor might generate after performing a comprehensive facility audit. This report includes the "Red Flags" from every audit session performed at a single facility.

Sample Summary Report

Question Heading	Question Text	References	Red Flag	Comment
SPILL PREVENTION AND CONTROL				
INVENTORY - GENERAL	Does the facility have a quantitative inventory, measured or estimated, of oil or hazardous substances stored and of releases to air, water and land? GMP–NOT SPECIFICALLY A REGULATORY REQUIREMENT	SP19	R	
OIL - RELEASE - GENERAL	The following questions concern spills or releases of Oil or petroleum products. A series of eleven spill inventory questions will repeat for each spill event. This information can be used to document recent and past releases and as a reference.	"Discharge" "Harmful" 40CFR110.3 33CFR153.203	R	
OIL - RELEASE - GENERAL - NOTIFICATION	For spill or discharge #1, was a determination made as to whether any outside entities (e.g. federal, state and local officials, emergency response teams) needed to be notified?	33CFR153.203/A	R	
AIR QUALITY				
NAAQS/TITLE I - MONITORING - GENERAL REQUIREMENTS	Does the site monitor emissions in accordance with the conditions of the permit?	40CFR60/F	R	
NAAQS/TITLE I - MONITORING - EQUIPMENT	Is all monitoring equipment working properly?	40CFR60.13	R	
ACID RAIN PROGRAM - APPLICABILITY - GENERAL	Is the facility listed in Table 1, Table 2, Table 3 of 40CFR73.10(a) or a utility unit as listed in 40CFR72.6(a)(3)?	"Commercial" 40CFR72.6(a)(3) 40CFR72.6(a)(3)(i) 40CFR72.6(a)(3)(ii)	P	
PROTECTION OF STRATOSPHERIC OZONE - OZONE DEPLETING SUBSTANCES - REFRIGERANT RECYCLING	Do you use, service or dispose of any ozone-depleting refrigerants (including equipment that uses ozone depleting refrigerants such as air conditioning or refrigeration equipment?	58FR28660/A 58FR28660/B 58FR28660/C	R	

Printed using Audit Master and a word processor

An auditor might print a report like this one which is a record of every Question answered, the Response given, and whether or not the response caused the item to be a "Red Flag."

Question Number	Question Heading	Question Text	Response	Red Flag
SCREENQ1	SCREENING QUESTION 1 - PCB ARTICLES	Does the facility have any PCB Articles?	Y	
SCREENQ2	SCREENING QUESTION 2 - PCB CONTAINERS	Does the facility have any PCB Containers?	Y	
SCREENQ3	SCREENING QUESTION 3 - PCB ARTICLES CONTAINERS	Does the facility have any PCB Article Containers?	Y	
SCREENQ4	SCREENING QUESTION 4 - PCB-CONTAMINATED EQUIPMENT	Does the facility have any PCB-Contaminated Electrical Equipment?	Y	
SCREENQ5	SCREENING QUESTION 5 - BULK PCB WASTE	Does the facility have any bulk PCB waste?	Y	
SCREENQ6	SCREENING QUESTION 6 - PCB TRANSFORMER	Does the facility have any PCB Transformers?	N	
SCREENQ7	SCREENING QUESTION 7 - CAPACITORS	Does the facility have any capacitors?	N	
SCREENQ8	SCREENING QUESTION 8 - HYDRAULIC MACHINES	Does the facility have any hydraulic machines that contain PCBs?	Y	
SCREENQ9	SCREENING QUESTION 9 - HEAT TRANSFER SYSTEMS	Does the facility have any heat transfer systems that contain PCBs?	Y	
SCREENQ10	SCREENING QUESTION 10 - MANUFACTURING, PROCESSING, DISTRIBUTING AND USE OF PCBs	Does the facility use, distribute, process, manufacture or transport PCBs or PCB Items? NOTE: PCB Items are PCB Articles, PCB Article Containers, PCB Containers, and PCB Equipment.	Y	
SCREENQ11	SCREENING QUESTION 11 - STORAGE AND DISPOSAL	Does the facility store for disposal, or dispose of any PCBs or PCB Items?	Y	
SCREENQ12	SCREENING QUESTION 12 - PCB SPILL CLEANUP POLICY	Has there been a release of materials containing PCBs at concentrations of 50 ppm or greater? NOTE: For complete coverage of spill regulations, please refer to the SPILL PREVENTION - FEDERAL module. Questions relating to PCB spills are accessible in the SPILL PREVENTION - FEDERAL module through a screening question and through focused audit.	N	

Printed using Audit Master and a word processor

Figure 9-1. Typical Audit Master Screening Questions.

This **Corrective Action Report** shows the auditor the text of the Questions, the Comments recorded while performing the audit, and the Corrective Action Plans created for each "Red Flag" item.

QUESTION HEADING	QUESTION TEXT	COMMENT	CORRECTIVE ACTION
S/N - STORM WATER GENERAL PERMIT - POLLUTION PREVENTION	Is the facility aware of the deadlines for plan preparation and compliance? NOTE: Please review deadlines for plan preparation and compliance in the references listed above.	The facility wasn't aware of the deadline dates outlined below. The plan for a storm water discharge associated with industrial activity that is existing on or before October 1, 1992 shall be prepared on or before April 1, 1993 (and updated as appropriate). It shall be implemented on or before October 1, 1993.	George Johnson, Pollution Prevention Coordinator will call to verify implementation dates for the Pollution Prevention Plan.
S/N - STORM WATER GENERAL PERMIT - POLLUTION PREVENTION	Does the plan include the following components and all of their subcomponents: 1) Pollution prevention team; 2) Description of potential pollutant sources; and 3) Measures and controls? NOTE: The components and subcomponents can be found in more detail in the references listed above.	There isn't currently a team in place though a pollution prevention coordinator has been assigned responsibility and has started his plan to submit to the team for approval.	George Johnson, pollution prevention coordinator, has the action item of ensuring the pollution prevention team is formed. His plan should be submitted prior to implementation date given by DEC.
S/N - STORM WATER GENERAL PERMIT - POLLUTION PREVENTION	Will a qualified person conduct site compliance evaluations at appropriate intervals (but not less than once a year)?	No one at the facility was assigned responsibility for these evaluations. That has rectified and evaluations will be conducted at appropriate interval.	The staff Industrial Hygienist will pick up responsibilities for ensuring site compliance evaluations are conducted at appropriate intervals.
S/N - DRP - VALID PERMIT	Is the N/SPDES permit currently valid (i.e., not expired)?	An extension has been applied for but has not yet been approved. I've sited the standard for valid permits below. 6NYCRR751.1; 40CFR122.1(b) With few exceptions, no person can discharge or cause a discharge of a pollutant to waters of the United States, (and many states "state waters") (including waters of the contiguous zone or the ocean) through an outlet or point source, without a currently valid NPDES or SPDES permit (N/SPDES permit).	We are unable to meet permit requirements due to unexpected increase in production. To handle this George Johnson, pollution prevention coordinator, will be reviewing chemicals for substitution. There seems to be no forseeable decrease in production.

Printed using Audit Master and a word processor

Figure 9-1. (*Continued*)

In July 1989, Utilicom released Federal Audit Master with six modules for federal regulatory compliance only. It did not include modules for air pollution and pretreatment. Version 2.0 (released in January 1990) allowed users to add information specific to their state or county.

As the environmental software industry began to grow, developers recognized the value of integrating their software. Utilicom expanded the capabilities of Audit Master by linking it with several software packages. In August 1989, Utilicom began distributing Chem Master (developed by Envirogenics/Softtouch Systems, Inc. of St. Mary's, Pa.). This software package manages inventories and chemical information, provides labeling capabilities, and generates reports for compliance under the Superfund Amendments and Reauthorization Act (SARA) Title III, Secs. 311, 312, and 313. In November 1990, Utilicom announced plans to provide links to ENFLEX DATA and ENFLEX INFO (developed by ERM Computer Services of Exton, Pa.). The link to ENFLEX DATA would help track corrective actions, analyze results, and identify trends, training needs, and common problem areas. The link to ENFLEX INFO would provide the complete text of environmental laws and regulations.

When the economic recession of the early 1990s began to slow the growth of the environmental industry, many software developers took the time to update

```
MATERIALS TO BE GATHERED                                         Page    1
         Standard Audit Master Report        SOLID AND HAZARDOUS WASTE MODULE
-----------------------------------------------------------------------------

*********************************************************
    MATERIALS TO BE GATHERED (MATERIALS)

MATERIALS TO BE GATHERED

In order to complete the following screening questions
and checklist all of the following materials which are
relevant to the site being audited should be gathered.
The materials listed below include all documents which
EPA requires for its Multi-Media Records Evaluation.
Please note that EPA's procedures call for a review of
records dating back four years, as well as in depth
review of documents for specific time periods.

    1. Any data for solid waste, which supports the
       facility's determination that the wastes are
       hazardous (including EPA hazardous waste codes) or
       nonhazardous; land disposal restriction analyses,
       notifications and certifications;

    2. List of all locations where hazardous wastes are
       generated, including type, quantity and EPA
       hazardous waste code;

    3. Written procedures for hazardous waste storage
       prior to off-site shipment for treatment,
       recycling, reclamation and/or disposal;

    4. List of all storage areas, including on-site
       accumulation areas, and the quantity stored at
       each location; for storage areas where waste is
       stored for more than 90 days, provide a description
       of the storage area, dimensions, waste management
       practices, shipping and receiving practices;

    5. A list of all hazardous waste storage areas,
       including tanks, drum storage areas (including
       on-site accumulation areas), pits, ponds, lagoons,
       waste piles, etc., that have operated at any time
       between November 1980 and the present;

    6. All hazardous waste manifests for the past
       four years, including notifications and
       certifications for land disposal restricted
       wastes;

    7. Waste Analysis and Waste Profile Sheets for each
       TSDF used by the facility;

    8. TSDF inspection forms;

    9. Exception Reports;

   10. Biennial Reports;
```

Figure 9-1. (*Continued*)

and improve their software. Utilicom focused on refining Audit Master's capabilities and adding additional modules. The refinements gave the user more control over the audit process (e.g., by allowing questions on a single compliance type and adding a *pending response* answer to screening questions). A worldwide licensing contract with General Electric in October 1989 called for Utilicom to modify Audit Master to develop several new modules. The subjects of these new modules, released in 1991, were hazardous material transporta-

```
SCREENING QUESTIONS                                         Page    1
     Standard Audit Master Report         SOLID AND HAZARDOUS WASTE
 - - - - - - - - - - - - - - - - - - - - - - - - - - - - - - - - - - -


************************************************************* Activity Codes **
1  SCREENING QUESTION 2 (SCREENQ2)                           * DR PI
                                                            ** References ******
                                                             *
SCREENING QUESTION 2 - SCREENING SECTION SCOPE -             *
LEVEL 1                                                      *
                                                             *
Level 1 of the screening section will take you to           *
direct questions which determine the categories for         *
the audit session.  In order to answer the questions        *
in this level, you must already know if hazardous           *
waste is generated, whether used oil generated is           *
hazardous waste, and whether recyclable materials are       *
hazardous waste.  LEVEL 1 IS RECOMMENDED FOR EXPERIENCED    *
AUDITORS ONLY.                                               *
                                                             *
To choose Level 1, answer "YES."                            *


     YES    NO    PENDING    DON'T KNOW    NOT APPLICABLE


************************************************************* Activity Codes **
2  SCREENING QUESTION 3 (SCREENQ3)                           * DR PI
                                                            ** References ******
                                                             *
SCREENING QUESTION 3 - SCREENING SECTION SCOPE -            *
LEVEL 2                                                      *
                                                             *
Level 2 of the screening section takes you to               *
questions which help you determine whether any used         *
oil generated is hazardous waste, and whether               *
recyclable materials are hazardous waste.  It does not      *
include questions which help you determine whether          *
solid waste is also hazardous waste (those questions        *
are in Level 3).                                            *
                                                             *
To choose Level 2, answer "YES."                            *


     YES    NO    PENDING    DON'T KNOW    NOT APPLICABLE
```

Figure 9-1. (*Continued*)

tion, industrial safety, and occupational medicine. Utilicom also developed an air quality module and an expanded toxic substances control module for the federal version and offered additional California regulations.

Utilicom originally (1988) priced Audit Master at $3500 per module; the federal version was $750 per module. After initially decreasing prices in the early 1990s to be more competitive, many environmental software developers increased prices. The cost of Audit Master's modules rose from $595 per module in 1991 to a range of $595 to $995 each in 1994. The price increase reflected increased regulatory coverage and more modules and training.

```
SCREENING QUESTIONS                                              Page   2
         Standard Audit Master Report          SOLID AND HAZARDOUS WASTE
-------------------------------------------------------------------------

*************************************************************** Activity Codes **
3   SCREENING QUESTION 5 (SCREENQ5)                            * DR
                                                              ** References ******
                                                              * "Solid waste"
SCREENING QUESTION 5 - WASTE IDENTIFICATION - SOLID           * 40CFR261.2
WASTE                                                         * 40CFR261.2(b)
                                                              * 40CFR261.2(c)
Does the facility generate ANY solid waste (liquid,           * 40CFR261.2(d)
gaseous, as well as solid) - discarded material that          *
is abandoned, recycled, or considered inherently waste-       *
like?                                                         *
                                                              *
NOTE:  Please view the references to see what                 *
materials are considered abandoned (40CFR261.2(b)),           *
recycled (40CFR261.2(c)) or inherently waste-like             *
(40CFR261.2(d)).                                              *


        YES    NO    PENDING   DON'T KNOW    NOT APPLICABLE


*************************************************************** Activity Codes **
4   SCREENING QUESTION 6 (SCREENQ6)                           * DR
                                                              ** References ******
                                                              * 40CFR261.4
SCREENING QUESTION 6 - WASTE IDENTIFICATION - SOLID           *
WASTE - EXCLUSIONS                                            *
                                                              *
Is ALL solid waste which you generate excluded from           *
regulation as a solid waste, for example:                     *
                                                              *
   1)  Domestic sewage;                                       *
                                                              *
   2)  Mixtures of domestic sewage and other wastes that      *
       pass through a sewer system to a POTW;                 *
                                                              *
   3)  Industrial wastewater discharges into a POTW           *
       permitted under the Clean Water Act; or                *
                                                              *
   4)  Specific wastewaters, certain mining wastes,           *
       samples meeting certain conditions, certain            *
       petroleum-contaminated media, media and debris         *
       subject for corrective action from underground         *
       tanks, and other items?                                *
                                                              *
NOTE:  Please check the first reference (40CFR261.4) in       *
the Requirements Guide for a complete list of                 *
exemptions.                                                   *


        YES    NO    PENDING   DON'T KNOW    NOT APPLICABLE
```

Figure 9-1. (*Continued*)

Question Number	Question Heading	Question Text	RES	References
SCREENQ1	SCREENING QUESTION 1 - NPDES PERMITS	Is there a discharge (including storm water and combined sewer overflows) to surface water from (or within) the facility?	Y	"Discharge
SCREENQ2	SCREENING QUESTION 2 - STORM WATER DISCHARGE	Does the discharge consist of storm water ONLY?	N	"Discharge
SCREENQ3	SCREENING QUESTION 3 - DISCHARGES TO POTWs	Does the site discharge to a Publicly-Owned Treatment Works (POTWs), also referred to as a municipal sewage treatment plant?	Y	"POTW"
SCREENQ4	SCREENING QUESTION 4 - UIC PROGRAM	Does the facility have (or has it ever had) any underground injection wells (UIW), including: 1) Parking lot or roof drain "dry wells" or dry sumps; 2) Other dry wells or sumps; or 3) Any structure used for disposal of liquid wastes which is deeper than its largest surface dimension?	N	"Undergrou
SCREENQ5	SCREENING QUESTION 5 - DRINKING WATER	Does any portion of the site's drinking water supply come from on-site wells or site controlled surface water sources?	Y	"Non
SCREENQ6	SCREENING QUESTION 6 - DRINKING WATER	Does the site's drinking water systems have 15 or more service connections or regularly serve at least 25 people for 60 days or more per year?	Y	"Non
SCREENQ7	SCREENING QUESTION 7 - DEPARTMENT OF ARMY PERMITS	Have ANY of the following occurred at the facility: 1) Discharge of dredged or fill material into a surface water; 2) Construction of a temporary or permanent artificial barrier (including a dock) across a water body; 3) Application for a federal license or certificate for a project which may result in a discharge to a navigable water and which may affect water quality; or	Y	"Discharge

Figure 9-1. (*Continued*)

Checklist Report
  Sample Audit Master CUSTOM Report   WATER MODULE

Question Number	Question Heading	Question Text	RES	References
		4) Discharge, spill and/or release of oil or hazardous substances to the environment?		
SCREENQ8	SCREENING QUESTION 8 - PRIVATE TREATMENT WORKS	Does the site discharge to a private treatment works?	Y	WQ81
SCREENQ9	SCREENING QUESTION 9 - GROUNDWATER MONITORING	Do ANY of the following categories apply to the facility: 1) Treatment, storage or disposal facility (TSDF); 2) Underground Storage Tank (UST); 3) Underground Injection Well (UIW); 4) Non-transient non community water system (NTNCWS); or 5) A site permitted to use radioactive material?	Y	"Treatment
QUSWQINV.00.02.Q	WATER QUALITY - INVENTORY - GENERAL	Do you wish to answer preliminary questions about the facility's water quality program? NOTE: These questions concern GMPs, water source(s), wastewater generation, and discharge locations. GMP--NOT SPECIFICALLY A REGULATORY REQUIREMENT.	NA	WQ2
QUSWQINV.00.12.Q	WATER QUALITY - INVENTORY - GENERAL - STATE PROGRAMS	Does the state have an EPA approved program (SPDES)?	Y	BNA -
QUSWQPPV.00.01.Q	WATER QUALITY - POLLUTION PREVENTION - GENERAL	Is there a procedure in place at your facility which assures pollution prevention is a consideration in decision making? GMP--NOT SPECIFICALLY A REGULATORY REQUIREMENT. NOTE: If you want to skip the two pollution prevention questions, please answer "Not Applicable."	NA	PPA6602(b)
QUSWQSTW.00.01.A	NPDES - STORM WATER - GENERAL - MUNICIPALITY	Does the discharge fall into ONE of the following categories: 1) A discharge from a large municipal separate storm sewer system (population greater	Y	"Discharge

Figure 9-1. (*Continued*)

```
RED FLAG QUESTIONS                                            Page    1
      Standard Audit Master Report          SOLID AND HAZARDOUS WASTE
--------------------------------------------------------------------------


                    Solid and Hazardous Waste - Federal


                         Audit Session File Name:
                                 FACILITY


        Prepared by Audit Master V4.9A        Printed on 09/10/1993 at 02:24pm




                               Facility Information:
September 10, 1993

General Manufacturing
2000 General Road
Kingstown, NY 14450

John Smith, Facility Manager
Joe Crawford, Admininstrator
Jean Burke, Plant Quality Engineer

The timeframe for this audit is four years from 1989 to 1993.

The scope is a comprehensive audit for Solid and Hazardous Waste.
```

Figure 9-1. (*Continued*)

Other Auditing Software Packages

CompQuest Pro+ by Semcor, Inc. (Moorestown, N.J.) is another environmental auditing software package for IBM-compatible microcomputers. Originally developed for a U.S. Department of Defense facility, CompQuest Pro+ includes a database of more than 5000 audit questions linked to more than 600 preaudit screening questions. It generates audit checklists and a final report with tables that cross-reference findings to applicable regulations (Fig. 9-2). Program areas covered in CompQuest Pro+ include:

- Air
- Asbestos
- Hazardous materials
- Hazardous waste generation
- Hazardous waste interim storage area

```
RED FLAG QUESTIONS                                               Page    2
          Standard Audit Master Report          SOLID AND HAZARDOUS WASTE
------------------------------------------------------------------- ----------

****************************************************************** Activity Codes **
20 QUESTION (Q1001)                                  * DR
                                                    ** References ******
                                                     * SH36
POLLUTION PREVENTION - DEVELOPMENT OF POLLUTION      * SH37
PREVENTION PLAN                                      * SH38
                                                     * SH39
Has a Pollution Prevention Plan been developed?      * SH40
                                                     * SH41
GMP -- NOT SPECIFICALLY A REGULATORY REQUIREMENT.    *

  RESPONSE : PENDING

Please comment on the question:
------------------------------

    SH37
    Some states have mandated and begun implementing pollution prevention
    programs, and require that facilities develop a Pollution Prevention
    Plan.


Describe the Situation:
-----------------------
In the process of developing a pollution prevention plan to meet the new
state mandates.


Corrective Action Planned:
--------------------------
A team has been assigned the task of researching the cost of developing a
pollution prevention plan. Plan and cost analysis to be completed and
submitted for approval•by September 30, 1993.

    CORRECTIVE ACTION DATA
         QUESTION  : Waste Q1001
         BRIEF_DES : Pollution Prevention
                   :
                   :
         LOCATION  :
         PRIORITY  : 1
         RESP_PER  : Pollution Prevention Task Team
         COMP_DATE : 09/30/93   Completed (Y if action is complete)  N
         COST      :
         FIELD_7   :
         FIELD_8   :
         FIELD_9   :
         FIELD_10  :
```

Figure 9-1. (*Continued*)

- Hazardous waste minimization
- Hazardous waste Part B storage area
- Hazardous waste transporter
- Infectious waste
- Installation restoration

- Laboratory
- Natural resources
- National Environmental Policy Act (NEPA)
- Noise
- Contingency planning
- PCBs
- Pesticides
- Potable water
- Radiation
- SARA Title III
- Solid waste
- Spill prevention, containment, and countermeasures (SPCC)
- Toxic Substances Control Act
- Underground storage tanks
- Wastewater

In 1993, a single module cost $1000, with discounts for multiple modules; the package of all 25 modules was $4000. Other companies developed their own environmental auditing software for internal use. At least one of these companies was expected to release a commercial product in 1994.

Environmental Legislative/Regulatory Software

An essential part of environmental auditing is knowing the environmental laws and regulations. Environmental legislative/regulatory software or electronic code books allow environmental professionals to quickly locate, review, print, and save regulatory text. Electronic code books store environmental, health, and safety (EHS) regulatory information on diskettes, compact disc read-only memory (CD-ROM), or on-line systems. The user can access these regulations using the system's *search engine*, a software application that searches and retrieves selected information. The user sets parameters for the system's search, then reviews a "hit list" describing all the regulations that meet those parameters. The user can then view or print those documents, continue to search, or export the text to a word processor. While all electronic code books share the same basic components (a database and a search engine), the electronic media, scope of coverage, and search methods vary.

General Company

Air Preaudit Checklist

04/93

QUICK REFERENCE TABLE

#	REFERENCE
1	COMAR Title 26.11
2	General Company 5090.1A
3	40 CFR

REF.	DRIVING REG.	REF.	COMPLYING REG.	QUES. NUM.	QUESTION	YES/NO	COMMENT
1	01.04(A)(1)	1	01.04(A)(1)	1.1	Has the State requested the facility to perform testing to determine its compliance with applicable air pollution control regulations?	Yes	Inspectors were at the site in August of 1992 to identify testing requirements.
1	02.04(A)	1	02.04(A)	1.2	Does the facility operate any of the installations listed in COMAR 26.11.02.04A?	No	
1	05.04(A)	1	05.04(A)	1.3	Has the State requested the facility to prepare a	Yes	
1	09.07	1	09.07	1.4	Does the facility operate any fuel-burning equipment and/or stationary internal combustion	Yes	There are generators in each of the 4 manufacturing buildings.
1	11.06	1	11.06	1.5	Is waste oil burned in any installation with approval from the MDE?	Yes	
1	12.01	1	12.01	1.6	Does the facility perform any batch-type hot-dip galvanizing?	No	
1	20.01	1	20.01	1.7	Does the facility operate any ships?	No	
2	6-5.12	2	6-5.12	1.8	Does the facility have buildings that are occupied for four hours or more?	Yes	There are 2 office buildings and 4 large manufacturing areas.
3	54.1	3	54.1	1.9	Has the facility received a notice of civil complaint?	Yes	A former employee has issued two complaints.
3	57.102	3	57.102	1.10	Does the facility operate any primary non-ferrous smelters?	No	
3	60	1	01.01(M)	1.11	Has the facility constructed any new stationary sources or maintained any facilities subject to air pollution permit requirements?	Yes	A new generator was completed in Building #4 in May of 1992.
3	60.10(a)	1	.06.03(D)	1.12	Is road or building maintenance performed at the	Yes	
3	60.30(a) & 60.50(a)	3	60.30(a) & 60.50(a)	1.13	Does the facility operate any municipal waste combustors?	No	
3	60.30(b)	1	01.01(M5)	1.14	Does the facility operate any sulfuric acid production units?	No	
3	60.40	1	01.10	1.15	Does the facility operate any fossil fuel-fired steam generators for which construction commenced after 8/17/71?	No	
3	60.40(a)	1	01.10	1.16	Does the facility operate any electric utility steam-generating units (>73MW) for which construction commenced after 9/18/78?	No	

Figure 9-2. Typical CompQuest Pro+ checklist.

257

General Company

Air Audit Checklist

04/93

QUICK REFERENCE TABLE	
#	REFERENCE
1	General Company 5090.1A
2	40 CFR
3	COMAR Title 26.11

REF.	DRIVING REG.	REF.	COMPLYING REG.	QUES. NUM.	QUESTION	YES/NO	COMMENT	FINDING #
1	6-4.6	1	6-4.6	1.1	Has the facility been screened or is it scheduled to be screened for radon concentrations?	Yes		
1	6-6.4(l)	1	6-6.4(l)	1.2	If the facility is being assessed, have the radon detectors been placed in accordance with the building schedule forwarded by the Department of Energy's contractor and installed in accordance with the detector manufacturer's instructions?	No	Not all sites have radon detectors.	AIR-93-001
1	6-6.4(m)	1	6-6.4(m)	1.3	If any structure is confirmed to have a radon concentration of more than four (4) picocuries per liter (pC/l), has the facility mitigated, or made plans to mitigate, the structure?	No	One building has been tested and determined to be above the limit. A mitigation plan, however, has not yet been completed.	AIR-93-001
1	6.5.12	1	6-5.12	1.4	Have all buildings at the facility that are occupied for four hours or more been tested for radon gas?	Yes		
2	54.1	3	06.08	1.5	Does the facility have any installations operated or maintained in such a manner that they create a nuisance or air pollution against which a record of complaints has been compiled?	No		
2	60.7(a)	2	60.7(a)	1.6	Has the facility furnished the EPA with the following information:	No answer required		
2	60.7(a)(1)	2	60.7(a)(1)	1.6.1	Notification of date that construction or reconstruction commences, postmarked no later than 30 days after such date?	No		AIR-93-002
2	60.7(a)(2)	2	60.7(a)(2)	1.6.2	Notification of anticipated initial start-up, postmarked no more than 60 days nor less than 30 days prior to such date?	No		AIR-93-002
2	60.7(a)(3)	2	60.7(a)(3)	1.6.3	Notification of actual initial start-up date, postmarked within 15 days after such date?	No		AIR-93-002

Figure 9-2. (*Continued*)

General Company
Air Program Findings/Questions Cross-Reference Matrix
April 06, 1993

FINDING	TITLE	QUESTION NUMBERS	FINDING	TITLE	QUESTION NUMBERS
AIR-93-001	Radon Detectors	1.2, 1.3	AIR-93-006	Bulk Material Storage	1.19, 1.27
AIR-93-002	Emission Source Notifications	1.6.1, 1.6.2, 1.6.3, 1.6.4	AIR-93-007	SERP	1.20, 1.28
AIR-93-003	Continuous Monitoring	1.7	AIR-93-008	Opacity of Fuel Emissions	1.22, 1.30
AIR-93-004	Excess Emissions	1.8, 1.9, 1.11	AIR-93-009	TAPs	1.23
AIR-93-005	Sample Ports	1.15	AIR-93-010	Recordkeeping	1.26

General Company
Air Program Findings and
Recommended Corrective Actions
April 06, 1993

FINDING #	TITLE	FINDING	RECOMMENDED CORRECTIVE ACTION
AIR-93-001	Radon Detectors	The facility has not installed radon detectors at sites where the tests have indicated radon levels above 4 picocurries/liter.	The facility should provide all the sites where radon levels exceed 4 picocurries/liter with radon detectors. Exhaust systems should be installed to reduce radon concentrations.
AIR-93-002	Emission Source Notifications	The facility has not made proper notifications of the date that construction or reconstruction commences with regard to new emission sources IAW State notification requirements.	The facility should complete proper notifications of the date that construction or reconstruction commenced with regard to new emission sources IAW State notification requirements.
AIR-93-003	Continuous Monitoring	The facility does not maintain records of the occurrence and duration of any startup, shutdown, or malfunction of continuous monitoring equipment.	The facility must maintain records of the occurrence and duration of any startup, shutdown, or malfunction of continuous monitoring equipment IAW State recordkeeping requirements.
AIR-93-004	Excess Emissions	The facility has not prepared written quarterly reports containing information outlined in 60.7(c)(1-4) of excess emissions measured by continuous monitoring system(s).	If the CMS data are to be used directly for compliance determination, the facility should submit written quarterly reports containing information outlined in 60.7(c)(1-4) of excess emissions measured by continuous monitoring system(s).
AIR-93-005	Sample Ports	Not all registerable air emission installations constructed after 1/4/71 have openings in the exhaust gas ductworks to enable the collection of effluent samples going into the atmosphere.	All registerable air emission installations constructed after 1/4/71 that do not have sample ports should be identified and sample ports should be installed in the exhaust ductworks.

Figure 9-2. (*Continued*)

GENERAL COMPANY
ENVIRONMENTAL AUDIT DEFICIENCY LOG

AIR

Log Number: AIR-93-005

Finding: Sample Ports.   Not all registerable air emission installations constructed after 1/4/71 have openings in the exhaust gas ductworks to enable the collection of effluent samples going into the atmosphere.

Class of Finding: MEDIUM

Recommendation: All registerable air emission installations constructed after 1/4/71 that do not have sample ports should be identified and sample ports should be installed in the exhaust ductworks.

Responsible Code: Primary - Sharon Johnson

Status: Behind Schedule

Driving Reference: 40 CFR 60.8(e)(1)
Complying Reference: COMAR Title 26.11.01.04 (A4)

Log Number: AIR-93-006

Finding: Bulk Material Storage.   The facility allows the storage of bulk materials of earth on paved streets and does not require that open-bodied vehicles be covered when transporting bulk materials.

Class of Finding: HIGH

Recommendation: The facility should revise its instructions and contractor documentation to stipulate that the bulk materials of earth stored on paved streets and open-bodied vehicles transporting materials likely to create air pollution be covered.

Responsible Code: Primary - Dave Smith

Status: Unscheduled

GENERAL COMPANY
ENVIRONMENTAL AUDIT FINDINGS SUMMARY

AIR				
LOG NUMBER	CLASS	ORGANIZATION	STATUS	ESTIMATED COMPLETION DATE
AIR-93-001	Medium	John Doe	On Schedule	04/27/93
AIR-93-002	High	Sharon Johnson	On Schedule	05/06/93
AIR-93-003	Medium	Sharon Johnson	On Schedule	08/11/93
AIR-93-004	Medium	Sharon Johnson	Ahead of Schedule	07/15/93
AIR-93-005	Medium	Sharon Johnson	Behind Schedule	07/30/93
AIR-93-006	High	Dave Smith	Unscheduled	09/01/93
AIR-93-007	High	Sharon Johnson	Behind Schedule	05/01/93
AIR-93-008	Medium	Chris Davis	Scheduled	06/01/93
AIR-93-009	Low	Dave Smith	Resolved	04/07/93
AIR-93-010	High	Sharon Johnson	Behind Schedule	08/01/93
UST-93-001	Critical	Fred Perkins	Scheduled	04/08/93
UST-93-002	Medium	Fred Perkins	Behind Schedule	06/01/93
UST-93-003	High	Fred Perkins	Ahead of Schedule	02/01/94
UST-93-004	Low	Fred Perkins	Resolved	04/07/93
UST-93-005	Medium	Fred Perkins	On Schedule	04/01/93
UST-93-006	Low	Fred Perkins	Behind Schedule	05/31/93

Figure 9-2. (*Continued*)

PLAN OF ACTIONS AND MILESTONES CHART

Air Emissions Study

Origin = January 01, 1994

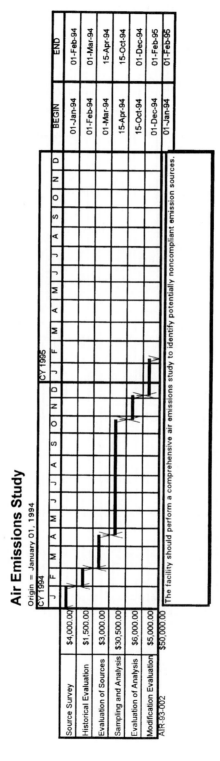

		CY 1994												CY 1995												BEGIN	END
		J	F	M	A	M	J	J	A	S	O	N	D	J	F	M	A	M	J	J	A	S	O	N	D		
Source Survey	$4,000.00																									01-Jan-94	01-Feb-94
Historical Evaluation	$1,500.00																									01-Feb-94	01-Mar-94
Evaluation of Sources	$3,000.00																									01-Mar-94	15-Apr-94
Sampling and Analysis	$30,500.00																									15-Apr-94	15-Oct-94
Evaluation of Analysis	$6,000.00																									15-Oct-94	01-Dec-94
Modification Evaluation	$5,000.00																									01-Dec-94	01-Feb-95
AIR-93-002	$50,000.00																									01-Jan-94	01-Feb-95

The facility should perform a comprehensive air emissions study to identify potentially noncompliant emission sources.

Figure 9-2. (*Continued*)

FINDINGS TRACKING LOG

Finding No.	Class Finding	Finding Title	Organization	Status	Date Completed	Finding No.
		AIR				
AIR-93-001	High	Radon Detectors	Environmental	On Schedule	May-93	AIR-93-001
AIR-93-002	Critical	Emission Source Notifications	Environmental	On Schedule	May-93	AIR-93-002
AIR-93-003	Medium	Continuous Monitoring	Environmental	Behind Schedule	Mar-93	AIR-93-003
AIR-93-004	Low	Excess Emissions	Environmental	Unscheduled	N/A	AIR-93-004
AIR-93-005	Medium	Sample Ports	Environmental	Ahead of Schedule	Jul-93	AIR-93-005
		HAZARDOUS WASTE GENERATOR				
HWG-93-001	Critical	Waste Analysis and Recordkeeping	Environmental	On Schedule	Jun-93	HWG-93-001
HWG-93-002	High	Exception Reports	Environmental	On Schedule	Jun-93	HWG-93-002
HWG-93-003	High	POA&M Waste Minimization	Liquid Waste	Ahead of Schedule	Aug-93	HWG-93-003
HWG-93-004	High	HW Contingency Plan	Environmental	Behind Schedule	Mar-93	HWG-93-004
HWG-93-005	Medium	Plating Process	Plating Shop	On Schedule	Jul-93	HWG-93-005
HWG-93-006	Medium	HW Training	Machine Shop	Unscheduled	N/A	HWG-93-006
HWG-93-007	Medium	Recyclable Materials	Machine Shop	Unscheduled	N/A	HWG-93-007
HWG-93-008	Medium	Arrangements With Local Authorities	Environmental	Ahead of Schedule	Aug-93	HWG-93-008
		PCBs				
PCB-93-001	Critical	PCB Elimination Plan	Environmental	On Schedule	Jul-93	PCB-93-001
PCB-93-002	High	PCB Prioritization	Environmental	On Schedule	Jul-93	PCB-93-002
PCB-93-003	Critical	PCB Plan	Environmental	Ahead of Schedule	Oct-93	PCB-93-003
PCB-93-004	High	PCB Content	Fabrication Shop	Behind Schedule	Feb-93	PCB-93-004
PCB-93-005	Medium	Removal of PCBs	Safety	Behind Schedule	Mar-93	PCB-93-005
		SARA				
SARA-93-001	Critical	EPCRA	Environmental	On Schedule	Aug-93	SARA-93-001
SARA-93-002	High	SARA Community Representative	Safety	On Schedule	Aug-93	SARA-93-002
		UST				
UST-93-001	Critical	UST Management Plan	Fuel Storage	On Schedule	Sep-93	UST-93-001
UST-93-002	High	UST > 15 Years	Environmental	On Schedule	Sep-93	UST-93-002
UST-93-003	Critical	Release Detection	Environmental	Ahead of Schedule	Dec-93	UST-93-003
UST-93-004	Medium	Closure	Liquid Waste	Unscheduled	N/A	UST-93-004

Figure 9-2. (*Continued*)

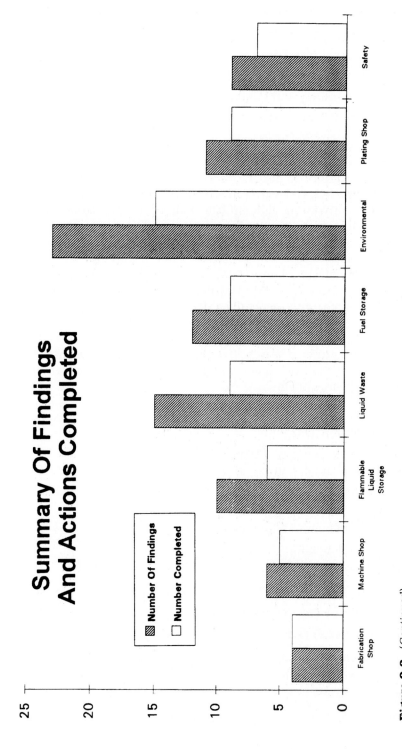

Figure 9-2. (*Continued*)

Federal regulations are the backbone of electronic code books. Most electronic code books include the full text of sections from three titles from the *Code of Federal Regulations* (CFR):

- Title 40 [Environmental Protection Agency (EPA)]
- Title 29 [Occupational Safety and Health Administration (OSHA)]
- Title 49 [Department of Transportation (DOT)/Hazardous Materials]

Code book developers are increasing the scope and type of regulatory information covered and offering consumers more database options. State regulations are the most common addition to the CFR because a facility's regulatory compliance does not stop at the federal level. Many code book developers also offer the *Federal Register,* which provides a daily update of all federal regulations. It covers proposed and final regulations as well as meeting and hearing notices. Some electronic code books also include the U.S. Code, legislative information, and international regulations.

The basis of the value of a regulatory database is not only the scope of its coverage but also its ability to locate and retrieve data. To locate text in the database, the user must select search terms, generally a string of characters (i.e., a word or phrase). The type of search terms the user can enter varies little from system to system. There is more variety in the different methods (e.g., boolean searches, proximity searches, or narrowed searches) for handling the search terms. These methods help the user narrow or expand a search. Following are some environmental code book software systems:

- BNA's Environment Library on CD, distributed by The Bureau of National Affairs Inc. (Washington, D.C.)
- Computer-aided Environmental Data System (CELDS), distributed by the University of Illinois (Urbana, Ill.)
- Counterpoint Publishing Federal CDs, distributed by Counterpoint Publishing (Cambridge, Mass.)
- EarthLaw, distributed by Infodata System, Inc. (Fairfax, Va.)
- Easy Regs, distributed by ERM Computer Services (Exton, Pa.)
- ENFLEX products, distributed by ERM Computer Services
- EnviroText Retrieval System (ETRS), distributed by the University of Illinois
- FastRegs, distributed by OSHA-Soft Inc. (Amherst, N.H.)
- FastSearch, distributed by DMSA Corporation (Minneapolis, Minn.)
- Federal Environmental and Safety Authority, distributed by Virtual Media Corporation (Scottsdale, Ariz.)
- IHS Environmental Safety Library, distributed by IHS (Englewood, Colo.)
- Legi-slate, distributed by Legi-slate Inc. (Washington, D.C.)

- RegScan, distributed by Regulation Scanning Technology Corporation (Williamsport, Pa.)
- Solutions Environmental Database, distributed by Solutions Software Corporation (Enterprise, Fla.)
- Super-Reg, distributed by Business & Legal Report Inc. (Madison, Conn.)
- TEXT-Trieve Products, distributed by TEXT-Trieve Inc. (Bellevue, Wash.)
- WESTLAW, distributed by West Publishing (Eagan, Minn.)

At the beginning of 1994, the cost of environmental code book software packages varied from $99 to $7500. On-line system use cost $90/hour to $195/hour. Distributors offer multiple price options based on electronic format, regulatory coverage, and update frequency. Customers can purchase an entire system, choose individual regulatory titles, order packages of titles, or select databases with supplementary information.

Related Software

There are more than 1000 EHS software packages, databases, and on-line systems available. Most of these software systems target a specific task or category of information, a reflection of the refinement of environmental protection goals. Since an audit can involve a variety of categories (e.g., water pollution, wastewater, solid waste, or air pollution), these packages can help an environmental professional manage the data required to conduct an environmental audit as well as the data generated by an audit or by corrective action. A brief description of some of this environmental information management software is given below.

The focus of environmental information management software is generally on hazardous substance management. This software, developed in response to regulatory requirements, helps maintain chemical inventories, manage data for regulatory reporting (e.g., training records or chemical safety information), and prepare for and manage data required for emergency response. Many systems address requirements under such federal laws as the Resource Conservation and Recovery Act (RCRA); the Hazardous and Solid Waste Amendments (HSWA); the Comprehensive Environmental Response, Compensation, and Liability Act (CERCLA); SARA; and TSCA. Some allow the user to set up and print reports required by SARA Title III [e.g., Form R, Tier I/II reports, and material safety data sheets (MSDSs)]. Many systems also include modeling and statistical analysis to determine the distribution of contaminants and predict trends.

Other software systems provide information on hazardous substances. Subjects include MSDSs, chemical properties, storage and handling, hazards, toxicity, regulatory requirements, biodegradation rates and fate, emergency response, and safety and health. Some of these databases also help train employees.

Software systems are also available that help with hazardous or solid waste management or remediation. Some of these systems help with waste tracking, generating manifests, and waste minimization. Other software provides information on treatment technologies or help with site remediation, landfills, incineration, and other types of waste disposal.

In addition, software is available for environmental risk assessments and hazard assessments, including hazard and operability studies. These systems help evaluate risks from such activities as transporting hazardous substances, exposure to site contamination, and chemical or radioactive spills or releases.

Environmental software for wastewater or water systems helps manage wastewater collection and treatment systems or drinking water systems. The software systems help with system design, data collection, testing, record keeping, and regulatory compliance. Some software packages help manage sludge and wastewater outflows. When a spill containing hazardous constituents could affect human health or the environment, software targeted at groundwater and/or soils helps predict migration and contaminant levels. Software for groundwater modeling and data management includes systems that address wells, soils, aquifer characterization, and contaminant transport.

Software is also available to address surface water. These software packages provide information on the aquatic environment, surface waters, and watersheds and for surface-water data management and modeling. Some systems provide hydrological data and perform calculations for storm-sewer system design.

If there is a release or the potential for a release of hazardous constituents to the air, air pollution models, including regulatory, screening, and air quality models, help define the plume and potential impact to the environment. Other systems help with emissions inventories or provide air quality data.

To help interpret environmental data, geographic information systems (GISs), geologic information systems, and mapping software provide graphical representations of a site and its attributes. Software is available with graphics and plotting software for two- and three-dimensional contour and surface maps.

To see a snapshot of the environmental history of a particular site, the site information software industry developed systems that provide environmental information on specific sites in the United States, including records on sites where there have been spills, sites proposed for cleanup, regulated sites, and sites that are the subject of EPA enforcement orders.

Other related software systems focus on ecology, biology, wildlife, forestry, agriculture, and meteorology. In addition, software systems used by environmental professionals that are not specifically environmental include energy, statistical, and graphics packages.

The Future of Environmental Auditing Software

In the 1980s many lawyers initially recommended against using auditing software to document a facility's environmental practices, fearing regulators could use this documentation against the facility. However, attitudes changed as environmental regulations increased and regulators began basing the size of fines for noncompliance on the facility's efforts to maintain compliance. Many consultants, environmental managers, lawyers, and environmental associations now recognize the value of environmental auditing and the use of environmental auditing software. As environmental auditing becomes more popular, the need for tools to simplify this process will increase. The number and types of software systems to help with environmental audits, like other environmental software, should increase dramatically. There is a need for industry-specific software and a continued need for custom, site-specific applications. This includes adapting software to help with regulatory compliance at the state and local levels. In addition, the hardware for environmental audits should improve. Pen-based and portable computers and automated data acquisition systems (e.g., sensor systems) should simplify the environmental auditing process.

The environmental code book software industry exploded with activity in 1993 and continues to be a growth area. As this software becomes even more popular, the number of integrated systems should increase. For example, a useful addition to a code book software system would be auditing software or an expert system to help the users easily determine if new or existing regulations affect their facility.

In the mid-1980s, environmental software systems emerged as comprehensive systems. Developers tried to make their software fit all users' needs. However, they soon realized that the task was beyond their capabilities, and more-targeted software packages began to emerge. Now in the mid-1990s, the trend is beginning to return to more comprehensive systems. However, the approach is more logical. Developers are merging their capabilities and resources by creating links to other targeted software systems. This helps to build more comprehensive systems but retains the expertise of the more-targeted software packages. This development is due in part to the refinement of the environmental industry, but it is primarily due to advances in the computer hardware and software industry. With the proliferation of client/server technology (e.g., Microsoft Windows), data translation software, and advanced data storage technology (e.g., CD-ROMs), environmental software developers can work together toward providing software that truly meets a facility's needs.

References

(The author used quotations from some of these references with permission from the publisher, Donley Technology, Box 335, Garrisonville, VA 22463).

1. Donley, Elizabeth, and Veronica Deschambault, editors. *Environmental Software Directory,* Donley Technology, 1993.
2. Donley, Elizabeth, and Veronica Deschambault, editors. *Environmental Software Report,* Donley Technology, 1988–1994.
3. Deschambault, Veronica, and Elizabeth Donley, editors. *Environmental Code Book Software Report,* Donley Technology, 1994.

10
Audit Team Selection and Training

Judith C. Harris, Ph.D., and Maryanne DiBerto

Arthur D. Little, Inc.

As in the case of other types of auditing, the quality of an environmental audit is only as good as the quality of the individuals conducting the audit. The challenge for the audit program manager is to select, orient, train, and manage the audit team members in a way that ensures that the audit program objectives will be met.

Necessary Knowledge and Skills to Be an Effective Auditor

In selecting audit team members, it is useful to begin by identifying their desired areas of expertise, as follows.

Regulatory Considerations. First, and most important, the audit team members need to know the laws, regulations, and permits that apply to the facility being audited. It is essential to know the federal, state, and local regulatory standards against which you are auditing. In addition, it is essential to know what corporate policies and procedures apply if you are auditing against them.

Auditing. Expertise in the discipline of environmental, health, and safety auditing has become particularly important as audit programs and the discipline for auditing have matured. This is more important than ever as senior management has begun to depend on the audit program to provide assurance about the company's environmental matters. To provide this kind of assurance, auditors need training in verification techniques, audit procedures, and other aspects of the internal auditing discipline, and a good understanding of the techniques and practices used by other companies.

Management Systems. An understanding of management systems—both environmental, health, and safety management systems and also facility management systems—can be especially important as an auditing program places increased emphasis on verification. Management systems include information on how work is planned, implemented, controlled, and reviewed; the roles and responsibilities of various managers and functional disciplines; and reporting relationships and information flow both within the facility and from the facility upward in the organization. Gaining an understanding of how the facility manages itself and how it is managed by the corporation has an impact on the effectiveness with which the team determines the extent to which the management systems are in place and working.

Facility Operations and Procedures. Even if the overall process is the same among select facilities, each facility has some differences. It is generally possible to search for and use people within the company who have some familiarity with the kinds of operations that are going on in the facility being reviewed. It is especially helpful to include on the team one or more people who understand the type of facility being audited. Knowledge of facility operations makes it easier for the auditor to relate to facility personnel and provides a basis for better understanding what facility personnel say and show to the auditor.

Relevant Environmental Control Technologies. Understanding the environmental, health, and safety technologies that fit the facility's operations and the kinds of process effluents and waste that the facility generates is an important component of the expertise of the team. Having expertise in the functional area being audited is important.

Scientific Knowledge. It is crucial to have on the audit team technical experts who recognize the potential for hazard in a particular situation or operation. This could range from the potential for groundwater contamination or the potential for pressure buildup in a tank or the impact on nearby foliage or wildlife to the presence of a particular substance not considered a threat in the past but that, given today's knowledge, could have major significance.

Knowledge of Peer Facilities. An understanding of what similar companies and facilities are doing to manage for compliance and to control hazards can be very helpful to the audit team.

Team Member Characteristics

Four broad team member characteristics, in particular, are important to the success of the audit:

Training. Have team members participated in, been exposed to, an actual environmental audit, especially at a facility similar to the one to be audited? Training or familiarity with not only the company's audit objectives but also the company's established auditing practices and approaches can be very important. Either training should be conducted prior to the audit, or extra time and resources should be provided during the audit to train new members of the audit team. Team members should be prepared to conduct an audit efficiently and effectively while at the facility. This is especially true if the team members are on a rotating basis. As crucial as preaudit training is for the auditors, actual on-the-job experience can be the best teacher, as with so many other kinds of training situations. Therefore, after team members have done some audits, they will become better auditors and more effective at the facility (and likely could become even more accomplished auditors with additional training by more experienced auditors).

Experience. The composition of the team is very directly related to the function that the team member is to perform during the audit. If it is desired that the team assess environmental performance and identify hazards, then technically trained people (engineers, chemists, safety professionals, industrial hygienists, and so on) who have that experience and expertise are desirable team members. If the focus of the audit is to verify records and find that procedures and policies are being followed, then people with experience in auditing, and not necessarily as much experience in environmental control techniques, are good team members.

Sensitivity. While an understanding of the regulatory requirements is of utmost importance, it is equally crucial for audit team members to be sensitive to the issues and concerns about identifying potential violations of laws and regulations or findings about inherent potential hazards that have not yet been controlled or for which there are no plans for control. An auditor should not set up a "smoking gun" situation where a regulatory exception is identified but not followed up properly. In addition, each individual auditor has an obligation to the auditing program to not duck any tough issues. Moreover, it is as much a disservice to the corporation to state that the facility is in violation of regulatory requirements when in fact no law or regulation yet exists as it is to state that the facility is only in violation of corporate policy when, in fact, it is violation of a regulation.

Independence. The degree of audit independence that is desired by the person for whom the audit is being conducted will influence who is on the team. If an inside, independent verification is sufficient, no outsiders need to be involved. This has been the view of many of the programs currently being conducted. However, if the board, corporate management, or environmental management decides that they would like to have an outsider's view of the compliance status of the corporation's facilities, it is necessary to go outside the corporation, for no one with the corporation can provide the independence sought.

What Constitutes an Effective Training Program?

The most successful environmental audit training provides hands-on experience to the auditor—either through an actual audit or in a role-playing situation. The ideal likely combines these two and is based on an exchange of information from a knowledgeable, experienced auditor and teacher to the trainee. In particular, it is essential that the individual being trained come to understand the following areas:

- What auditing is, its context within environmental, health, and safety management, and the basic audit process—including team members' roles and responsibilities
- General understanding of EHS management and control systems
- Audit tools to "guide" the auditor during the audit
- Using working papers and other record-keeping tools effectively
- Effective interviewing, evidence gathering, and sampling techniques
- Evaluating audit findings
- Reporting audit findings

Understanding What Auditing Is

As environmental auditing has grown around the world in the last twenty years, it has taken on various definitions—even different names. It is helpful for the auditor to recognize these various definitions and their differences, and then to understand the definition of auditing within the company or facility being audited. Also important is a thorough understanding of the EHS auditing framework (Fig. 10-1), in addition to the role of assessment and verification in the audit process.

Assessment.   Provides expert judgment and opinion on EHS hazards, associated risks, and management and control measures. Assessment activities can identify knowable hazards and estimate the significance of risks; identify programs, practices, and capabilities that are missing or weak; and provide the basis for recommendations to improve the organization's approach to EHS management.

Verification.   Determines and documents performance by evaluating the application of, and adherence to, EHS policies and procedures; certifies the validity of EHS data and reports; and evaluates the effectiveness of management

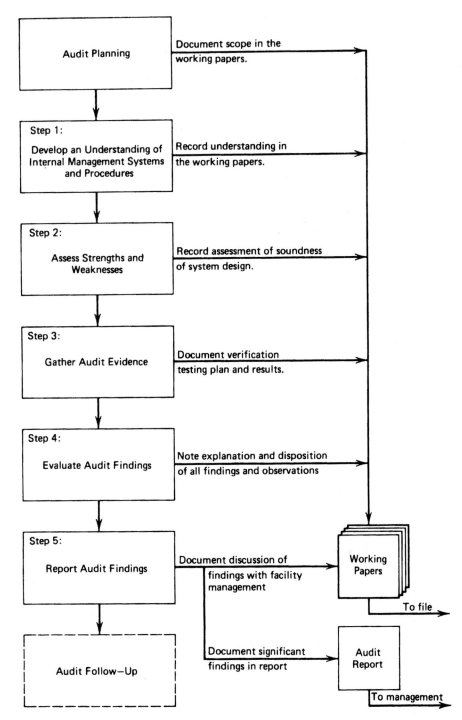

Figure 10-1. Basic steps in the typical audit process. (*Source: Arthur D. Little, Inc.*)

systems. Verification activities can substantiate that regulations are being adhered to; assist in identifying gaps in implementation of policies and procedures; and confirm that management control systems are functioning effectively.

The auditor needs to understand this mix and emphasis in the context of the company's audit program in order to help fulfill the company's EHS auditing goals and objectives.

Understanding Internal Management and Control Systems

The first step of the environmental audit is to develop an understanding of the facility's internal management and control systems. Internal control refers to the actions taken within an organization to help the organization regulate and direct its activities. Most managed activities, by definition, have some sort of internal control. Every facility audited has an internal environmental management and control system. While this "management system" may not be very explicit or even thought of as a system by the facility staff, it will have at least some of the following:

- *Policies.*   Information, directives, guidelines, and standards concerning operations and performance
- *Procedures.*   Instructions intended to ensure that operations are carried out as planned
- *Practices.*   Ways that people typically and routinely carry out operations
- *Controls.*   Checks and balances built into facility operations that may have an impact on environmental compliance.

A management control system provides a framework for guiding, measuring, and evaluating performance. Management control systems can include both managerial controls and equipment controls. In addition, as shown in Fig. 10-2, both the formality and the complexity of these systems can and do vary tremendously. In general, the more critical to the organization a desired action or outcome becomes, the more appropriate or desirable a formal environmental management control system may be.

It is a goal of the audit team to develop a working understanding of how the facility intends to manage those activities that can influence environmental, health, and safety performance. This understanding provides the auditors with a benchmark by which they can measure how accurately they understand the facility's management approach and processes. An understanding of the internal control system is generally developed by several means, including review of information, questionnaire, plant tour, and detailed review.

Then, once an understanding of the facility's internal controls has been developed and recorded, the next basic step of the audit process is to evaluate the

Nonformal				Formal
A	B	C	D	E
Implicit policies and procedures	Formal policies; implicit procedures	Formal policies; limited written operating procedures	Formal policies; written operating procedures; limited control procedures	Formal policies; full written operating procedures; formal control procedures

Figure 10-2. Range of management systems likely to be encountered. (*Source: Arthur D. Little, Inc.*)

soundness of the design of those controls. Principles of satisfactory control include clearly defined responsibilities, adequate system of authorizations, division of duties, trained and experienced personnel, documentation, protective measures, and internal verification. Each of these seven characteristics can require significant *judgment* on the past of the auditor, which must be developed through a combination of training and experience. There simply are not many widely accepted standards, especially in the environmental, health, and safety management areas, that an auditor can use for comparison or as a guide to what is acceptable internal control. Accordingly, the auditor must keep referring to the goals and objectives of the audit program as well as to the corporation's basic environmental philosophy for guidance about what is satisfactory.

Understanding What Audit Tools to Use as "Guides"

Most environmental audit programs use some form of written document to help guide the auditor through the on-site audit process. Such a guide, or *audit protocol,* adds consistency to the audit approach and is particularly helpful to help train the audit team. In addition, written audit guides are sometimes used to assist the facility in preparing for the audit.

Some companies use detailed guidelines that spell out the specific requirements and a step-by-step procedure for auditing against these requirements. Other companies prefer to use simple checklists or topical outlines that depend upon the auditors' knowledge of the regulatory and corporate requirements.

An audit protocol lists the step-by-step procedures that are to be followed during the audit to gain evidence about the facility's environmental practice. The protocol also provides the basis for assigning specific tasks to individual members of the audit team, for comparing what was accomplished with what

Approach, Style, Format	Characteristics
Basic protocol	A sequenced set of instructions outlining or specifying what is to be reviewed or examined, how the examination is to be conducted, and what is to be documented in the audit working papers; usually formatted to provide for quick identification of audit team assignments, deviations from audit plan, and cross-referencing of audit steps with working papers.
Topical outline	Listing of topics to be included in the audit; generally leaves the manner in which each topic is audited up to the experience and discretion of the auditor.
Detailed audit guide	Materials to familiarize the auditor with the basic requirements and standards against which the audit is to be conducted. The audit guide places substantial emphasis on requirements; often is in flow chart format.
Yes-or-no questionnaire	Translation of the basic requirements against which the audit is to be conducted into a series of yes-or-no questions. Format can be expanded to provide for notation of the basis from which the answer is derived (inquiry, observation, testing) and/or cross-referencing of audit working papers. Tends to result in a questionnaire or inquiry-driven rather than review or examination-driven audit.
Open-ended questionnaire	A questionnaire designed and formatted to include both selected explanations of yes-or-no responses and open-ended questions that cannot usually be answered with a yes or no. Focuses the audit on fact- finding and recording.
Scored questionnaire or rating sheet	A questionnaire where responses are scored against criteria developed for that purpose resulting in either a numerical score or a satisfactory or unsatisfactory result. Tends to shift the audit team from review and fact-finding to a grading role.

Figure 10-3. Alternative approaches to audit protocols or guides. (*Source: Arthur D. Little, Inc.*)

was planned, and for summarizing and recording the work accomplished. In addition, a well-designed audit protocol can also be used to help train inexperienced auditors and thus reduce the amount of supervision required from the audit team leader. The audit protocol or guide can be organized and formatted in a variety of ways as listed in Fig. 10-3.

Many environmental audit programs supplement the audit protocol or guide with a questionnaire specially designed to assist in efficient collection of specific background information about the facility's internal environmental management systems. Such questionnaires are commonly known as *internal control ques-*

tionnaires. Internal control questionnaires are a vehicle to assist the auditors in identifying and reviewing internal management procedures and systems. They are used to supplement the audit protocol. Internal audit questionnaires should enhance the audit protocol rather than duplicate or substitute for it. An example of a questionnaire is provided in Fig. 10-4.

Does the facility treat or dispose of any hazardous wastes, including containers, by means of incineration?   Yes_____ No_____

	Answer		
	Yes	No	N/A
1. Does the facility analyze wastes before burning them for the first time to establish proper operating conditions?	___	___	___
If yes, does the analysis include at least the following?			
(a) Heating value of the wastes	___	___	___
(b) Halogen and sulfur content	___	___	___
(c) Concentrations of lead and mercury in the wastes	___	___	___
2. Are waste analysis results made part of the operating record of the facility?	___	___	___
3. Does the facility incinerator have instruments that relate to the following aspects of combustion and emission control?	___	___	___
(a) Measurement of waste feed rate	___	___	___
(b) Measurement of auxiliary fuel flow rate	___	___	___
(c) Measurement of air flow rate	___	___	___
(d) Measurement of combustion temperature	___	___	___
(e) Measurement of scrubber flow	___	___	___
(f) Measurement of scrubber pH	___	___	___
(g) Pressure measurements throughout system	___	___	___
4. Are the above instruments monitored at least every 15 minutes when hazardous wastes are being incinerated?	___	___	___

Figure 10-4. Example of a basic "yes-or-no" questionnaire. (*Source: Arthur D. Little, Inc.*)

Learning Effective Data-Gathering Methods

During the audit, each audit team member carries out a variety of procedures to determine the environmental compliance status of the facility. In the most basic sense, audit procedures are methods of acquiring audit evidence. Auditors should be trained in the three basic approaches to data gathering: inquiry, observation, and verification testing.

Inquiry. This is perhaps the most frequently used means of collecting environmental audit evidence. Here, the auditor asks facility personnel questions—both formally (e.g., via a questionnaire) and informally (e.g., through discussions). Inquiry often provides satisfactory explanations of unclear items in the records.

Observation. Often one of the most reliable sources of audit evidence is observation or physical examination. Where specific operations are material to the environmental performance, it may be desirable for the auditor to observe such operations.

Testing. The wide variety of verification activities that can be employed to increase confidence in the audit evidence and the facility's internal controls is "testing." This can be a very powerful technique in assisting the environmental auditor to achieve the objectives of the audit, and thus is an important part of the training process.

During the conduct of an audit, each team member will spend considerable time asking questions and engaging in many informal discussions. Whether the auditor is gathering evidence through inquiry, observation, or testing, he or she will be interacting with facility personnel. To gain the maximum benefit from those interactions, the auditor can be trained to adopt certain practices. Three particularly important techniques for gaining useful information in interviews or discussions are:

Concreteness or Specificity of Response. Focusing on the adequacy of the data included in a response and getting the interviewee to be as concrete as necessary.

Respect. Developing and maintaining rapport.

Constructive Probing. Focusing on resolving ambiguities or contradictions. (Figure 10-5 provides interviewing examples of "constructive probing.")

Using Working Papers and Record-Keeping Tools

A particularly critical area for auditors to learn is the use of working papers. The working papers in many ways represent the core of the environmental audit—the documentation of the work performed, the techniques used, and the conclusions reached. Working papers help the auditor to achieve the audit objectives and provide reasonable assurance that an adequate audit was conducted consis-

Question: How are you tracking movement of hazardous waste from the plant to treatment, storage, and disposal sites?

Interviewee: I have my secretary call the contractor every so often to check on our manifests—we are very committed to complying with the federal regulations.

Poor   *Interviewer:* I see, then she's pretty conscientious about following up on those things.

(*Note:* This will usually lead to premature closure rather than the eliciting of more detailed information.)

Fair   *Interviewer:* How do you ensure that she follows up on all shipments?

(*Note:* This does not confront the questions of the basic adequacy or inadequacy of the control process.)

Good   *Interviewer:* I'm not sure about your overall process for tracking manifests and how your secretary's efforts fit in. Could you explain the process to me?

(*Note:* This restates the question in a more focused way and respectfully challenges the notion that the secretary's efforts are adequate.)

Figure 10-5. Interviewing example: constructive probing. (*Source: Arthur D. Little, Inc.*)

tent with program goals and objectives. Because the working papers document the information gathered by each audit team member during the audit, the information included should substantiate the conclusions reached about the areas of both compliance and noncompliance.

Working papers also can:

- Supplement the audit protocols by providing audit planning details (such as the time allotted by the audit team to each audit step) and specific documentation of the internal control systems
- Provide the rationale for the auditors' approach to testing, a record of tests conducted, and the evidence accumulated
- Provide data that support the audit report and that may be useful in subsequent action planning and follow-up activities
- Provide a basis for review of the audit by the team leader, the audit program manager, or some other individual

Generally speaking, the environmental audit working papers contain three basic elements: (1) a description of the environmental management systems in place for managing various aspects of compliance; (2) a description of the specific audit methods or actions taken to complete each step of the protocol (including documentation of the tests conducted, sources of information

Item

7a) Of the sources identified in 3 and 5, all met applicable standard limitation or compliance standards with the exception of the following:

Source A-2: Recent process changes have resulted in emitting volatile organic compound B instead of the permitted volatile organic compound A. Since the latter was given a limitation under NESHAP at about the same time as the proven change, no data have been obtained to establish compliance, nor has a revised permit application been submitted to the Air Pollution Control District (APCD) for County R.

Source B-4: Permit requires semi-annual submittal of test data on performance of Air Pollution Control System for particulate removal -- at the time of survey. These tests had not been carried out for the latest reporting period, nor was there any evidence that failure to meet the specified reporting had been communicated to the APCD for County R.

7d) Selected charts from all automatic recording instruments required under permits (see Item 5) were reviewed and no evidence was found of out-of-compliance periods. A non-isokinetic sample for particulate sampling was used on all sample locations--however the permit for Source D-3 requires isokinetic sampling. Although it is doubtful that there will be a significant change in results, especially since the measurements are well below limits, approval for this sampling method should be sought.

Records of instrument maintenance, calculations, and analytical techniques were inspected and found to be in accordance with all applicable regulations.

IMC-10

Figure 10-6. Example of working paper documentation. (_Source: Arthur D. Little, Inc._)

obtained, and evidence accumulated); and (3) a summary of the auditors' findings and observations along with the conclusions reached at each audit step.

Figure 10-6 provides an example of working paper documentation. Environmental auditors who have experienced the pressures of an audit may doubt their ability to generate high-quality working papers consistently. But these guidelines, in particular, should be kept in mind: Write down your observations while performing your fieldwork. Do not wait until the end of the audit—or even the end of the day—to document your understanding and findings. Experience suggests that understandable, complete, and factual working papers can be prepared as the audit is conducted. Keep in mind that the working papers are not a report that the auditor is to prepare from notes after the audit is completed. Rather, working papers are the auditors' field notes to keep track of audit procedures undertaken, results achieved, and items requiring further review or investigation.

Evaluating Audit Findings

Auditors must learn basic techniques and options to consider in evaluating audit findings and then reporting those findings. After the audit fieldwork is completed, the auditors must wrestle with how best to organize the findings so that they are reported to appropriate levels of management (so that any deficiencies are corrected). During this step of the audit process, the findings and observations of each auditor are evaluated and their ultimate disposition determined. Each auditor prepares his or her own list of findings. In preparing this list, it is crucial that the auditor ask the question: Have I gathered sufficient evidence to substantiate this finding? Each auditor should feel comfortable that, if challenged, he or she has the data to substantiate the finding.

Moveover, the auditors must become adept at examining the lists of audit findings and observations of each auditor in order to look for situations where the sum may be greater than the parts. Several common findings (such as informal preventive maintenance systems for different types of equipment) when viewed as a group may have greater significance.

The next step of the audit process is to report audit findings. While the first activity is to apprise facility personnel during the audit, the first formal reporting generally occurs at the audit's exit interview with an oral report to facility management. Auditors should learn to carefully plan and prepare for the exit interview and oral report. A handwritten summary of the audit findings, which may take a variety of forms, is often used to provide a framework for the oral report and to document what the auditors told facility management.

Developing Audit Reports

All audit reports seek to provide clear and appropriate disclosure of the audit findings by:

- Providing management information
- Initiating corrective action
- Providing documentation of the audit and its findings

Given these objectives, the audit report must give due consideration to a number of concerns that are inherent in environmental audit reporting. Among those most frequently raised are whether certain disclosure requirements are being met, the risk of raising issues perhaps not previously recognized, and the confidentiality of reports. When the report is written, these concerns must be addressed with sufficient clarity to ensure that undue liability is avoided.

Key principles to master in audit report writing include these:

- *Report facts clearly and concisely.*   Every statement should be based on sound evidence developed or reviewed during the audit.

- *Establish a context for the report.*   The auditor should provide enough background information so that the reader clearly understands who conducted the audit and what the audit did or did not include. The reader should be able to understand why the report was written and who should take corrective action.

- *Report in a timely manner.*   Since management expects deficiencies to be corrected promptly, the report must be communicated in a timely fashion.

- *Provide for reviews and follow-up of audit reports.*   Audit findings must do more than paint an accurate picture of the environmental status of the facility. Any identified deficiencies must receive prompt consideration, action planning, and follow-up.

- *Manage the collection, dissemination, and retention of audit data.*   The audit program manager will establish policies on how data collected during the audit as well as the audit reports are to be managed.

The content of the report divides naturally into four basic sections: background, audit findings pertaining to environmental statutes, audit findings pertaining to company policies and procedures, and an evaluation of the environmental management and control system. The audit report should be prepared with an appropriate form and content. At a minimum, audit reports should present the purpose, scope, and results of the audit. However, audit reports that are highly valued within a corporation, relative to their role in providing assurance, go beyond the basics. They provide the right level and mix of information and contextual background to ensure that the managers receiving the report understand the implications of what is being reported.

It is best that auditors come to understand the different report formats and how to understand whether or not the report approach is consistent with the needs of management and the culture of the organization.

Team Leader Characteristics and Training

In addition to the team member characteristics, questions arise about the special characteristics and training the team leader should have and what kind of experience he or she should bring to bear on the audit program. Typically, team leaders are technically trained personnel, very frequently engineers, who have had experience in environmental management and extensive experience in at least one functional area and possibly more. Because of experience, most if not all team leaders have a general familiarity with facility operations.

The team leader needs to have the support and respect of management, the audit team, and facility personnel. If the audit program is designed to serve the needs of top management, those individuals need to be assured that the team leader will conduct a comprehensive audit. To maximize the effectiveness of the on-site review, the team leader needs to be someone who can command the respect of the other team members. In addition, facility personnel, quite understandably, may be very concerned about being audited, what the audit team is going to do in the facility, and what they are going to say about the facility to management. In short, the team leader should be believable to the facility management and should have the capability and willingness to help the facility understand why the facility is being audited, what the audit is all about, and what the audit team is doing. And the team leader must recognize the fine balance between not interfering with the facility's primary mission, of producing a product, and not being so supportive that the team does not in fact do the rigorous audit that generally needs to be done.

Audit team leaders must be accomplished in these areas (in particular in addition to their technical and regulatory experience and knowledge):

Effective Communications. The ability of the team leader to motivate and work well with the audit team members as well as the facility management and staff is critical to the success of the audit. In particular, the extent to which the team leader encourages shared information within the team and with the facility personnel will largely determine the success of the audit. One particularly helpful skill to develop is that of "constructive feedback." Soliciting feedback from team members can make the individual a better team leader. Moreover, an effective team leader is adept at conducting meetings.

Personal Leadership Style. Several kinds of understanding on the part of the team leader are necessary for achieving effective leadership, including an understanding of his or her own leadership style and an appreciation that other people may have very different styles, the nature of his or her working relationships with other individuals, team dynamics, and how to adapt to the situation at hand.

The team leader must also have the ability to guide the team in the development of an effective audit report in order to document the audit findings clearly, accurately, and in a timely manner. The audit should be designed to include the type of information and level of detail appropriate for the problems identified, the individuals who have to be notified, and the scope of the audit.

11

Environmental Auditing and the Law

Carol Clayton

P. Kathleen Wells
Wilmer, Cutler & Pickering

Jeffrey D. Watkiss
Bracewell & Patterson

Outside environmental counsel work with a variety of clients on many different types of "environmental audits," a term loosely applied to a host of environmental assessment activities ranging from preacquisition due diligence reviews to internal corporate investigations of suspected violations to routine and systematic regulatory compliance audits. This chapter addresses routine audits conducted primarily to ascertain a facility's regulatory compliance status and evaluate the effectiveness of its environmental compliance program. Each client has unique needs and concerns; nevertheless, most clients seeking advice and assistance with respect to auditing initially ask the same basic questions: What are the legal benefits of auditing? What are the legal risks? Specifically, will the audit generate evidence that can be used against the company in enforcement proceedings? How can we minimize this and other risks? We seek to answer these recurring questions in this chapter.

The Benefits of Environmental Auditing

Environmental auditing, when performed with care and deliberation, offers many benefits. Some of these benefits fall outside the legal realm and therefore are not explored in detail here.[1] From a legal perspective, the major benefits of environmental auditing are twofold: (1) auditing allows a company to detect conditions or activities that may violate environmental laws and regulations, so that the company can correct them before they cause harm or lead to the imposition of penalties; and (2) even if a violation does occur, implementation of an auditing program (at least one that is part of a structured overall environmental compliance program) may help mitigate the penalties sought or imposed.

The cost of environmental violations can be exceedingly high. Virtually all of the major federal environmental statutes authorize the imposition of severe civil and criminal penalties on both individuals and corporations. Inadvertent violations typically are punishable by civil fines of up to $25,000 per day.[2] Knowing violations (and, under the Clean Air Act and Clean Water Act, even *negligent* violations) are punishable by substantial criminal fines and prison terms.[3] Violations can also lead to temporary suspension and permanent debarment from government contracts.[4]

The federal and state governments have placed increasing emphasis on criminal enforcement in recent years, making the specter of criminal liability daunting for both corporations and their individual officers and employees. It is well established that a corporation may be held criminally liable for the acts of an employee performed within the scope of his or her employment.[5] The courts have held that, under "public welfare" statutes such as the environmental laws, the government need not prove that the offender had specific intent to violate the law; a lesser (although as yet somewhat ill defined) general intent or knowledge standard applies.[6] A corporation may be deemed to have the collective knowledge of its employees as a group, so that the corporation may be held liable even though no one employee knew all the facts constituting the violation.[7]

Ideally, routine auditing will allow a company to avoid environmental violations altogether. Even if a violation does occur, however, the existence of a systematic auditing program may serve as a mitigating factor that either prevents a government agency from bringing an enforcement action or reduces the penalties sought. In evaluating the benefits of auditing in the context of enforcement, it is important to distinguish auditing programs in existence *before* a violation occurs from programs implemented afterward.

Previolation Auditing

Both the United States Environmental Protection Agency (EPA) and the U.S. Department of Justice (DOJ) have policies for encouraging and rewarding auditing and other voluntary compliance efforts.[8] EPA has explained that, when responding to a violation, it will "take into account, on a case-by-case basis, the honest and genuine efforts of regulated entities to avoid and promptly correct violations and underlying environmental problems."[9] "[H]onest and genuine efforts," according to EPA, are evidenced by "reasonable precautions to avoid noncompliance, expeditiously correct[ing] underlying environmental problems discovered through audits or other means, and implement[ing] measures to prevent their recurrence."[10] EPA's enforcement response policies for certain statutes make a violator's "`good faith' efforts to comply" with the statute and related regulations—of which an effective environmental auditing program should be evidence—a factor in determining EPA's response to violations of the statute.[11]

The Pollution Prosecution Act of 1990[12] significantly increased the capability of EPA's Office of Criminal Enforcement (OCE) to detect and criminally prosecute environmental violations. In a 1994 internal guidance memorandum, OCE recognized that central to an effective criminal investigation strategy is a case-selection process that "distinguishes cases meriting criminal investigation from those more appropriately pursued under administrative or civil judicial authorities."[13] Consistent with EPA's stated policy of encouraging self-auditing, the OCE Guidance explains that "a violation that is voluntarily revealed and fully and promptly remedied as part of a corporation's" compliance auditing program will "not be a candidate for the expenditure of scarce criminal investigative resources."[14]

DOJ similarly encourages voluntary environmental compliance activities by treating them as "mitigating factors in the Department's exercise of criminal enforcement discretion."[15] DOJ published in July 1991 a description of factors that affect the degree of DOJ's enforcement response to environmental violations, including factors dictating against criminal prosecution.[16] One such factor is the existence of an environmental auditing program sufficient "to identify and prevent future noncompliance [that]…was adopted in good faith in a timely manner."[17] Audits undertaken after an environmental problem is known or suspected are *not* "timely."[18] "[V]oluntary, timely and complete disclosure" of matters under investigation, taking into account the "degree and timeliness of cooperation," may also influence DOJ enforcement.[19] In many if not most cases, timely disclosure and cooperation will be possible only where a routine auditing program is already in place.

Some states have similarly identified the existence of a routine auditing program as a factor that will influence environmental enforcement discretion. New Jersey considers the existence and scope of an environmental audit program an important factor in deciding whether to bring a criminal prosecution for violating an environmental statute.[20] Likewise, in selecting enforcement options, environmental prosecutors in California are directed to "consider the existence and scope of any regularized, intensive, and comprehensive environmental compliance program."[21] Like DOJ, both New Jersey and California credit violators who voluntarily disclose violations in a timely manner.[22]

If a federal or state agency *does* decide to bring an enforcement action, the existence of an auditing program may warrant a reduction in civil penalties. For example, both RCRA and the Clean Air Act require EPA to consider "good faith efforts to comply" when assessing civil penalties.[23] EPA has also explained that good faith compliance efforts may reduce penalties under the Toxic Substances Control Act (TSCA),[24] the Comprehensive Environmental Response, Compensation, and Liability Act (CERCLA),[25] and the Emergency Planning and Community Right-to-Know Act (EPCRA).[26] The existence of compliance and management auditing programs should be strong evidence in most instances of a company's good faith efforts to comply with all statutory requirements addressed by these programs.

The existence of an auditing program (again, at least as part of a comprehensive overall compliance program) may also lead to a reduction in criminal fines. On November 16, 1993, the U.S. Sentencing Commission's Advisory Group on Environmental Sanctions issued its revised final draft of recommended sanctions for criminal violations of federal environmental laws by organizational defendants (Draft Sentencing Guidelines).[27] Part B of the Draft Sentencing Guidelines establishes a point system under which six primary categories of environmental offenses would be assigned a "base offense level" from 2 to 24, with 24 indicating 100 percent of the maximum penalty permitted under the relevant law. Each base offense level would then be either increased by the presence of aggravating "culpability factors" or decreased by the presence of mitigating "culpability factors." On the aggravating side of the equation is the complete absence of a compliance program or other organized effort to ensure compliance, such as an auditing program.[28]

On the mitigation side, base offense levels could be reduced up to eight levels for companies that

> prior to the offense…had committed the resources and the management processes that were reasonably determined to be sufficient, given its size and the nature of its business, to achieve and maintain compliance with environmental requirements….[29]

Reductions of up to six levels would be available for companies that promptly report violations and cooperate in the investigation and remediation of violations.[30]

It is particularly noteworthy that the Advisory Group members would deny altogether recognition of any mitigating factors to any organization failing to "substantially satisfy" seven "minimum factors demonstrating a commitment to environmental compliance."[31] Central to those seven minimum factors is that the company has designed, funded, and staffed a program for "frequent auditing" of principal operations and pollution control facilities.[32]

Postviolation Auditing

Even if a company did not have an auditing program in place before a violation occurred, it may be able to mitigate penalties by implementing a program afterward. EPA may agree to settle an action brought in response to an environmental violation if the violator "can assure the Agency that [the violator's] noncompliance will be (or has been) corrected."[33] EPA will sometimes agree to a reduced settlement fine if a company undertakes "projects remediating the adverse public health or environmental consequences of the violations at issue."[34] These projects, called "Environmentally Beneficial Expenditures" or "Supplemental Environmental Projects," may include initiation of environmental auditing programs. EPA will not accept, however, a program that simply represents "generally good business practices" or that deals only with "similar, obvious violations at other facilities."[35] Rather, the agency may reduce penalties where an audit program goes beyond the minimum required by good business practices and is

> designed to seek corrections to existing management and/or environmental practices whose deficiencies appear to be contributing to recurring or potential violations. These other potential violations may encompass not only the violating facility, but other facilities owned and operated by the defendant/respondent, in order to identify, and *correct as necessary*, management or environmental practices that could lead to recurring or future violations of the type which are the basis for the enforcement action.[36]

As noted above, a criminal conviction under the Clean Water Act or Clean Air Act can lead to debarment from government contracting. Before EPA will remove the bar, a violator must demonstrate that it has taken steps to prevent future violations, including creation and enforcement (which EPA suggests may be by means of "a monitoring and auditing system") of "an effective program to prevent and detect environmental problems and violations of the law."[37] Furthermore, once a violation has been detected, the violator must take steps to prevent further similar violations. One recommended step is to conduct a compliance audit—"an independent environmental audit to ensure that there are no other environmental problems or violations at the facility."[38]

Environmental Auditing: The Risks and How to Minimize Them

We have seen that environmental auditing can offer important benefits from a legal perspective; however, auditing is not without risks. A major—and legitimate—concern of every company that is considering initiating an auditing program is that audit reports and other documentation generated through the audits may be used as evidence against it in enforcement proceedings. Consequently, clients often ask whether EPA, DOJ, or other government entities will seek to obtain copies of audit reports and, if so, whether the reports can be protected from disclosure under the attorney-client or other legal privileges. This section explains EPA's and DOJ's positions on the use of audit reports in enforcement proceedings, summarizes the relevant law of privilege, explains how to maximize the chances that a claim of privilege will be upheld, and discusses potential drawbacks of structuring an audit program with a primary aim of establishing a legal privilege for audit reports.

At the outset, it is important to recognize that an audit report is most likely to become evidence against a company in an enforcement case when the company fails to correct a violation identified through the audit. Indeed, ignoring the information in the report may transform a relatively insignificant infraction into a *criminal* violation. EPA's OCE Guidance explains that "[c]orporate culpability may...be indicated when a company performs an environmental compliance or management audit, and then knowingly fails to promptly remedy the noncompliance and correct any harm done."[39] A company obviously can minimize the risk that an audit report will be used against it in an enforcement case by expeditiously rectifying violative conduct or conditions brought to its attention through the audit process.

Government Policies on Audit Reports

Despite their policies to encourage auditing, neither EPA nor DOJ has been willing to promote this objective by making a commitment to forgo use of voluntary audit reports as evidence in enforcement efforts. EPA recognizes, however, that "routine Agency requests for audit reports could inhibit auditing in the long run, decreasing both the quantity and quality of audits conducted"; accordingly, the agency has stated that it "will *not* routinely request environmental audit reports."[40] Nevertheless, the agency will exercise its "broad statutory authority...to request an audit report" if it deems the report "material to a criminal investigation" or determines that disclosure is necessary "to accomplish a statutory mission."[41] In an attempt to balance the chilling effect of disclosure with its need for reports under certain circumstances, EPA has indicated that it will "most likely" limit its requests to "particular information needs rather than the entire report" and will "usually" make such requests only "where the information needed cannot be obtained from monitoring, reporting or other data otherwise available to the Agency."[42] The California EPA has adopted an identical policy.[43]

DOJ's enforcement policy notes the importance of voluntary audit programs while cautioning that DOJ is not legally required "to forego [sic]...the use of any evidentiary material."[44] In fact, DOJ's definition of "cooperation" contemplates full "disclosure of any documents relating either to the violations or to the responsible employees."[45]

In addition to government attempts to obtain environmental audit reports, private parties may seek them during the course of litigation instigated under either common law or federal or state statutes that permit citizens' suits to enforce environmental laws. In federal courts, the Federal Rules of Civil Procedure establish a strong bias in favor of the disclosure of any information "relevant" to the subject matter of litigation. Even information not admissible in court may be discovered if it "appears reasonably calculated to lead to the discovery of admissible evidence."[46] When a company is alleged to have violated environmental standards, the contents of its environmental audit reports will almost invariably be deemed relevant and discoverable, if not otherwise protected by an applicable privilege.

Protection against Disclosure

Despite their strong bias in favor of disclosure of relevant information, the Federal Rules of Civil Procedure provide some protection for "privileged" material.[47] Among the several privileges that the courts have articulated, three are potentially applicable to environmental audit reports: (1) the attorney-client privilege, (2) the work-product doctrine, and (3) the self-evaluation privilege.

The Attorney-Client Privilege.   The attorney-client privilege is intended to facilitate effective legal representation by protecting from discovery certain communications between a client (or potential client) and his or her lawyer. When the client is a corporation, the Supreme Court has held that the privilege applies to communications (1) by a corporate employee[48] to counsel (acting as such) or counsel's agent,[49] (2) made at the direction of corporate superiors, (3) regarding matters within the scope of the employee's duties, (4) for the purpose of enabling counsel to render legal advice to the corporation, and (5) that were considered confidential when made and have been kept confidential by the corporation.[50] Significantly, the privilege protects only the communications between the attorney and the client; the *facts* underlying the communication are *not* protected if evidence of them is available in another form.[51] Further, failure to maintain the confidentiality of the communication can result in the complete waiver of the privilege.

If a company seeks to afford its environmental audit reports protection under the attorney-client privilege, it should involve either outside or in-house legal counsel directly in the actual auditing process, rather than simply sending copies of the audit reports to counsel. Audits should be conducted by or under the supervision of counsel and for the express purpose of obtaining legal advice from counsel.[52] The reports themselves should be written in the form of legal advice and should be labeled "privileged and confidential." The company should also restrict distribution of the audit reports to attorneys, management personnel, and others having responsibility for legal compliance decisions, and establish internal controls to ensure that the audit reports are not inadvertently disclosed to others.

Companies should recognize that it may be difficult to establish the privilege for audit reports prepared in the regular course of a routine auditing program—and that there may be drawbacks in attempting to do so. In order for the privilege to attach, the report would need to be for the purpose of seeking legal advice; the focus of routine auditing may instead be on engineering, operations, and management matters. Establishing the privilege requires limiting distribution of the audit report, which may prevent those who most need the information contained in the report from receiving and responding to it (such as relatively low-level employees handling recordkeeping required under the environmental laws, or operators of pollution control equipment). Consequently, measures taken to establish the privilege could undermine the basic compliance management goals of auditing.

The Work-Product Doctrine.   The Federal Rules of Civil Procedure provide a qualified immunity from discovery for documents and tangible things that were (1) prepared in anticipation of litigation or for trial, (2) by or for another party or by or for that other party's representative.[53] This protection, known as the "work-product doctrine," is intended to protect the thought processes and mental impressions of the attorney. Like the attorney-client privilege, however, the work-product doctrine does not shield from discovery facts contained in work-product documents.[54]

Because work-product protection applies only to documents prepared in anticipation of litigation, the privilege does not extend to documents created in the regular course of business, even if prepared by an attorney.[55] Routine audits conducted as part of an ongoing environmental management program are therefore unlikely to be covered by the privilege because they are performed in the ordinary course of business, and not "in anticipation of litigation." However, an audit performed as part of an internal corporate investigation of a specific known or suspected violation should ordinarily be deemed to have been prepared in anticipation of litigation (i.e., enforcement proceedings) and therefore should be protected by the work-product doctrine.[56]

Work-product protection is not limited to materials prepared directly by a company's counsel. Recognizing that "attorneys often must rely on the assistance of investigators and other agents in the compilation of materials in preparation for trial," the Supreme Court has held that the protection extends to materials prepared by agents of a party's attorney.[57] Accordingly, the work product of environmental consultants, if prepared for the company in anticipation of litigation, should also be protected from discovery. In such cases, a company should create a strong documentary record concerning the likelihood or expectation of litigation. Investigating the subject of possible litigation separately, rather than as part of a routine audit conducted under the company's environmental auditing program, may give further credence to a claim that the audit was conducted in anticipation of litigation.

Even if a report qualifies for work-product protection, the Federal Rules of Civil Procedure permit an opposing party to gain access to a work-product document by showing a substantial need for the materials in preparing the case, and an inability to obtain substantially equivalent materials without undue hardship.[58] Some courts distinguish between pure compilations of factual information and materials containing "mental impressions, conclusions, opinions, or legal theories of an attorney or other representative of a party concerning the litigation," providing a higher level of protection (nearly absolute protection, according to some courts[59]) for the latter, and permitting deletion of pure statements of opinion before a document is disclosed.[60]

Like the attorney-client privilege, the work-product doctrine is waived if the communication is voluntarily disclosed in a manner that "substantially increase[s] the opportunities for potential adversaries to obtain the information."[61] Some courts have held that voluntary disclosure of internal audit reports to government agencies waives the privilege,[62] although explicit written agreements in which the agency agrees not to disclose the materials to third parties have sometimes preserved the work-product protection.[63]

The Self-Evaluation Privilege. A handful of courts have recognized a "self-evaluation" or "self-critical" privilege to protect the confidentiality of internal discussions and investigations regarding compliance with legal or other requirements. The privilege has been described as applying where the public interest in encouraging confidential self-analysis and critical evaluation outweighs the need of litigants and the judicial system for access to the information.[64] Although not consistently defined, the privilege has been applied where (1) the information results from a critical self-analysis undertaken by the party seeking protection from disclosure; (2) there is a strong public interest in preserving the free flow of the type of information sought; (3) disclosure would curtail the flow of that type of information; and (4) the documents in question were prepared with the expectation that they would remain confidential, and they have in fact been kept confidential.[65]

According to one court, the self-evaluation privilege applies to certain subjective or evaluative materials, but not to objective data.[66] To the extent the self-evaluation privilege is held to apply only to subjective analyses of factual information that is required to be disclosed to governmental entities, it has limited value in protecting sensitive information from discovery since environmental violations typically turn on raw facts, such as what was emitted, where, and in what quantity. This same court, however, rejected the view espoused by other courts that self-critical analyses warrant privilege only to the extent that they are performed pursuant to a legal requirement.[67]

Some courts have held that the self-evaluation privilege cannot protect documents from disclosure to the government.[68] The privilege was raised and rejected as a defense to government efforts to obtain "self-evaluative documents" in an action to enforce the Clean Water Act.[69] After reviewing the purposes of the act, the court found that recognition of a "'self-critical' privilege...would effectively impede the Administrator's ability to enforce the [Act]," and held the privilege inapplicable where "the documents in question have been sought by a governmental agency."[70] A couple of courts have gone further. One held that even when the government is not a party, the "public interest in preventing and remediating environmental pollution" creates a "need for disclosure of documents relating to environmental pollution and the [related circumstances that] outweighs the public's need for confidentiality in such documents."[71] Another court found that reasoning persuasive when it refused to apply the privilege in a suit by an insured seeking indemnification from its insurers for environmental liability, holding that "the self-evaluation privilege does not apply *a fortiori* to environmental reports, records, and memoranda."[72] Similarly, an Ohio court rejected a claim of "self-critical privilege" by the operator of a hazardous waste facility on the ground that the "heavy regulation of the hazardous waste industry" by the state required disclosure of company records to state regulatory officials.[73]

In another recent case, a court distinguished between voluntary, routine safety reviews and reviews undertaken in response to an accident. The court held that voluntary, routine safety reviews are not protected by the self-evaluation privilege because they are designed to prevent litigation and are therefore unlikely to be curtailed by the possibility of discovery.[74] This reasoning (if followed by other courts) arguably would deny the self-evaluation privilege to all routine and systematic audits.[75] The usefulness of the privilege is further limited because, like both the attorney-client privilege and the work-product protection, any factual information in the self-evaluative analysis regarding a violation remains discoverable.[76]

A few states recently enacted legislation creating a self-evaluation privilege for environmental audit reports, and several other states are considering similar measures.[77] Oregon adopted legislation in 1993,[78] and Colorado, Indiana, and Kentucky followed suit in 1994.[79] The state laws vary in some respects,[80] but Oregon's statute has served as a model for the others and is an instructive example of these statutory privileges. Part of a larger bill that established criminal penalties for air, water, and hazardous waste violations, the Oregon law would extend a privilege to (a) audit reports, (b) analyses of such reports, and (c) implementation plans that improve compliance, address noncompliance, and prevent future noncompliance. To attain privileged status, these communications must be labeled when prepared "Environmental Audit Report: Privileged Document."

The Oregon privilege is not absolute. It permits adverse parties in civil cases to obtain access where *in camera* inspection shows that the privilege has been waived by the owner or operator, the privilege was asserted for a fraudulent purpose, or the audit report demonstrates that the company had discovered a violation and failed to attempt to rectify the violation with reasonable diligence. Government prosecutors are also allowed access in criminal cases where they can demonstrate (a) a compelling need for the audit information, (b) that the audit information is not otherwise available, and (c) an inability to obtain the substantial equivalent without unreasonable cost and delay.[81] Notwithstanding these limitations, the Oregon statute confers much greater protection than is afforded by the common law privilege currently evolving in the courts. Specifically, the Oregon law protects both subjective analysis *and* underlying facts. Further, the Oregon law protects self-evaluative reports irrespective of whether they are prepared pursuant to governmental mandate.

The American Bar Association has also prepared legislation that would protect from discovery or use at trial "all activities"—including audit reports—that are part of environmental and other compliance programs.[82] According to this proposal, the protection would apply only to a corporation that had acted in good faith to implement an effective compliance program, and there would be severe financial penalties for any company or individual claiming the protection in bad faith. Unlike the attorney-client privilege and the attorney work product protection, the ABA's proposed privilege would not be waived by disclosure of documents related to the compliance program to courts, government bodies, or any third party.

Managing Risks Associated with Disclosure

A company's most important protection against the risk of disclosure of audit results is to ensure that it has an effective system in place for addressing violations as soon as they are detected. In this way, even if an audit report disclosing a violation becomes public, the company can demonstrate to the government and the public that it promptly remedied the problem. The possibility that audit reports could become public or be disclosed to potentially adverse parties should also shape the language auditors use in their reports. It is usually unwise and unnecessary to opine in an audit report that a given activity or condition is "illegal" or a "violation." Rather, auditors should identify applicable requirements and summarize conditions noted in the field. Thereafter, in a separate written or oral communication, legal counsel can assess whether a violation has occurred.

Developing and Implementing an Environmental Audit Program

A company can maximize the benefits and minimize the risks of environmental auditing through careful advance planning, thoughtful implementation, and rigorous follow-up on auditor's recommendations. This section sets out a few basic suggestions for designing and conducting effective environmental audits.

The first question we ask when a client seeks our assistance on an audit is "Why are you conducting the audit?" The answer sometimes is unclear, particularly if the client is performing an audit for the first time and has not considered how the audit "fits" into the company's overall compliance strategy. Independent counsel can be of great assistance at this early stage of the process, helping the client define its environmental compliance goals and develop a plan to achieve them.

It bears repeating that the greatest downside risk of auditing occurs when a company produces a written report revealing a violation of law and then takes no (or insufficient) action to correct the violation. If a company does not have an adequate management structure in place to respond to negative information in an audit, the fundamental purpose of auditing, which is improving the company's compliance status, cannot be achieved. Auditing in these circumstances is particularly likely to generate a "paper trail" for prosecutors showing the existence of a violation and no meaningful action to correct it. For this reason, companies that lack at least rudimentary compliance programs may be well advised to postpone conducting comprehensive audits, and instead focus their initial efforts on creating a basic compliance management structure, clear and workable compliance policies, and an internal compliance training program.

If it appears that auditing will be a useful compliance tool in light of a company's individual needs and existing compliance procedures, the company should invest adequate time and resources into audit planning. Sending a hastily assembled audit team into the field with a standard "off the shelf" checklist would likely produce little more than check marks that even the auditors would be hard-pressed to explain. To avoid this result, the company must expressly define the goals and scope of the audit. For example, will the audit cover compliance under all applicable federal, state, and local environmental laws and regulations? If so, substantial effort may be required to ascertain in advance exactly what those requirements are with respect to a particular facility or operation. Are the auditors expected to assess the effectiveness of the company's compliance program and the performance of compliance personnel? If so, they should be told as much at the inception of the audit, and given explicit direction on what aspects of the program they are to evaluate.

The company should ensure that the audit is conducted by a team with the necessary legal, technical, and operational knowledge. The auditors should also be sufficiently independent of the facility and personnel whose performance they are assessing that their objectivity will not be compromised. In EPA's words:

> The status or organizational locus of environmental auditors should be sufficient to ensure objective and unobstructed inquiry, observation and testing. Auditor objectivity should not be impaired by personal relationships, financial or other conflicts of interest, interference with free inquiry or judgment, or fear of potential retribution.[83]

The most thorough audit will be of little value and will carry little weight with enforcement agencies if the independence of the auditors is uncertain.

It is crucial for the company to determine in advance *who* will be responsible for evaluating the information generated by the audit and ensuring that appropriate measures are taken in response to that information. Because of the complexity of environmental statutes and regulations, advice of legal counsel often will be required to determine whether questionable conditions or activities identified in the audit in fact constitute a violation of law, and if so, how the company should respond. As previously explained, auditors should not draw legal conclusions in their report but should instead identify the requirements they believe are applicable and the conditions they have observed, leaving counsel to evaluate the legal implications of the information presented.

The company can help ensure the accuracy of audit reports by instructing auditors to prepare draft reports for review by legal counsel, corporate compliance managers, and facility-level compliance managers. Even the most skilled and experienced auditors may misunderstand information they have obtained; errors can best be avoided by allowing those who know the audited facility or operation best to comment on the draft. This procedure will also help establish the open dialogue among all the audit participants that is so critical to the audit's success.

The company should plan in advance *how* it will respond to audit recommendations. There should be an established protocol and schedule for evaluating audit recommendations and taking necessary corrective measures. From an enforcement defense counsel's perspective, it is always preferable to have a written remedial plan and documentation demonstrating that the plan in fact was implemented in a timely fashion.

It is important to consider in advance what types of information contained in the audit report would be legally required to be reported to the federal or state government. The company should also be aware that even if information isn't legally required to be reported, there may be legal and strategic reasons for *voluntarily* disclosing the information. The question of what information must or should be reported merits discussion here.

Environmental laws and regulations, as well as individual environmental permits issued to the company, may mandate reporting of information identified during the audit. Many environmental regulations and individual permits establish a continuing duty to report certain events or to correct prior reports that have become inaccurate. Furthermore, some environmental statutes (and EPA's penalty assessment policies under those statutes) authorize the imposition of fines on a *daily* basis for violations of reporting requirements; penalties accrue from the date of the violation until the required report is made. Even though the initial failure to make a timely report cannot be "cured" by making a late report, continuing exposure to daily penalties can be stopped.[84]

Even in the absence of a legal duty to report, there are a number of reasons why a company may be well advised to report noncompliance voluntarily. As previously discussed, both DOJ and EPA encourage voluntary reporting and may reduce the penalties sought for an alleged violation based on a company's good faith compliance efforts, which include voluntary disclosure of the violation. EPA's RCRA penalty policy indicates that a company acts in good faith by "promptly identifying and reporting noncompliance or instituting measures to remedy the violation" before EPA detects it.[85] EPA's penalty policy under TSCA calls for a reduction in the penalty of up to 50 percent if the violator immediately discloses the violation. EPA also considers prompt identification and reporting of violations to be a positive factor in penalty assessment under the Clean Air Act,[86] CERCLA,[87] and EPCRA.[88] Since the core element of these interpretations of good faith is prompt disclosure, it warrants reiteration that prompt disclosure requires prompt detection, which is more likely to occur when a company has an auditing program in place.

There may be other (although less tangible) benefits resulting from voluntarily disclosing a violation or environmental problem. By bringing a matter to the regulator's attention voluntarily, a company exhibits a responsible and cooperative attitude that may cause regulators to treat the company more favorably in later dealings. Of course, a company should develop a plan for addressing the violation or environmental problem *before* approaching regulatory agencies so that the company can both demonstrate its responsiveness and take the lead in defining and implementing any remedial measures.

Conclusion

Companies should not wait until they have discovered a violation of environmental laws to implement a systematic compliance program, including auditing. Commitment to such a program is essential to complying with the growing body of complex and increasingly punitive environmental laws and regulations. Routine auditing is a valuable element of a compliance program because it can detect existing violations—permitting prompt remediation—or help prevent imminent violations before they happen. If violations do occur, the existence of an environmental auditing program may help establish a foundation for lenience by enforcement authorities. Finally, in an era of increasing public concern for environmental protection, auditing programs are evidence of environmentally responsible management. Although environmental auditing does entail some risk, the risk can be minimized through careful planning and thoughtful implementation.

References and Notes

1. These benefits may include obtaining information useful for general business planning or applying for insurance coverage, satisfying conditions for membership in a trade association (e.g., the Chemical Manufacturers Association's "Responsible Care" program), meeting standards under various private environmental initiatives (such as the CERES principles, formerly known as the Valdez principles), providing a tool for evaluating a facility's progress in achieving internal corporate objectives (such as pollution prevention goals), and serving as a mark of corporate good citizenship that will enhance the company's public image.

2. E.g., Clean Air Act, 42 U.S.C. § 7413(d); Resource Conservation and Recovery Act (RCRA), 42 U.S.C. § 6928(d).

3. E.g., RCRA, 42 U.S.C. § 6928(d); Clean Air Act, 42 U.S.C. § 113(c)(4); Clean Water Act, 33 U.S.C. § 1319(c)(1).

4. The Clean Water Act, 33 U.S.C. § 1368, and the Clean Air Act, 42 U.S.C. § 7606, provide for the automatic, mandatory debarment of those convicted of criminal violations of the statutes. EPA's regulations provide for discretionary debarment based on "continuing" or "recurring" noncompliance with environmental standards. 40 C.F.R. § 15.11(a). In addition, the Federal acquisition regulations applicable to all federal government contracting allow discretionary debarment of persons convicted of violating any environmental statute, as well as temporary suspension pending disposition of criminal charges. 48 C.F.R. §§ 9.400–9.409; see also 40 C.F.R. §32.

5. *C.I.T. Corp. v. United States*, 150 F.2d 85, 89-90 (9th Cir. 1945) (criminal intent of individual agent imputed to corporation).

6. See, e.g., *United States v. Park*, 421 U.S. 658, 672-74 (upholding conviction of corporate president under Food, Drug and Cosmetic Act based on his "responsible relation to the situation" and failure to prevent violation); *United States v. Frezzo Bros., Inc.*, 546 F. Supp. 713, 720 (E.D. Pa. 1982) (criminal convictions under Clean Water Act did not require proof of specific intent to violate law, but only that defendants

intended to do acts for which convicted), aff'd, 703 F.2d 62 (3d Cir.), cert. denied, 464 U.S. 829 (1983).

7. E.g., *United States v. T.I.M.E.-D.C., Inc.,* 381 F. Supp. 730, 740-41 (W.D. Va. 1974) (convicting corporate defendant of violations of Interstate Commerce Act based on collective knowledge of employees).

8. See EPA Environmental Auditing Policy Statement, 51 Fed. Reg. 25,004 (July 9, 1986) [hereinafter EPA Auditing Policy]; U.S. Department of Justice, Factors in Decisions on Criminal Prosecutions for Environmental Violations in the Context of Significant Voluntary Compliance or Disclosure Efforts by the Violator (July 1, 1991) [hereinafter DOJ Enforcement Policy]. EPA is reevaluating its policy on auditing, and will address alternative options for enforcement relief granted to those conducting audits, legal privilege issues, use of audit results in criminal cases, and other aspects of its policy. EPA Notice of Public Meeting on Auditing, 59 Fed. Reg. 31,914 (June 20, 1994); EPA Restatement of Policies Related to Environmental Auditing, 59 Fed. Reg. 38,455 (July 28, 1994) [hereinafter EPA Restatement of Audit Policy].

9. EPA Auditing Policy, *supra* note 8, at 25,007.

10. Ibid.

11. See, e.g., U.S. Envtl. Protection Agency, Enforcement Response Policy for Section 313 of the EPCRA 14 (Dec. 2, 1988).

12. 42 U.S.C. § 4321.

13. Memorandum from Earl E. Devaney, Director, OCE/EPA, to All EPA Employees Working in or in Support of the Criminal Enforcement Program 1 (Jan. 12, 1994) [hereinafter OCE Guidance].

14. Ibid. at 6.

15. DOJ Enforcement Policy, *supra* note 8, at 1.

16. Ibid.

17. Ibid. at 4.

18. Ibid. at 8-9.

19. Ibid. at 3.

20. New Jersey Envtl. Prosecutors Office, New Jersey Voluntary Environmental Audit/Compliance Guidelines at A (May 15, 1992) [hereinafter N.J. Guidelines].

21. California Envtl. Protection Agency, General Cal/EPA Policy on Envtl. Auditing 10 (Mar. 8, 1993) [hereinafter Cal. Audit Policy].

22. N.J. Guidelines, *supra* note 20, at L; Cal. Audit Policy, *supra* note 21, at 9-10.

23. RCRA § 3008, 42 U.S.C. § 6928(a)(3); Clean Air Act § 113, 42 U.S.C. § 7413(e)(1).

24. EPA Polychlorinated Biphenyls (PCB) Penalty Policy (Apr. 9, 1990), reprinted in [Administrative Materials] 20 Envtl. L. Rep. (Envtl. L. Inst.) 35,235 (Aug. 1990).

25. U.S. Envtl. Protection Agency, OSWER Directive 9841.2, Final Penalty Policy for EPCRA §§ 302-304, 311-312 and CERCLA § 103 (1990) [hereinafter EPCRA/CERCLA Penalty Policy].

26. Ibid.

27. Proposals to U.S. Sentencing Commission by Advisory Group on Envtl. Offenses Issued Nov. 16, 1993, reprinted in Daily Env't Rep. (BNA), Nov. 17, 1993, at E-1.

28. Section 9C1.1(f) of the draft would increase the base offense level by four levels if the organization were found to lack "a genuine organized effort to monitor, verify, and bring about compliance with environmental requirements." Ibid. at E-6.

29. Ibid. § 9C1.2(a) at E-6.

30. Ibid. § 9C1.2(b) at E-7.

31. Ibid. § 9D1.1(a) at E-7.

32. Ibid. § 9D1.1(a)(3)(i)-(v) at E-8.

33. U.S. Envtl. Protection Agency, OSWER Directive 9891.3, EPA Policy on the Inclusion of Environmental Auditing Provisions in Enforcement Settlements 2 (1986).

34. U.S. Envtl. Protection Agency, EPA Policy on the Use of Supplemental Environmental Projects in Enforcement Settlements 1 (Feb. 12, 1991).

35. Ibid. at 4.

36. Ibid. at 3-4 (emphasis in original). California has a similar policy of crediting certain "supplemental environmental projects," including environmental auditing programs that go beyond simple good business practice and are "designed to seek corrections to existing management and/or environmental practices whose deficiencies" contribute to recurring or potential violations. See California Envtl. Protection Agency, Supplemental Environmental Projects 3 (Mar. 22, 1993).

37. EPA Policies Regarding the Role of Corporate Attitude, Policies, Practices, and Procedures, in Determining Whether to Remove a Facility from the EPA List of Violating Facilities Following a Criminal Conviction, 56 Fed. Reg. 64,785, 64,787 (Dec. 12, 1991).

38. Ibid.

39. OCE Guidance, *supra* note 13, at 6.

40. EPA Audit Policy, *supra* note 8, at 25,007.

41. Ibid.

42. Ibid. Examples of such situations include instances in which "audits are conducted under consent decrees or other settlement agreements; a company has placed its management practices at issue by raising them as a defense; or state of mind or intent are a relevant element of inquiry, such as during a criminal investigation."

43. Cal. Audit Policy, *supra* note 21, at 2.

44. DOJ Enforcement Policy, *supra* note 8, at 15.

45. Ibid. at 9.

46. Fed. R. Civ. P. 26(b)(1).

47. Fed. R. Civ. P. 26(b). This discussion of privileges focuses predominantly on federal law. State laws may define privileges or their applications differently. Even in federal court, state law may govern the application of privileges if the federal court is applying state law. See Fed. R. Evid. 501. "Although the ancient common-law origins of the privilege suggest that the applicable rules will generally be the same in different jurisdictions, precise application of the rules may differ from one court to another." Stephen F. Black & Robert M. Pozin, Internal Corporate Investigations, 20 Bus. L. Monograms (MB) at 6-3 n.14 (May 1993) [hereinafter Black & Pozin].

48. Some states adhere to a stricter "control group" test in determining whether a communication from a corporate employee will be deemed a communication from the "client." The test focuses on whether the employee making the communication is in a position to control or take a substantial part in a decision about any action which the corporation may take upon the advice of the attorney. If a communication is circulated beyond those individuals with substantive responsibility for the subject matter at issue (i.e., the control group), the privilege is lost. Although the Supreme Court has rejected the strict control group test, *Upjohn Co. v. United States*, 449 U.S. 383 (1981), state law governs on issues of privilege where state law provides the rule of decision in the case. In states that apply the control group test, it would be difficult to establish the privilege for audit reports because the employees communicating with counsel likely would include technicians and other nonmanagement employees who would have no role in deciding the course of action that the company would take.

49. Communications from corporate personnel to experts retained by counsel may fall within the attorney-client privilege if the experts are retained to assist in the provision of legal advice. See *United States v. Cote*, 456 F.2d 142, 144 (8th Cir. 1972); *Bailey v. Meister Brau, Inc.*, 57 F.R.D. 11, 13 (N.D. Ill. 1972).

50. *Upjohn*, 449 U.S. at 394-95.

51. Ibid. at 395.

52. The court in *Diversified Indus. v. Meredith*, 572 F.2d 596, 608, 610 (8th Cir. 1977) (en banc), concluded that the participation of independent lawyers in an internal investigation provided prima facie evidence that the purpose of the investigation was to secure privileged legal advice. The law is less clear with respect to reports prepared by in-house counsel, who often provide business as well as legal advice. See *In re Grand Jury Subpoena*, 599 F.2d 504, 511 (2d Cir. 1979) ("Participation of the general counsel does not automatically cloak the investigation with legal garb.").

53. Fed. R. Civ. P. 26(b)(3) (codifying the doctrine of *Hickman v. Taylor*, 329 U.S. 495 (1947), see 8 Charles A. Wright and Arthur R. Miller, Federal Practice and Procedure § 2024 at 196-97 (1970) [hereinafter Wright & Miller].

54. See, e.g., *Protective Nat'l Ins. Co. v. Commonwealth Ins. Co.*, 137 F.R.D. 267, 280-81 (D. Neb. 1989).

55. See, e.g., *Bristol-Myers Co. v. F.T.C.*, 598 F.2d 18, 28 (D.C. Cir. 1978).

56. It is not necessary that litigation be an absolute certainty. The Supreme Court has held materials to be protected that were voluntarily generated by the company because of its suspicion, resulting from an internal audit, that a foreign subsidiary might have made illegal payments. See *Upjohn*, 449 U.S. at 386-87. According to several courts, all that is necessary for the protection to apply is that the documents can "fairly be said to have been prepared or obtained because of the prospect of litigation." *Martin v. Bally's Park Place Hotel & Casino*, 983 F.2d 1252, 1258 (3d Cir. 1993) (quoting *United States v. Rockwell Intern.*, 897 F.2d 1255, 1266 (3d Cir. 1990)); see *Diversified Inds. v. Meredith*, 572 F.2d at 604.

57. *United States v. Nobles*, 422 U.S. 225, 238-39 (1975); Fed. R. Civ. P. 26(b)(3).

58. Fed. R. Civ. P. 26(b)(3); see also 4 James W. Moore et al. & Jo D. Lucas, Moore's Federal Practice ¶ 26.64[3.-1] (2d ed. 1991).

59. See, e.g., *In re Martin Marietta Corp.*, 856 F.2d 619, 626 (4th Cir. 1988), cert. denied, 490 U.S. 1011 (1989); *In re Murphy*, 560 F.2d 326, 336 (8th Cir. 1977); *Duplan Corp. v. Moulinage et Retorderie de Chavanoz*, 509 F.2d 730, 734 (4th Cir. 1974), cert. denied, 420 U.S. 997 (1975).

60. Wright & Miller, *supra* note 53, § 2026 at 231-32.

61. *Niagara Mohawk Power Corp. v. Stone & Webster Eng. Corp.*, 125 F.R.D. 578, 587 (N.D.N.Y. 1989); see Wright & Miller, *supra* note 53, § 2024 at 210.

62. See *In re Martin Marietta Corp.*, 856 F.2d at 623 (disclosure of internal audits to governmental agencies waived attorney-client privilege and nonopinion work product privilege).

63. Black & Pozin, *supra* note 47, § 6.04[1] at 6-11.

64. *Reichhold Chemicals, Inc. v. Textron, Inc.*, No. 92-30393-RV, 1994 WL 532165, at *4 (N.D. Fla. Sept. 20, 1994); *Granger v. National R.R. Passenger Corp.*, 116 F.R.D. 507, 508 (E.D. Pa. 1987); *Webb v. Westinghouse Elec. Corp.*, 81 F.R.D. 431, 433-35 (E.D.Pa. 1978) (discussing evolution of privilege in the context of internal analyses of employer's equal employment opportunity goals and policies).

65. See *Dowling v. American Hawaii Cruises, Inc.*, 971 F.2d 423, 425-26 (9th Cir. 1992) (analyzing decisional precedents and literature on privilege); *Reichhold*, No. 92-30393-RV, 1994 WL 532165, at *4 (applying *Dowling* factors to protect environmental self-evaluation); see also *CPC Int'l, Inc. v. Hartford Accidents & Indemn. Co.*, 620 A.2d 462, 467 (N.J. Super. Ct. 1992); *Ohio ex rel. Celebrezze v. CECOS Int'l, Inc.*, 583 N.E.2d 1118, 1120 (Ohio Ct. App. 1990).

66. *Hoffman v. United Telecommunications, Inc.*, 117 F.R.D. 440, 442-43 (D. Kan. 1987).

67. Ibid. at 443; contra *Webb*, 81 F.R.D. at 434 (employment discrimination case finding "materials protected have generally been those prepared for mandatory governmental reports").

68. See, e.g., *Emerson Elec. Co. v. Schlesinger*, 609 F.2d 898, 907 (8th Cir. 1979) ("[T]he rationale underlying the qualified privilege—encouraging complete and candid self-evaluations—is less persuasive...[when] the reports are disclosed only to federal agencies, not to third parties.").

69. *United States v. Dexter Corp.*, 132 F.R.D. 8, 10 (D. Conn. 1990).

70. Ibid. at 9-10 (quoting *Federal Trade Comm'n v. TRW, Inc.*, 628 F.2d 207, 210 (D.C. Cir. 1980)); see also *Celebrezze*, 583 N.E.2d at 1120 ("In the presence of a clear legislative directive that the hazardous waste industry be subject to public scrutiny," the court refused to apply a self-evaluative privilege because to do so would permit companies "to skirt [disclosure] obligations created by law.").

71. See *CPC Int'l*, 620 A.2d at 467.

72. *Koppers Co., Inc. v. Aetna Casualty and Surety Co.*, 847 F. Supp. 360, 364-65 (W.D. Pa. 1994).

73. *Celebrezze*, 583 N.E.2d at 1121.

74. *Dowling*, 971 F.2d at 427.

75. See, e.g., *Reichhold*, No. 92-30393-RV, 1994 WL 532165, at *5 ("privilege in this case applies only to reports which were prepared after the fact for the purpose of candid self-evaluation and analysis of the cause and effect of past pollution").

76. See *Hoffman*, 117 F.R.D. at 442.

77. Table of State Legislation on Environmental Audit Privilege, 18 Chem. Reg. Rep. (BNA) 447 (July 15, 1994).

78. Or. Rev. Stat. § 468.963 (1993).

79. Act of June 1, 1994, 1994 Colo. Sess. Laws (to be codified at Colo. Rev. Stat. §§ 13-25-126.5, 13-90-107, 25-1-114.5); Act effective July 1, 1994, 1994 Ind. Acts (to be codified at Ind. Code § 13-10); Act effective July 15, 1994, 1994 Ky. Acts, Ch. 430 (to be codified at Ky. Rev. Stat. Ch. 224).

80. For instance, Colorado's statute is the only one that provides for a presumption against the imposition of penalties when a violation detected through an environmental self-evaluation is voluntarily disclosed.

81. Or. Rev. Stat. § 468.963(3).

82. Joseph E. Murphy, Organizational Compliance Program Improvement Act, attached to Report on the Self-Evaluative Privilege (First Draft July 16, 1991) (unpublished manuscript on file with the Corporate Counseling Committee, ABA Antitrust Section).

83. EPA Auditing Policy, *supra* note 8, at 25,009; see also Draft Sentencing Guidelines, *supra* note 27, § 901.1(a)(3) at E-8 (requiring frequent audits with "appropriate independence from line management"). The California EPA has instructed that "[a]uditor objectivity should not be impaired by personal relationships, financial or other conflicts of interest, interference with free inquiry or judgment, or fear of potential retribution." Cal. Audit Policy, *supra* note 21, at 7.

84. See, e.g., EPCRA/CERCLA Penalty Policy, *supra* note 25, at 21.

85. EPA's penalty policy under RCRA states: "The violator can manifest good faith by promptly identifying and reporting noncompliance or instituting measures to remedy the violation before the Agency detects the violation. Assuming self-reporting is not required by law and the violations are expeditiously corrected, a violator's admission or correction of a violation prior to detection may be cause for mitigation of the penalty, particularly where the violator institutes significant new measures to prevent recurrence. Lack of good faith, on the other hand, can result in an increased penalty." RCRA Civil Penalty Policy (Oct. 29, 1990), reprinted in [Federal Laws] 1 Env't Rep. (BNA) 21:5091, 21:5102 (Oct. 30, 1992).

86. EPA Penalty Policy for Stationary Source Clean Air Act Violations as Corrected (Jan. 17, 1992), reprinted in Daily Env't Rep. (BNA), Jan. 31, 1992, at E-8 (prompt reporting "where there is no legal obligation to do so" is evidence of violator's "degree of cooperation").

87. EPCRA/CERCLA Penalty Policy, *supra* note 25, at 26-27 (prompt identification and reporting is evidence of good faith, justifying mitigation of a penalty by up to 25 percent).

88. Ibid.

PART 5

Safety
Auditing

12

Safety Management Systems Audits in the Chemical Process Industry

Ray E. Witter

For the Center for Chemical Process Safety,
American Institute of Chemical Engineers

The American Institute of Chemical Engineers (AIChE) has a 30-year history of process safety and loss control in the chemical, petrochemical, and hydrocarbon processing industries. The Center for Chemical Process Safety (CCPS), a directorate of AIChE, was established in 1985 to develop and disseminate technical information for use in preventing major chemical process incidents. In today's society, the public, customers, in-plant personnel, and governmental regulatory agencies all demand that companies take necessary actions to reduce the possibility of episodic hazardous materials incidents. In fact, the majority of the 100 largest property losses in the chemical and petroleum industries occurred during the past 20 years. Incidents such as Flixborough, Bhopal, and Louisiana's Norco refinery need no further discussion. Reversing this trend requires new initiatives. This chapter focuses on a key means to meet these initiatives—auditing process safety management systems. It provides fundamentals and guidance, as developed by CCPS,[1] for developing and implementing an audit program to help ensure that the elements of a process safety management (PSM) system are in place and functioning. A sound auditing program for a process safety management system will enhance PSM effectiveness. Those requiring additional detail and discussion on these audits should consult Ref. 1.

Management Systems

Management systems are comprehensive sets of policies, procedures, and practices designed to ensure that the goals of an organization are met in an effective and efficient manner. These management systems seek to appropriately allocate activities between levels of management and to eliminate barriers to the effective cooperation of various organizational groups.

Process safety management is one such set of systems that has and will become increasingly important to enterprises manufacturing, storing, and handling hazardous materials. Process safety management systems require management to provide sound design, construction, operation, and maintenance, along with systems to ensure that resources are made available and used properly. These PSM systems should provide checks and balances to ensure that the various functions are carried out as intended.

At top-management levels, PSM systems should establish and review the overall process safety goals and policies of the organization. At the middle-management level, process safety management systems should provide information and decision support to assure that process operations are conducted safely. And at the first-line supervisory level, PSM systems should control the regular, ongoing activities.

Twelve distinct but interactive elements of an effective PSM have been identified and are enumerated in Table 12-1. Catastrophic accidents can occur when one or more of these 12 elements are not designed or functioning properly. Auditing—one of these elements—is critical since it contributes to management control of all the other elements. A sound process safety management auditing program will improve the effectiveness of an entire process safety program.

Auditing includes different types of activities, and process safety management is still a relatively new term. (Definitions of these and other relevant terms are included in Table 12-2.) Process safety auditing differs from PSM systems auditing. The former identifies and evaluates specific hazards, while the latter assesses the management systems that ensure ongoing hazard control. Both types of reviews are important in enhancing process safety. This chapter addresses the PSM systems audit.

PSM systems audits determine the existence, completeness, and functioning of management systems. Corrective actions are recommended to ensure that operating facilities and process units are designed, constructed, operated, and maintained such that the safety and health of employees, customers, communities, and the environment are being protected properly.

It is the auditor's task to investigate, gather, and evaluate relevant information for each of these elements to determine if PSM systems are in place, are adequate, and are functioning as designed. The first part of this chapter addresses an auditing program organization, and auditing fundamentals and techniques. The remainder is intended to inform auditors what to look for, where to look, and how to obtain information for each of the 12 elements defined in CCPS's process safety management system.

Table 12-1. Elements and Components of Process Safety Management

1. Accountability: objectives and goals

   Continuity of operations
   Continuity of systems (resources and funding)
   Continuity of organizations
   Company expectations (vision or master plan)
   Quality process
   Control of exceptions
   Alternative methods (performance vs. specification)
   Management accessibility
   Communications

2. Process knowledge and documentation

   Process definition and design criteria
   Process and equipment design
   Company memory (management information)
   Documentation of risk management decisions
   Protective systems
   Normal and upset conditions
   Chemical and occupational health hazards

3. Capital project review and design procedures (for new or existing plants, expansions, and acquisitions)

   Appropriation request procedures
   Risk assessment for investment purposes
   Hazards review (including worst credible cases)
   Siting (relative to risk management)
   Plot plan
   Process design and review procedures
   Project management procedures

4. Process risk management

   Hazard identification
   Risk assessment of existing operations
   Reduction of risk
   Residual risk management (in-plant emergency response and mitigation)
   Process management during emergencies
   Encouraging client and supplier companies to adopt similar risk management practices
   Selection of businesses with acceptable risks

5. Management of change

   Change of technology
   Change of facility
   Organizational changes that may have an impact on process safety
   Variance procedures
   Temporary changes
   Permanent changes

6. Process and equipment integrity

   Reliability engineering
   Materials of construction
   Fabrication and inspection procedures
   Installation procedures
   Preventive maintenance
   Process, hardware, and systems inspections and testing (prestart-up safety review)
   Maintenance procedures
   Alarm and instrument management
   Demolition procedures

7. Human factors

   Human error assessment
   Operator-process and equipment interfaces
   Administrative controls vs. hardware

8. Training and performance

   Definition of skills and knowledge
   Training programs (e.g., new employees, contractors, technical employees)
   Design of operating and maintenance procedures
   Initial qualification assessment
   Ongoing performance and refresher training
   Instructor program
   Records management

9. Incident investigation

   Major incidents
   Near-miss reporting
   Follow-up and resolution
   Communication
   Incident recording
   Third-party participation as needed

10. Standards, codes, and laws

    Internal standards, guidelines, and practices (past history, flexible performance standards, amendments, and upgrades)
    External standards, guidelines, and practices

11. Audits and corrective actions

    Process safety audits and compliance reviews
    Resolutions and close-out procedures

12. Enhancement of process safety knowledge

    Internal and external research
    Improved predictive systems
    Process safety reference library

Reprinted from Guidelines for Technical Management of Chemical Process Safety, with permission. © 1989 by the American Institute of Chemical Engineers.

Table 12-2. Definition of Terms

An *audit* is a systematic, independent review to verify conformance with established guidelines or standards. It employs a well-defined review process to ensure consistency, and to allow the auditor to reach defensible conclusions.

An *inspection* is physically examining a facility.

Process safety is protecting people and property from episodic and catastrophic incidents that may result from unplanned or unexpected deviations in process conditions. (This is an ideal condition, since working with materials having inherent hazardous properties can never be done in the total absence of risk.) Not included here are personnel safety (e.g., slips, trips, and falls), security issues, and chronic releases to the environment.

Process safety management (PSM) is the application of management systems to the identification, understanding, and control of process hazards to prevent process-related incidents and injuries.

Process safety management systems (PSM systems) are comprehensive sets of policies, procedures, and practices that ensure that barriers to episodic incidents are in place, in use, and effective.

Process safety auditing is a formal review that identifies process hazards relative to established standards; for example, examining plant and equipment, often using a checklist or other audit guide. This type of audit verifies to plant management that the management systems in place are effective in assuring that company and plant policies and procedures are being implemented, and it identifies areas in which management systems can be strengthened.

Process safety systems auditing is the systematic review of PSM systems and is used to verify the suitability of these systems and their effective, consistent implementation.

Planning an Audit Program

Process safety management audits, in general, involve collecting and analyzing information about the system being audited, comparing the systems to established criteria, and reporting the results and recommended corrective actions to management. Although simple in concept, a PSM audit program, to be effective, must be fully thought through and a number of decisions made before an audit is carried out.

Important areas for early decisions include:

- Objectives of the audit
- Audit scope
- Audit frequency
- Audit management responsibilities and staffing
- Administration placement
- Audit reports and reporting
- Follow-up of recommendations
- Audit quality assurance

Objectives of the Audit

As a first step in planning a PSM auditing program, senior management should set the program's objectives. Objectives vary, depending on the company's culture, management philosophy, and size. Some see them as tools for confirming that rules and regulations are met; others use them for spot checking. Typical objectives include:

- Helping to improve overall process safety
- Determining compliance with regulations
- Increasing the level of process safety awareness
- Improving the process safety risk-management system
- Meeting compliance reviews

As an example of what is involved in meeting objectives, consider compliance reviews, i.e., confirming that a facility's operations are in line with laws and regulations as well as company policies, procedures, and practices. Such an audit requires rigorous effort to determine and document performance by evaluating the application of and adherence to relevant process safety standards. Compliance reviews can certify the validity of process safety data and reports and can identify gaps in policies and standards—and in the facility's adherence to these standards. Also, such an audit may consider compliance with other parameters, such as industry or association standards. Compliance criteria should be set during the planning phase.

Audit Program Scope

The scope of an audit program must be clearly defined. Selecting the physical boundaries or scope of a PSM systems audit, however, can be complicated. The number of disciplines and area in the scope of the PSM audit vary—the scope can be very broad or very specific. When less than the whole facility is audited, boundaries may not be clear. Operating units or processes generally do not run in isolation. Thus it is challenging to set boundaries that fulfill program objectives without spreading limited resources too thinly.

The scope of an audit program specifies the facilities and units to be covered, the subject areas to be addressed, disciplines required, and the criteria against which the audit is to be conducted. Failure to set the scope can lead to misunderstandings among the groups being audited, the auditors, and the management recipients of the audit. It can also lead to inconsistent and inaccurate audit results.

In defining the scope of a PSM systems audit program, a number of factors should be considered. Among these are company policies, regulations, and the nature of the risks. Company policy may dictate audit frequency. Regulations will influence the design of the program. At minimum, the scope should be that required by company or plant policies and objectives.

Among the parameters considered in defining the scope of the audit program are:

- The type of facility (manufacturing, terminals, etc.)
- Geographical location
- Program content (all process safety management elements versus selected ones)

Specific aspects or functional areas can be included in the scope of the audit. Usually included are:

- Process risk management
- Process control
- Electrical hazards
- Fire protection
- Critical operating parameters
- Safety standards

The available time should also be considered. It is better to perform a thorough audit with a narrower scope than a hurried, incomplete audit with a broader scope. A comprehensive PSM systems audit in a large continuous process plant may require four to eight person-weeks of effort, including preparation and reporting.

The nature of a company's operations influences decisions on the scope of the audit program. In some companies, locationwide PSM systems apply to several process units. Where this is the case, it is practical and efficient to review PSM systems for the entire location. In other instances, a large location may have separate plants with independent PSM systems. In such cases, time may dictate performing separate audits of the individual units at different times.

Other management control systems (e.g., self-inspection or internal reporting) may also influence decisions on the scope of the PSM systems auditing program. For example, where there are many effective process safety management control systems, it is comparatively less important for the audit program to be frequent and broad in scope. There is no single correct approach to defining the scope of an audit program. Rather, decisions on audit scope should be made within the context of the overall process safety management program.

Audit Frequency

The frequency of an audit depends on the objectives of the program and the nature and hazards of the operations involved. Frequency also depends on when the feedback information is needed to attain process safety management goals. It may be necessary to establish different frequencies for different categories of facilities. As part of planning the audit program, a procedure is needed for selecting the facilities to be audited. If all facilities are not visited on some cycle, a strategy for selecting a sample of the facilities to audit should be developed and documented. In addition, provisions should be included for special requests from specific locations for audits. Some major factors to be considered when establishing audit frequency are:

Degree of Risk. This is the most important factor in deciding on the frequency. Generally the frequency will be higher for operations that pose higher levels of risk. For example, it may be necessary to change a hazardous catalyst every 20 days, which suggests auditing catalyst change procedures quarterly or semiannually. On the other hand, process changes, if done infrequently, may require audits only every 2 to 3 years.

Process Safety Management Program Maturity. Frequency is likely to be higher for operations that have new or evolving process safety management programs, as compared with operations that have established, well-developed programs. For the former, there is a greater chance for PSM systems to require greater management attention. In a location with a more mature process safety management program, it is more likely that PSM systems have been integrated into the normal operations, generally requiring less frequent reviews. Changes in either the process safety management program or the audit criteria may prompt reconsideration of established audit schedules.

Results of Prior Audits. When such results indicate significant gaps in process safety management system implementation, the next audit may have to be done sooner than normal.

Other Factors. A location's incident history, company policies, and government regulations all may help dictate the frequency of the audit.

Audit Management and Staffing

In organizing a program to audit process safety management systems, major considerations are:

- The basic program management responsibilities
- Criteria used to select auditors
- Proper use of internal versus external auditors

Whether or not a company has a full-time audit program manager, basic program responsibilities must be assigned, including:

- Resource management—managing the budget, personnel, and resources from other parts of the organization
- Staff selection and training—selecting, orienting, training, and continually developing the audit team staff

- Program development—continuing to develop, refine, and advance the audit program
- Keeping current—staying up to date on program activities

Selecting effective team members ultimately determines program success. Most PSM systems auditing is performed by a team. The ideal team consists of no fewer than two and no more than six members. A single individual can conduct an audit, but the one-person approach lacks the benefit of bringing a variety of insights to the process. On the other hand, a team of more than six can be used, but teams of that size are difficult to coordinate.

The ideal team for a PSM systems audit will include individuals familiar with the process and having experience in process safety management and audit techniques. Team members should have expertise in process engineering and design (control technologies), safety disciplines (fire protection, electrical hazards, and materials handling), chemistry, facility operations, maintenance, and, of course, auditing.

Staffing an audit team exclusively with individuals from the process unit being audited is generally not desirable. This sacrifices the benefits often derived from having fresh eyes looking at a unit. In addition, the use of such a staff can make it difficult to avoid potential conflicts of interest, that is, instances where an auditor is reviewing things for which he or she has at least some responsibility or involvement, or where the auditor reports to the manager whose activities are being audited.

A variation is using interfacility exchanges to staff audits—staffing with individuals from other company locations where similar operations are performed. This provides high process familiarity but without direct involvement in the operations of the plant being audited. Rotation of assignments for audits provides valuable cross training of persons in activities in which they normally are not involved. First-level supervisors and hourly employees will benefit and contribute to the audit if given appropriate tools. Each worker manages some amount of activity—the scope varies.

Sometimes an audit must be done quickly or with particular impartiality, as when issues are under litigation. Here, it is often preferable to staff the audit with persons not directly associated with the site.

Some companies employ a staff of dedicated auditors. Sometimes they comprise the audit team, and other times they are used as team leaders. Dedicated auditors are best able to develop strong auditing skills and a broad perspective on the topics being audited. However, they may not have in-depth process knowledge for all processes.

Sometimes outside consultants are used in staffing audits. They provide independence to the process and may help supplement scarce internal resources. However, during an audit there is an opportunity to gain valuable knowledge about and appreciation for PSM systems. If outside consultants are used exclusively, the company may fail to capitalize fully on, and to enhance further, the knowledge base of the internal staff. Whichever staffing strategy is chosen, the audit team must be trained in both process safety management and audit skills and techniques.

Program Location

The appropriate location of the program is a question of where it should be housed to ensure that the proper skills are brought to bear. The audit program should have enough organizational influence to ensure (1) that the required resources are made available to achieve program objectives and (2) that findings will receive serious attention and proper consideration. By selecting a separate internal audit department to house the process safety management system auditing program, the program may be seen as a corporate (rather than process safety) management tool.

Audit Reporting

The audit's findings must be documented. The report should be issued in a timely fashion to expedite initiation of corrective actions. An audit is similar to an accident or incident investigation in that it identifies deficiencies in process safety management. Thus the report should develop recommendations to prevent or reduce process risks. The goal is to bring to plant management's attention findings identified by the audit and to provide appropriate personnel with information in order to respond to and formulate action plans.

In designing the reporting process and executing the actual preparation of reports, there is a series of issues to consider:

Report Content. This should be decided on as part of the audit program design and should be consistent with the objectives of the audit program. There is no single correct way to determine the content of an audit report. However, once the content is set, all audits should produce consistent reports.

The audit report should document the results of the audit, indicating where and when the audit was done, who performed the audit, the audit scope, and the audit findings. The specific content of the report can vary. Some companies prefer an exception report, addressing only areas of deficiency; others use one commenting on every subject reviewed. The report may include recommendations, or it may prioritize or simply list the findings, among other options.

Distribution of Reports. When a PSM systems audit report has been prepared, it must go to appropriate parties for follow-up. Failure to properly distribute the audit report may compromise the audit.

Distribution may be set by corporate policy. Typically, the appropriate recipients include the manager of the facility being audited and at least one level of supervision above that manager. Report distribution can be sensitive since the report typically documents deficiencies, which may be seen by attorneys as poor practice. As a result, some corporate attorneys prefer to have report distribution managed by the legal department. This sensitivity sometimes leads to reporting findings orally. This is not recommended as the sole means of reporting.

Language of Reports.   Great care needs to be taken to use appropriate wording. An audit report must clearly communicate its findings and observations. However, the report should be worded carefully so as not to imply findings or observations that were not found. In the report, facts should be reported clearly and concisely. Every statement should be supportable; speculation must be avoided. Also, avoid using words such as "alarming," "dangerous," "violation," and other inflammatory terms.

Report Retention.   A company or location should develop a policy on the retention of audit reports and backup records (including working papers and follow-up correspondence). Some companies retain all records permanently, while others do so for a limited time. Retaining audit reports for at least one full audit cycle allows comparison with prior audits. In addition, regulations may mandate record retention.

Audit Follow-Up

In some organizations, corrective action is a formal part of the audit program; in others, it is merely closely linked to the process.

Timely implementation of corrective actions enhances process safety. Following issuance of the audit report, an action plan should be developed. This plan should include the timetable for implementing follow-up actions (serving as a project schedule) and the person responsible for each action (as a management-control document). In cases where no action is necessary, this should be explained and the reasons given, so future reviewers know that the finding was not ignored.

Management determines what actions (if any) are appropriate or desirable. Other management decisions include corrective action priorities, timetables, resource requirements, and assignment of responsibilities. Corrective action planning can include such items as interim, or stopgap, measures for immediate corrections to, or mitigation of, especially hazardous conditions.

Management review and determination of a course of action are key when audits include issues or recommendations for areas where established criteria or clear requirements are unavailable. Note that often a corrective action also serves to prevent or further reduce the likelihood that the undesirable situation will recur.

The action plan should be developed by the manager(s) responsible for the facility or operation that has been audited. There should be an established system for review and approval of the plan by appropriate levels of management. Copies of the plan should be distributed to those assigned responsibility under it, the audit team, and the next higher level of management.

On an established basis, the action plan should be reviewed and updated to indicate completed items and the status of other items. As items are completed, the specific action taken should be documented and kept on file. Quarterly updating of action plans is often used, but more or less frequent updates may be chosen. To assist in tracking corrective actions, several types of reports are used:

- Periodic status reports (e.g., quarterly or monthly)
- Milestone (summarizing accomplishments)
- Exception reports (other major accomplishments)

Responsibility should exist in the organization for tracking action plan status so that corrective actions are completed. In some organizations this is part of the audit program. In others, this is a line function and is performed by the facility management. Also, regardless of who actually conducts the tracking, line management must assume responsibility for the execution of corrective action plans.

A final step in the overall audit process is to verify that corrective actions effectively address the audit deficiencies. This may be done by the audit teams. In some audit programs, verification is part of the next regularly scheduled audit. In others, verification is done sooner as a separate, special review.

Quality Assurance

The development of performance criteria for the program is one method of helping to assure quality of audits. Parameters chosen might include the team composition, the nature and number of facility staff interviewed, and the availability of key records to the audit team (past incidents, relief valve test records, etc.).

Another method is independent review of the audit process. In some programs, an independent quality assurance person accompanies the audit team on some fraction of the audits. In others, the audit working papers and report are reviewed by someone not involved in the audit. Periodic critiques and evaluations of the audit program done by those not in the program help in identifying program weaknesses.

Audit Techniques

Some basic activities are common to most audit programs. A number of them are undertaken before the audit, some during the fieldwork, and others after the fieldwork has been completed. Virtually all involve gathering and analyzing information, evaluating PSM systems against criteria, and reporting the results to management. Reference 2 provides information on audits of process safety management systems and includes examples of auditing questions and forms. Figure 12-1 presents the steps that are typically employed by most companies in their audit process.

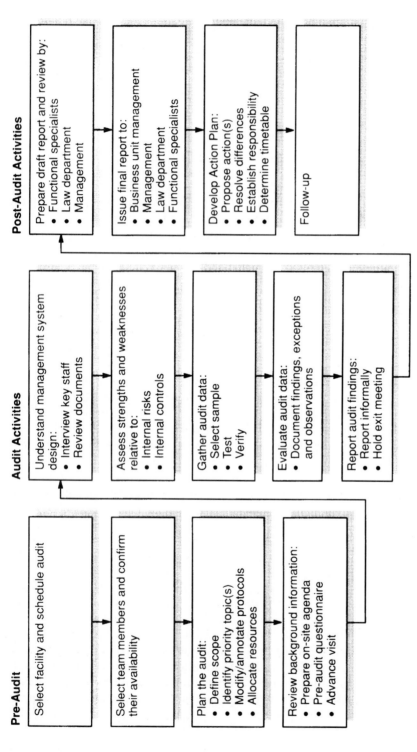

Pre-Audit

Select facility and schedule audit

Select team members and confirm their availability

Plan the audit:
- Define scope
- Identify priority topic(s)
- Modify/annotate protocols
- Allocate resources

Review background information:
- Prepare on-site agenda
- Pre-audit questionnaire
- Advance visit

Audit Activities

Understand management system design:
- Interview key staff
- Review documents

Assess strengths and weaknesses relative to:
- Internal risks
- Internal controls

Gather audit data:
- Select sample
- Test
- Verify

Evaluate audit data:
- Document findings, exceptions and observations

Report audit findings:
- Report informally
- Hold exit meeting

Post-Audit Activities

Prepare draft report and review by:
- Functional specialists
- Law department
- Management

Issue final report to:
- Business unit management
- Management
- Law department
- Functional specialists

Develop Action Plan:
- Propose action(s)
- Resolve differences
- Establish responsibility
- Determine timetable

Follow-up

Figure 12-1. Typical steps in the process safety management audit process. (*Source: Arthur D. Little, Inc. Reprinted from Guidelines for Auditing Process Safety Management Systems, with permission, © 1993 by the American Institute of Chemical Engineers.*)

Preaudit Activities

Selecting Facilities and Units to Be Audited.   The audit scheduling and review frequency depend on the goals of the program and the number of facilities and functional areas in its scope. Facilities or functional areas can be selected by a number of methods, for example, random selection, potential hazards, or importance in terms of business considerations.

Scheduling the Audit.   If this is the first audit at the facility or by the corporation, the required lead time will always be greater than normal. Scheduling generally begins with the audit team leader communicating to the facility manager that the facility has been selected for a PSM systems audit. In selecting dates, the auditor's major considerations are that key facility personnel will be available and that the facility will be in a normal mode of operation.

The facility should be asked to designate a contact person to coordinate the collection of background material and the scheduling of interviews. The team leader should inform all audit team members about specific safety rules that the audit team will need to comply with.

Gathering and Reviewing Background Information.   The audit team leader should identify the types of information needed or desired before the audit. Some of this may come from the facility, some from other sources. Typical background information includes the previous audit report, action plan, rules and regulations, and facility layout and organization. An audit questionnaire is sometimes administered beforehand to assist auditors and the facility in audit planning and preparation. This helps the audit team (1) identify and understand key elements of the facility's internal process safety management procedures and systems and (2) identify audit topics that are not applicable at a particular facility. Only that information which can be studied prior to an audit should be requested.

Advance Visit to the Facility.   Such a visit can increase the effectiveness of the audit, particularly early in the development of an audit program. The objectives of a preaudit facility visit are to inform the facility manager and staff about the audit program goals, objectives, and procedures, and to obtain information about the facility that allows the team to develop a more comprehensive audit plan. The costs and benefits of an advance visit should be considered in the context of the goals and objectives of the program. Whether or not there is a preaudit visit, all information requested should be reviewed by the team leader and team members, as appropriate, prior to arriving on site.

Developing the Audit Plan.   Commonly, some form of audit protocol serves as the outline. Before the audit, the team leader should identify priority topics for review, modify and annotate the audit protocols or checklists as necessary, and make an initial allocation of audit team resources. The audit team should determine an appropriate cross section of employees to interview. If there are specific things that the team will want to observe (e.g., an intermittent operation), arrangements need to be made in advance.

Audit Activities

Audits normally start with an opening meeting at which the facility staff provides an overview of site operations. This meeting also allows the audit team to find out the site safety rules and procedures; also, any special sensitivities can be aired. After the opening meeting, the audit team typically is given a facility tour. The on-site audit typically involves the following five steps:

Understand Management Systems.   Most on-site activities begin by developing a working understanding of the facility's internal PSM systems. The PSM system involves the formal and informal procedures and activities used to control and direct process safety. Where informal management systems are in place (e.g., systems with little or no documentation or procedures), the auditor must assess whether this form of management can be effective if operations or staff change. This step usually includes developing an understanding of facility processes, internal controls (both management and engineering), plant organization and staff responsibilities, compliance parameters and other applicable requirements, and any current or past problems. The auditor's understanding is usually gathered from multiple sources, including interviews, review of documents, and the advance visit. The auditor records his understanding in a written description against which it is possible to audit.

Assess Strengths and Weaknesses.   After understanding how process safety is intended to be managed, auditors evaluate the soundness of the facility's management systems to see if they will achieve the desired performance when functioning as intended. For each of the areas or topics assigned, the auditors should ask themselves three questions:

- "If the facility is doing everything the way they say, is that acceptable?"
- "Will the facility be in compliance with applicable requirements?"
- "Is the company adequately protected?"

Auditors typically look for such indicators as written policies and plans, clearly defined responsibilities, an adequate system of authorizations, capable personnel, administrative controls, and internal verification. Each of these indicators usually requires judgment by the auditor since there are no widely accepted standards. This step helps to determine how the balance of the audit will be conducted. Where internal controls are judged to be sound, the auditor will spend time confirming the existence of the control systems and testing to see if they function effectively on a consistent basis.

Gather Audit Data.   Data are collected to verify and validate the functioning of PSM systems. Such data form the basis on which the team determines whether the process safety management system has been implemented as designed. Auditors collect data by interviewing, observation, and verification.

Evaluate Audit Data. Once data gathering is complete, the data are evaluated to identify audit findings. Findings are reviewed in terms of process safety management system criteria to determine their significance. Negative findings are typically called "exceptions."

Audit teams usually make preliminary evaluations of their data throughout the audit and compare notes at the end of each day, developing tentative findings. The auditors should be careful in reaching conclusions based on a single data point.

Report Audit Findings. Process safety management system findings are discussed regularly by the auditors with facility personnel to avoid any surprises at the closing meeting. The formal reporting process usually begins with an exit or closeout meeting between the audit team and facility personnel. During the meeting, the audit team communicates all findings. Any ambiguities about the findings are then clarified.

Postaudit Activities

After the on-site audit work is complete, the audit team must complete its report and in some cases monitor the completion of an action plan to address audit findings. When the team has left the site, all audit findings should have been identified and communicated to the local staff at the closing meeting. The audit team usually prepares a draft report, has the report reviewed by appropriate company groups, and issues the final report. The location audited has an opportunity to review the report at the draft stage to assure that the report is clear, concise, and accurate.

Next, the audited facility or unit should prepare an action plan, indicating what is to be done, who is responsible, and what are the completion dates. All exceptions should be addressed and any action taken on the exceptions or the rationale for not taking action should be documented. The role of the auditor differs here among companies. For example, the auditor may be responsible for tracking the resolution of exceptions or simply have no further role.

Audit Guides

The PSM systems audit is most commonly supported by some important tools:

Protocol. An audit protocol is a written step-by-step guide for accomplishing the audit, developed as part of the audit program design. It typically includes a list of audit steps and procedures for gathering data about facility programs and their implementation.

A standard protocol can be discretionary or fixed. The discretionary protocol lists all audit procedures and verification tests that could be used to achieve the audit goals and objectives. The auditor uses it much like a menu, selecting those procedures appropriate to the specific audit and documenting the results. The fixed protocol, on the other hand, lists procedures that must be carried out in every audit unless there is good reason to deviate. Fixed protocols are used where audit goals are served by some degree of standardization from audit to audit.

Whether a discretionary or fixed protocol is selected, a number of steps should be taken in developing the protocol:

- Decide the scope of the audit and its specific areas of coverage; then list the selected audit topics.
- For each topic selected in step 1, identify and list the performance criteria against which the locations will be audited.
- Determine and identify the depth of review for each topic selected above.
- Determine the type and level of audit techniques (e.g., interviewing, observation, verification) to use for each topic selected, paying particular attention to audit resources and time constraints.

Finally, a draft of the protocol is written, reviewed, and revised and completed. Figure 12-2 provides a sample page from a PSM systems audit protocol.

Questionnaire.   Some PSM systems audit programs supplement their audit protocol with a questionnaire to assist in identifying and reviewing internal management procedures and systems. This internal controls questionnaire allows a large amount of background information to be collected quickly and efficiently. It identifies key elements of the facility's management systems and procedures (e.g., maintenance, record keeping, and internal reporting).

The audit protocol may be used as a guide in determining the types of information to request. The auditor must have a clear understanding of what information is needed when preparing the questionnaire and know the objective, scope, and focus of the audit. During the audit, the questionnaire can help determine items for follow-up. The information gathered in the questionnaire should be documented.

There are a variety of ways to administer the internal controls questionnaire. Most audit teams meet with facility personnel at the beginning of the audit, with the team leader administering the questionnaire to appropriate facility personnel. It may also be administered by the audit team leader to facility personnel in advance of the audit.

Topical Outline.   This is a summary or list of the major topics to be covered during the audit. Figure 12-3 shows a sample page from a process safety management topical outline. As a simple list of key subjects, the topical outline relies to a great extent on the experience and judgment of the auditor. The outline's principal advantage is that it is short and easy to use for someone who knows how to go about reviewing each topic; however, it does not provide any substantive guidance to the audit team.

Introduction
Background Information
1. Review background information obtained from the facility to develop a general understanding of process safety hazards, areas of process safety concern, chemicals and processes used, etc. Typical background information includes:

a. List of major hazardous chemicals used at the facility and applicable MSDS sheets for those chemicals;

b. Description of the facility including site plan, flow diagrams, and general arrangement diagrams;

c. Types of operations conducted;

d. Organizational charts;

e. Environmental, Health, and Safety Policy and any other applicable corporate policies;

f. Selected facility reports (i.e., insurance carrier reports, incident reports, etc.);

g. Copies of prior PSM audits or any other safety audits conducted at the facility;

h. List of process hazards analysis conducted; and

i. Number of contractors typically on-site.

2. Obtain and review applicable company and facility plans, policies, procedures, and standards.

3. Obtain and review Pre-Audit Questionnaire completed by facility.

Figure 12-2. Sample page from an audit protocol format, page 1 of 17. (*Source: Arthur D. Little, Inc. Reprinted from Guidelines for Auditing Process Safety Management systems, with permission. © 1993 by the American Institute of Chemical Engineers.*)

Data Gathering

During the on-site audit, each team member gathers data to evaluate the facility's PSM systems from a number of places through a variety of means and techniques. It is very important that the auditor maintain control of the data sample selection. In sampling a population of persons to interview or records to examine, auditors must obtain as representative a sample as possible in order to minimize bias and distorted results.

Interviewing is perhaps the most frequently used means of collecting audit data. Interviews can be either formal (e.g., via a questionnaire) or informal (e.g., through discussions). In evaluating information gained through interviews an auditor should consider:

- The level of knowledge or skill of the individual

- His or her objectivity

- The consistency of each response with other audit data

- The logic and reasonableness of the response

---

1. Process Safety Information

1.1 Written process safety information

1.1.1 Chemical hazard information
    1.1.1.1 Toxicity information
    1.1.1.2 Permissible exposure limits
    1.1.1.3 Physical, reactivity, and corrosivity data
    1.1.1.4 Thermal and chemical stability data
    1.1.1.5 Hazardous effects of inadvertent mixing
1.1.2 Technology of the process
    1.1.2.1 Block flow diagram or simplified process flow diagram
    1.1.2.2 Process chemistry
    1.1.2.3 Maximum intended inventory
    1.1.2.4 Safe upper and lower limits for temperatures, pressures, flows, and/or compositions
    1.1.2.5 Evaluation of the consequences of deviations, including those affecting safety and health
1.1.3 Equipment in the process
    1.1.3.1 Materials of construction
    1.1.3.2 Piping and instrument diagrams (P&IDs)
    1.1.3.3 Electrical classification
    1.1.3.4 Relief system design and design basis
    1.1.3.5 Ventilation system design
    1.1.3.6 Design codes and standards employed
    1.1.3.7 Material and energy balances
    1.1.3.8 Safety systems (e.g., interlocks, detection and suppression systems)
1.1.4 Documentation that equipment complies with recognized and generally accepted good engineering practices
1.1.5 Determination and documentation that equipment is designed, maintained, inspected, tested and operating in a safe manner for existing equipment that was designed and constructed in accordance with codes or standards that are no longer is use.

Figure 12-3. Sample page from a process safety management topical outline, page 1 of 12. (*Source: Arthur D. Little, Inc. Reprinted from Guidelines for Auditing Process Safety Management Systems, with permission. © 1993 by the American Institute of Chemical Engineers.*)

Observation, or physical examination, is one of the most reliable sources of audit data. Where knowledge of specific operations or equipment is important, it is desirable for the auditor to observe them.

Verification refers to activities that increase confidence in the audit data and the facility's internal controls. Verification can be a powerful tool in assisting the auditor to achieve the objectives of the audit.

Audit programs vary regarding the amount and balance of interview, observation, and verification. Many of the more sophisticated audit programs use a considerable amount of verification to determine if management systems perform as intended.

Interviews

Prior to conducting an interview, the auditor should identify those to interview, outline what is to be done, and determine where and how the interview will be handled and the steps necessary to maximize the effectiveness of the interview. Several points need to be covered:

Iron Out Logistics.   Obtain a brief understanding of the interviewee's title, responsibilities, and reporting relationships. Whenever possible, establish a specific time and place suitable for the interview and interviewee, and the duration for the interview. Identify the specific type of information desired and areas to be addressed, and define the desired outcome.

Organize Your Thoughts.   Interviews need not be long. However, the auditor should decide what information is to be sought and then organize, consolidate, and write down the questions to ask.

Carrying Out the Interview

Opening the Discussion.   This is perhaps the most crucial aspect of the interview. The quality of information gathered is closely related to the interviewee's sense of comfort. To build this sense, auditors should follow a few basic guidelines:

- Introduce yourself, explaining why you are at the facility, and briefly recap the purpose, scope, and desired outcome of the audit.
- Inform the interviewee of the estimated length of the interview.
- Explain how the information will be used and that the primary purpose of the discussion is to help develop a complete understanding of how the facility manages its process safety activities.
- Say that comments are confidential.

Conducting the Interview.   The auditor should next shift emphasis and begin obtaining specific information such as a brief overview of the interviewee's job, how it fits into the overall organization at the facility, and principal responsibilities. The questions previously written down should serve as the basis for this part of the interview. There are a number of guidelines that are normally used to conduct an interview, including:

- Ask for concrete or specific responses and avoid leading questions.
- Show respect, interest, and understanding by helping the interviewee clarify or deepen responses and confirm the responses by paraphrasing them.
- Use constructive probing when responses are inconsistent.

- Phrase inquiries to focus on the data, not on the respondent, and enlist the help of the respondent in clarifying information.
- Summarize the information learned as frequently as needed to ensure that the data gathered are correct.
- Provide feedback, as if requested by the interviewee and as appropriate under company policies.
- Take notes throughout in a manner not to interfere with the interview, but try not to indicate areas of concern through note-taking patterns.
- Keep to the agreed-upon time limit unless consented to by the interviewee.

Closing the Interview. Close in a concise, timely, and positive manner, thanking interviewees for their time. It is often useful to ask a question such as: "Is there anything else I should have asked you but haven't?"

Documenting Interview Results. This begins with notes taken during the interview. Immediately following the interview, review working papers to ensure that they accurately and completely reflect the information obtained. In documentation, there are four key points:

- Record the name of the interviewee, title or job responsibilities, the date and time of the interview, and the applicable protocol steps.
- Include notes taken.
- Flag the key items to ensure that important items receive attention.
- Summarize the outcome, reviewing the key points, and noting any changes or additions.

Sampling Strategies and Techniques

Although sampling is a well-established aspect of auditing, selecting appropriate sampling methods and sample size can be difficult. Thus the auditor must exercise caution when selecting a sampling method. Here are some steps to help ensure that each sample selected is appropriate and defensible:

Determine the Objective of the Protocol Step Being Conducted. This is vital since it will help the auditor to set boundaries of the population under review.

Identify the Relevant Population (Records, Employees) to Be Reviewed. Independent records should be used whenever possible to develop the sample. For example, in reviewing training do not start with a sample developed from a stack of training records provided by the facility coordinator, since these reflect only those with completed training records. Start with personnel department or departmental organization charts and develop a sample of employees who should have been trained. Then the training records can be reviewed to help determine whether each employee in the sample was trained.

Determine the Sampling Method. Samples selected by an auditor are judgmental but may be aided by a systematic selection strategy. A systematic sample is one selected by a systematic process chosen to represent the population that is being reviewed. Numerous methods will select a sample for review, but no single one is correct for all situations.

Determine the Sample Size. This can be done by either the auditor's judgment or statistics, depending on the goals and objectives of the audit program. In some audits, it may be desirable and adequate to review 10 to 20 percent of the population.

Conduct Sampling. To avoid bias, independent records should be used wherever possible to develop the sample, and records for sampling should be selected by the auditor, not by facility personnel.

Document the Sample, Strategy, and Methodology. To assure management that the audit was reasonable and ensure the quality control of the sampling process, the auditor should be prepared to indicate why the particular sample was selected.

Evaluating Fieldwork

It is necessary to determine the sufficiency of the information gathered during fieldwork to support the objectives of the audit and the conclusions of the auditor. Data can be defined as whatever influences the auditor's findings and opinions; others should be able to draw the same conclusions from the data.

Gathering Sufficient Information. An auditor has probably accomplished this if he or she understands both management system design (e.g., facility policies and procedures) and implementation (e.g., availability of compliance records); has interviewed key personnel in major functions or tasks and can summarize the basic programs, practices, and management systems; and understands the probable cause of differences between management and employee perspectives, or between those of the safety staff and the operating personnel.

Determining Adequacy of Information. In order to determine if the information collected is adequate to develop valid findings, three factors should be considered:

Relevance. Information gathered during a PSM systems audit should produce a flow of logic from the auditor's discoveries to the conclusions drawn. Thus a sample of process change authorizations would show that these changes were handled appropriately along with management change requirements.

Freedom from Bias. Information must be free from influence that makes one decision more attractive than another or that excludes information supporting an alternative decision. Bias can arise from the source of information or from the auditor's choice of sample or items to examine. It is human nature for facility personnel to want to describe facility practices in their best possible light. An auditor may need additional information whenever it is thought that an inter-viewee's response or records provided are biased or otherwise unreliable. In crucial matters, the auditor should not rely on a single source of data.

Objectivity and Persuasiveness.   Objective data that are persuasive should lead two auditors to reach the same conclusion.

Working Papers

These are the auditor's field notes and documentation, which usually do not become part of the permanent audit record. Working papers are used in prepar-ing the final report and are usually destroyed.

Accountability and Responsibility

For a process safety management program to be effective, responsibilities must be assigned clearly and those given responsibility must be held accountable. Accountability and responsibility must be addressed in reviewing every element of process safety management. They can be more difficult to audit, since these two elements are principles rather than activities. However, there are specific indicators for them.

Indicators of Accountability and Responsibility

Although each organization approaches accountability, goals, and objectives to fit its own culture, there are several indicators that can be audited.

Policy Statement.   The auditor should be able to identify a written statement of policy and assess its effectiveness. It should express an organization's com-mitment and approach to safe design and operation, indicating that safe opera-tion is a shared responsibility that requires participation from every employee. The policy should be disseminated broadly throughout the organization by posting or by inclusion in manuals and employee orientation programs. The statement should be current and signed by senior management. Management's commitment to the policy can be demonstrated on facility tours, references made in written communications, or in other actions.

Management Commitment.   The auditor should verify that management is involved appropriately in the process safety management program by determin-ing there are written performance criteria for the program, for either the corpo-ration or facility. Criteria should be set for both process safety performance and the programs designed to effect that performance. The auditor should confirm that management actively monitors progress toward achieving the performance criteria and provides feedback to those responsible for programs. This can be done in various ways such as via periodic written progress reports, presentations to management, or assessments by a safety committee. Providing sufficient resources to implement PSM management programs is another indicator of management's commitment.

Requirements for Procedures. Procedures help to assure the clear assignment of responsibility by management. The auditor should sample process safety management procedures to ensure that they are complete as written, up-to-date, and are implemented comprehensively and consistently. To do this, the auditor should determine that procedures are clear, concise, and comprehensive. The auditor should verify that each procedure clearly states who is responsible for implementing or approving it, who is to follow it, and whether affected employees are familiar with it and their roles. A system for writing, approving, and communicating revisions of procedures, especially to concerned employees, should be documented. The auditor should carefully review this system, probing for ways in which obsolete versions of a procedure could remain in place. A formal mechanism should be in place for obtaining a variance from procedures, and it should be verified.

Individual Performance Management. Performance appraisals relating to process safety should be confirmed by the auditor where possible.

Organizational Changes. When these occur, they should be incorporated into appropriate procedures, assignment of responsibilities, and performance criteria.

Culture. Organizational culture affects values and expected performance. The auditor should verify that PSM systems continue to function adequately when organizational and corresponding cultural change takes place.

Acquisitions. When a facility or business is acquired and integrated into the PSM systems of the acquiring company, there should be mechanisms to effect prompt and thorough process safety management system implementation. Finally, PSM programs of the acquisition candidate should be assessed for gaps and possible corrective actions prioritized. It is the auditor's role to confirm that this process takes place.

Process Safety Knowledge

Process safety *information* is the data describing the process, its chemistry, its design, and equipment required for its operation. Process safety *knowledge* includes process safety information plus the understanding, or interpretation, of the information.

Knowledge of the process is vital for the design and operation of a safe facility. Without adequate knowledge of the chemistry and design of the process and how it is intended to operate, it is difficult to adequately implement many of the other process safety management elements such as capital project safety reviews and management of changes, or to conduct hazard analyses and risk assessments.

Also, several of the elements of process safety management, including incident investigations, hazard analyses, and risk assessments, contribute to the body of process knowledge. This creates a cycle in which process safety information is used during hazard analysis, after which the results become part of the process safety knowledge.

Process Safety Audits

When making an audit of process safety knowledge, the auditor must verify the existence and functioning of a number of systems. These include systems to:

- Collect safety information
- Maintain and control changes to data
- Verify data prior to storage
- Disseminate up-to-date information
- Identify gaps in information and a means to fill the gaps

Data Collection.   The auditor should first verify that there is a mechanism to capture and maintain necessary process safety information throughout all stages of a facility's life cycle—from process development, design, construction, operation, and maintenance through decommissioning. Personnel responsible for gathering and updating process safety information should be interviewed. It is important in a functioning system that individuals involved should have received guidance on the types of information to be documented and where and how the information will be retained.

Systems also should be in place to ensure that stored process safety information is accurate and its source identified. The auditor should make sure that information being entered is verified prior to date storage. There also should be systems to identify gaps in process safety information and a means to fill the gaps.

Finally, the auditor should have discussions with project and process engineering managers and staff who generally are responsible for assembling the initial process design package to assess their understanding of the process safety information system.

Data Availability and Dissemination.   The process safety management system should address information dissemination to users. Staff involved in process safety should be aware of the types of process safety information available and where it can be found. In an audit, potential users of process safety information should be interviewed to establish their understanding of the availability of information. Whether the process safety information is maintained in a central location, in paper format, or in a multiuser automated database, there are distribution issues to be considered:

- Access
- Confidentiality
- Data integrity
- Dissemination of updates

For process safety information disseminated within an organization, there should be systems for assuring that only the most recently updated versions are in use. Approaches range from having every copy numbered and assigned to an individual to using regular staff meetings to disseminate information about changes.

Maintaining Information. Information must be kept current during process changes, equipment maintenance, and other normal activities. This requires appropriate links between the process safety information management system and the PSM systems for capital project reviews, management of change, process equipment integrity, and process risk management. The linkages are depicted in Fig. 12-4. The management system should include controls over the changing of process safety information to ensure that the integrity of these data is not compromised.

Types of Process Safety Information

The following is a discussion of the types of information important to process safety. The auditor should be sure that such information is available and maintained.

Chemical Data. Typical types of chemical data for raw materials, intermediates, catalysts, products, and waste streams are given in Table 12-3. Some of these data are available on material safety data sheets (MSDS); others might be found in technical literature or process development technical reports. In addition, information on required regulations is also important. Examples of such information are OSHA Standards, DOT Classifications and Placarding Requirements, and state or local regulations.

Design Data. Process chemistry, catalysts, reactive chemicals, and kinetic data are usually required to perform the process design and develop the front-end process design package. Such data are important for conducting a process hazards analysis. Examples include:

- Equations for the main chemical and side reactions (i.e., the process chemistry)
- Composition and hot spot potential for catalysts

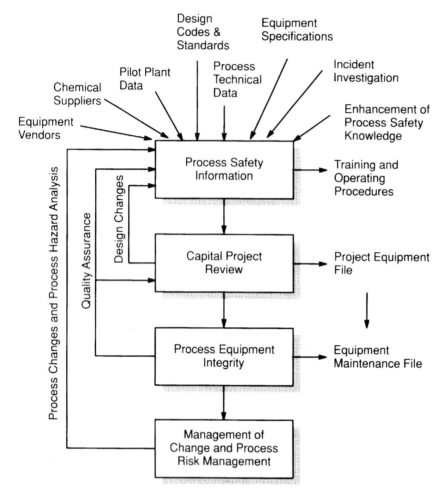

Figure 12-4. Process safety knowledge linkages. (*Reprinted from Guidelines for Auditing Process Safety Management Systems, with permission. © 1993 by the American Institute of Chemical Engineers.*)

- Compositions and process locations of reactive chemicals
- Reaction rates and heats of reaction (typical kinetic data)

The type of process will determine which of these are pertinent.

Design Basis.   The design basis (such as pressures, temperatures, and control philosophy) is used during process design to define the operating conditions and process limits. Also, the consequences of exceeding these limit should be defined. Much of this information is found in research reports or licensee data and is usually incorporated into the operating manuals. Copies should be available to appropriate site personnel.

Table 12-3. Typical Chemical Data

Identification data	Thermodynamic data
Chemical name	Latent heats of vaporization
Synonyms	Heat capacity
Chemical formula	Thermal conductivity
Code names	Heats of fusion
Trade names	Equilibrium constants
Physical property data	**Reactivity and stability data**
Molecular weight	Flash points
Density	Flammability data
Boiling point	Ignition energy for dust
Freezing point	Spontaneous decomposition conditions
Vapor pressure data	Spontaneous polymerization conditions
Viscosity	Shock sensitivity
Refractive index	Pyrophoricity
Surface tension	Hydrophoricity
Solubility	Reactivity with common materials of construction
Particle size and form	Incompatibilities with other chemicals
Azeotrope points	**Explosive properties**
Critical pressure, temperature, volume	Maximum rate of pressure rise
Vapor-liquid equilibrium	Peak explosive pressure

Reprinted from Guidelines for Auditing Process Safety Management Systems, with permission. © 1993 by the American Institute of Chemical Engineers.

Diagrams and Plots

Process Flow Diagrams. Much of the process design information just described is shown diagrammatically on process flow diagrams (PFDs). These can present information in varying levels of detail. A detailed diagram shows the major pieces of process equipment, along with the process lines into and out of the equipment, and principal utility demands. Operating conditions are also indicated. For a cyclical process operation or a batch process, there are likely to be supplemental sheets that list the sequence for the major process steps, plus a valve position chart.

Piping and Instrumentation Drawings. These show all process and utility piping, instrumentation, and protective systems, along with process and safety settings.

Plot Plans. A plot plan shows the layout of major process equipment and should note any special design and layout considerations, such as separation distances and underground pipeline locations.

Electrical Classification Plot Plan. This document is a plot plan of the unit or site showing the electrical classifications that are applicable to the different sections of the plant. These classifications are important where flammable or combustible materials are handled.

Plot Plan of Underground Services.   This plan helps avoid accidental rupture of underground services when excavating.

Electrical One-Line Diagrams.   These show power distribution of power from the source to all the power users in the unit, including major transformer and switching apparatus.

Special Design Considerations.   Special considerations must be met for some processes, such as:

- Nonlubricated equipment for oxygen
- Exclusion of water
- Heat-transfer salts
- Lethal-service requirements for vessels and piping
- Stress-relieving requirements

Where applicable, these special criteria should be well documented.

Equipment Specifications.   Data sheets should be provided for each piece of process equipment such as pumps, compressors, tanks, vessels, heat exchangers, and for each protection system (e.g., overpressure). These sheets typically show the design codes and standards used, the expected conditions, the design conditions, materials of construction, and other mechanical details. Other information should be available such as hydrostatic test reports, manufacturing inspection reports, weld radiographs, and other documentation on tests and inspections conducted during fabrication and installation.

Piping Specifications.   For each line, piping specifications include the type of pipe used along with data on the acceptable types of fittings, valves, and gasketing materials.

Safety-Critical Instrument Index.   This index lists all the instruments in the process important to maintaining process safety. These may include sensors, control valves, and pressure reducers. Each instrument should have its own complete specification sheet, listing design conditions, the required materials of construction for various portions of the instrument, as well as the range it must cover and its protective function.

Programmable Controllers and Computers.   For a plant that is controlled or monitored by a computer, there are two critical sources of documentation: (1) the hardware system specifications, including the computers, instrumentation wiring diagrams, and the uninterruptible power supply (UPS); and (2) the documentation of software, usually the service manuals from the computer system vendor. In the computer control system, two elements are subject to modification:

- The configuration of the instrument control loops, control actions, tuning constants, and alarm points
- The control software

These must be under secure management control and not accessible to random changing by operators. Documentation of the alarm points, control-loop configurations, and the control software should be stored away from the process.

Vendor Data.   Packaged items (e.g., deionizing systems) come with sets of drawings and operating and maintenance instructions or manuals. Such units should be identified in the process flow diagrams and the P&IDs. At least one set of the vendor documents should be maintained in a central file with controlled access.

Other Information.   The process safety information should also include reports documenting process hazard identification and analysis. These reports should be retained under an appropriate retention policy.

Operating Procedures.   Most process safety information is used off line. Operating procedures differ in that they are required and should be available for day-to-day operations. Accordingly, the operating procedures should be in place for new and modified facilities prior to start-up, and they should be clearly understandable. Process operators should have ready access to the operating procedures and should be familiar with them.

Operating procedures must be kept up-to-date and reviewed periodically for accuracy and completeness. They should provide clear safety instructions for activities involved in each process and for switching between products in a batch operation. Procedures should address each operating phase, including initial start-up, normal operation, emergency operations (including emergency shutdowns), normal shutdown, start-up following an emergency shutdown or a turnaround, and temporary or experimental operation.

Also, procedures should describe safety-related operating limits, making clear the consequences of a deviation and how to correct it. Operating procedures should also describe safety information related to the process, for example, the hazardous properties of chemicals used or produced. They should also cover special restrictions or procedures.

The auditor should examine operating procedures to verify completeness. He or she should ensure that recent process changes or capital projects are reflected in the procedures. Operators should be interviewed to see if they understand the procedures, particularly with new employees or on start-up of new projects. The auditor should also examine written procedures for other operations such as maintenance, emergency response, and training. There should be systems in place to assure the updating, dissemination, and accuracy of these procedures.

Enhancing Process Safety Knowledge.   Codes, standards, and guidelines for topics ranging from design to maintenance of equipment evolve over time. A facility must keep abreast of such advances. The auditor should determine if there is a system for continuing education in process safety information for technical personnel (attending conferences, etc.). The auditor should verify that this is taking place by exploring what the facility is doing relative to recent developments in process safety such as the AIChE Design Institute for Emergency Relief Systems (DIERS) technology, new regulations, and results of incident investigation reports from other facilities and companies.

Project Safety Reviews

When a plant is built, a facility expanded, or a process modified, most companies require formal approval of the major capital projects based on the project's dollar value. Small projects, on the other hand, may be approved by local management and implemented by facility staff. These small projects may not fall under the company's formal capital project procedure and should then have a review conducted under management of change procedures. The auditor should be aware of the company's criteria for defining a capital project that requires a process safety review to ensure that all projects in fact have a safety review.

Since capital projects may involve equipment or technology changes, one or more formal safety reviews should be performed. The number and type of project safety reviews should be appropriate to the scope of the project. Project safety reviews can employ different techniques (e.g., a hazard and operability study or a failure mode and effects analysis). Thus the person auditing this element should have a reasonable understanding of project stages and the applicability and effectiveness of the various hazard evaluation methods. The audit of this element should examine both the existence and quality of project safety review procedures and should determine if the procedure is being effectively implemented.

As indicated in Fig. 12-5, project safety reviews interface with many other process safety management elements. For example, process safety knowledge and documentation are essential for a credible and effective review. Procedures to ensure information is transferred among the various elements need to be examined.

Depending on the findings of the project safety review, recommendations may involve various process safety management activities. Because of this, the auditor may need access to information from several departments in order to verify follow-up and completion of recommendations. Interviews with the safety coordinator, process chemist, engineering department supervisor, project engineer, technical-service supervisor and engineer, operations manager, unit supervisor, or operators might be appropriate.

The auditor should determine how project safety reviews are scheduled and funded. The project's safety reviews should be part of the project schedule and project control plan, and the project should bear costs of review costs and auditors'

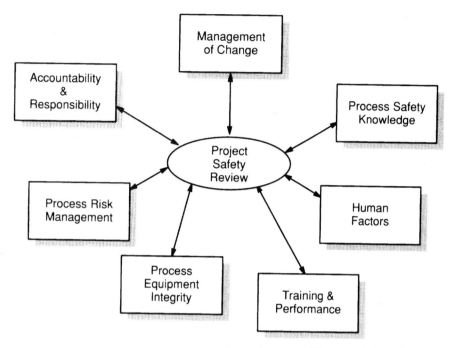

Figure 12-5. Project safety review interfaces. (*Reprinted from Guidelines for Auditing Process Safety Management Systems, with permission. © 1993 by the American Institute of Chemical Engineers.*)

time. The auditor should examine how project review costs are estimated and charged and if the project safety review is integrated into the project schedule.

Project Safety Review Procedures

The auditor should begin by reviewing the documentation that describes the project safety reviews. There should be written procedures indicating the schedule for the safety reviews, how they are initiated, who assigns the staff, the techniques used, and the documentation required.

Written protocols or procedures for project safety reviews should exist and be reviewed by the auditor. A copy of the project safety review procedures may be obtained during the preaudit. Reviewing the procedures in advance lets the auditor understand the scope of and approach to project safety reviews and date of the last revision of the procedures. A preaudit review can also provide some insight into management controls and responsibility for the system. The auditor should interview project safety review leaders, participants, and operators to understand how the review system is implemented.

Copies of the project safety review procedures should be readily available to the proper company staff. The auditor should determine if those participating in these reviews have access to the information.

The project safety review procedures should address documentation of the review. The auditor should study the reporting requirements and a sample of reports to determine compliance and quality of the project safety review. Some aspects to evaluate include report type (interim draft, final), content, and distribution. The auditor should also identify typical projects in progress or recently completed to audit the adequacy of the reports documenting the reviews.

The auditor should see if there is an effective site and facility procedure for periodically updating the review procedure. An interval of 1 to 3 years is typical. The auditor should ensure that there is a management system in place so that the procedure is reassessed in a timely manner.

With regard to the actual review and rewriting of the procedure, the auditor should ascertain that those who use and are affected by the procedure are involved in its reassessment and updating. Interviews with engineering and safety personnel should cover the updating process. Because the project safety review procedure may affect several departments, there should be a formal sign-off requirement for changes to the procedure.

Scope of Hazard Analysis

Safety reviews are often conducted during the conceptual engineering, detail design, and prestart-up phases, and the scope of each depends on the stage. While not every project passes through all phases, one way to characterize these phases, as given in Ref. 6, is:

- Phase I—Conceptual Engineering
- Phase II—Basic Engineering
- Phase III—Detail Design
- Phase IV—Equipment Procurement and Construction
- Phase V—Commissioning Prior to Start-up

Examples of safety reviews at various stages are:

Conceptual Engineering.   This phase can involve both laboratory and pilot-plant development work to determine the process definition. The purpose of safety reviews at this stage is (1) to ensure that the laboratory or pilot systems are designed and operated safely and (2) to see if the project will be safe and environmentally sound when scaled up. The auditor should verify that the safety review addresses both points.

Detail Design Phase.   In this phase, the auditor should determine if an adequate design review has been conducted by interviewing members of the design team and examining study reports. The auditor should verify that safety information from the conceptual engineering phase was transferred and used. As piping and instrument diagrams are now available, a more in-depth hazard analysis can be completed. The auditor should confirm that the safety review completed at this phase utilized the piping and instrument drawings.

Prestart-up Phase.   Prestart-up safety reviews are important because they are the final check before the process is placed in operation. These analyses include a review of the operating procedures and a walk-through of the facility. Typically, final checks verify that:

- Recommendations from previous reviews have been addressed.
- All safety information has been incorporated.
- The plant adheres to the intended design.
- Operating, maintenance, and emergency procedures are in place and personnel have been trained.

The auditor should verify that the prestart-up safety review system meets these objectives.

Personnel must know which type of safety review to conduct at a specified project phase. Therefore, along with assuring that a review is being conducted, the auditor should verify that the type of review is appropriate.

Hazard Analysis Techniques

In addition to considering generic requirements of project safety reviews, auditors must use specific review techniques properly. These techniques are described in Ref. 7. This reference will assist the auditor in assessing the adequacy of the technique.

Since review techniques are sometimes loosely interpreted, the auditor needs to determine how the hazard analysis is actually done—which method is being used and whether any steps are missing. Hazard analyses should consider the impact a project may have on existing facilities. The auditor should determine if there are provisions for quantitative risk assessment of major hazards and, if so, whether or not they are used.

Staffing

There are three criteria for selecting the review chairperson: experience, leadership skills, and objectivity. The project safety review procedure may address minimum skill requirements for review leaders, including process experience. Regarding objectivity, there is a trade-off between process knowledge and the perception of objectivity that must be considered. The project safety review procedures should address this potential conflict.

Team members should have the skills and experience to conduct an adequate review. The auditor should verify that the team is selected on this basis, not purely by job title. Also, the auditor should verify that the safety review procedures specify the proper roles and responsibilities for team members.

Recommendations, Follow-up, and Closure

A project safety review can result in various actions, such as requirements to meet standards and regulations, recommendations based on knowledge and experience, and issues requiring further study.

Follow-up and closure of recommendations are as important as conducting the review itself. Action items that must be addressed prior to start-up should be documented and clearly identified. The auditor should verify that the safety review procedure includes a formal tracking system, with assigned responsibility to ensure that all recommended items are addressed. The auditor should be aware that a successful tracking system comprises formal documentation, periodic reporting, easy updating, and information distribution. The auditor should also verify that the person assigned to handle corrective measures has the resources, authority, and knowledge to ensure the follow-up.

A list of action items with responsibility and status should be distributed to the project review team periodically. Such a tracking system ensures prompt attention to all safety concerns, as well as reassuring the project review team that their action items are being addressed. Other areas that should be covered in process safety review procedures include:

Resolution of Disagreements.   The project safety review process should provide a means for resolving differences. The auditor should determine if such provisions exist and whether they have ever been invoked.

Updating Process Safety Information.   The auditor should verify that there is a clearly defined initiating mechanism for such updating. This mechanism should recognize that the project safety review team will often disband at the end of the project.

Report.   The report should briefly describe the process, name those attending the review, and present findings and recommendations, and it should include a prioritized list of action items with responsibilities and follow-up notes.

Dissemination.   This practice varies from company to company. Findings should be made available to other company facilities that have similar process units. The auditor should determine whether the company has a policy on dissemination of project safety review reports, and whether it is followed.

Record Retention.   The final safety review report should become part of the process safety information documentation. The auditor should determine the provisions for retaining project safety review reports and sample project safety review reports to validate that the system is working.

Management of Change

Process changes range from large facility expansions or new plants to minor changes in chemistry, technology, equipment, or procedures. Even simple changes, if not managed properly, can have catastrophic consequences. Three types of changes should be managed at any location: technology, facility, and organization.

By far the most challenging aspect of managing change is identifying that a proposed modification is, in fact, a change. Once a change is identified, the next task is determining the level of review necessary prior to implementing the change. These elements should be the primary focus of an audit.

This process safety management element interfaces with other process safety management elements, including organization changes, process safety information, project safety reviews, and training.

Auditing Approach

A management of change audit can vary in scope and approach, depending upon company requirements. One approach is shown in Fig. 12-6. The procedure should be discussed with the appropriate technical, operations, maintenance, and purchasing personnel to fully understand how the system is supposed to work and to ensure that the procedure addresses the key elements of management of change. This interview may involve only one or two facility staff with knowledge of the procedure and its applicability. A list of recent large and small changes may be identified for later verification.

The auditor should review the change reports to verify compliance with written procedures and to ensure that changes follow the procedure. A random check to ensure that process safety information and other documentation are updated should also be performed. Additional verifications may be done by interviewing maintenance and operations personnel. Interviews may include a cross section of the facility, with personnel from separate process areas. The auditor should verify that changes were reviewed properly and that training was completed prior to implementation of the change.

Written Procedures

The auditor should ensure there is a written procedure for change authorization and that management of change procedures conforms to corporate or facility standards or guidelines.

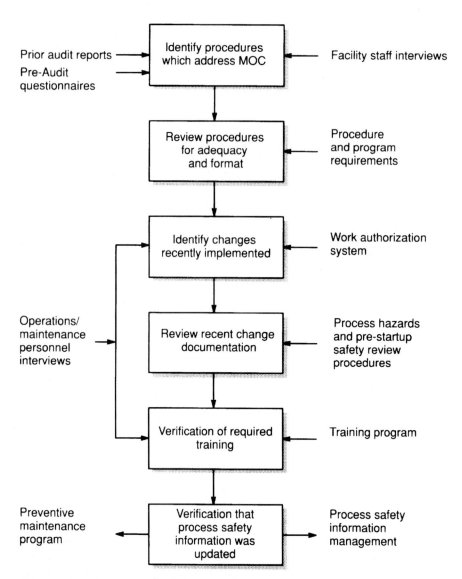

Figure 12-6. Management of change audit process. (*Reprinted from Guidelines for Auditing Process Safety Management Systems, with permission. © 1993 by the American Institute of Chemical Engineers.*)

The auditor should identify and review procedures that address management of change. If the procedures require an in-plant audit of the program, documentation supporting the completion of these audits and resolution and closeout of findings should also be verified by the auditor.

Definition of Change. The written procedure should specify what is a change. One definition is that all hardware changes must follow the procedures except for replacements-in-kind. A good definition of a procedural change is anything that requires modification of process safety information, such as new raw materials, operation outside safe operating limits, and decommissioning. When there are multiple procedures for different types of change, the auditor should verify that each procedure sufficiently overlaps others to include all technology and facility changes.

Identification of Change. There should be a clear mechanism for initiating a change authorization, including changes made for emergency reasons. This mechanism should differentiate capital changes from routine or preventive maintenance. The system should require documentation, such as by a work order. If changes are requested by a work order, the auditor should verify that the required information is documented, including naming the person responsible for obtaining the review and authorization.

The auditor should understand how facility changes can be made and ensure that there is written documentation and approval by those responsible for operation. The auditor should verify that appropriate facility personnel have been notified of the procedures for managing change and that appropriate training has been conducted. There should be a means for cross checking or for requesting additional approval by personnel who will implement the change to assure that the change is properly identified and reviewed. To verify that changes are identified and managed, the auditor might interview staff to review some changes that have occurred recently and sample appropriate documentation.

Description of Change. Any change should be described, including sketches or drawings, and the technical basis for the change should be discussed. The impact of the change on process safety should be described in sufficient detail to help define the required safety reviews. When reviewing a change request, the auditor should understand which process safety information must be subsequently updated, especially the piping and instrument drawings, operating procedures, and safe operating limits. The auditor also needs to ensure that information describing the technical basis and impact on process safety is adequate.

Temporary Changes. To avoid temporary changes from becoming permanent, a maximum duration for such a change should be specified, such as days or weeks, but not months. Once the approved time expires, the procedure should require that the temporary modifications be removed and the process returned to normal conditions. The auditor should verify that temporary changes have expiration dates and that changes are restored after that date. When temporary changes are extended beyond their scheduled times, or become permanent, the auditor should verify that they were properly authorized.

Authorization. The procedures should identify those who can authorize a change. The level of authorization should relate to the potential risks, rather than to the investment level or duration of the change. The level of authorization and/or number of approvals must be appropriate for those changes that may impact one or more process areas and change. Any alternative staff who are authorized to approve changes should be identified, and they should comply with the variance procedures discussed before. The auditor should verify that proper authorization is obtained for the type of change requested—especially for changes done during off shifts, on weekends, or on holidays.

Safety Review. All changes should have some level of prestart-up safety review. The auditor should verify that these requirements exist and are done following the safety review procedures. For changes with a low risk potential, a formal and systematic process hazards analysis procedure may not be required. A hazard analysis, however, should be conducted on any changes that require engineering. A sampling of the safety review documentation should be made to verify completion and adequacy of review prior to start-up.

Training. Before a change is carried out, all employees who are affected by the change must be trained. Any changes that involve safety, operating, maintenance, and emergency procedures should be included in the training.

The auditor should interview operations and maintenance staff (including contractors) to ensure they are adequately trained in the change and are able to operate or maintain equipment so affected. The auditor should also verify that the training is completed before the change is made. All relevant shifts, not just those who do the start-up, must be trained.

Documentation. The procedures just discussed must be documented. Typically, a management-of-change form documents the key requirements and authorizations for implementing change. Separate documents should be filed for the safety review findings and how they are addressed. Before the change file is closed, the process safety information must be updated.

The auditor should make sure that documentation of the change is complete, including a description, authorizations, and safety reviews. Principal documents to audit include piping and instrument diagrams, operating and emergency procedures, and safe operating limits.

Process Equipment Integrity

An audit of the process equipment integrity element begins with equipment design and continues through to fabrication, installation, operation, and maintenance. To be comprehensive, the audit of process equipment integrity should include:

- Review of standards and procedures for format and content
- Interviews with construction and maintenance (including contractors) to verify qualifications
- Sampling of equipment records for completeness and compliance with procedures
- Observation of new construction and maintenance in progress for compliance with procedures

The auditor should interview engineering, construction, and maintenance personnel, since they have the responsibility for these activities. Where a plant relies on corporate engineering and construction departments or on outside contractors, these groups should be interviewed.

In preparing for the audit, the auditor should identify key individuals and obtain the following documentation from them for review:

- Design standards, specifications, and records
- Fabrication specifications and records
- Installation, inspection, and maintenance procedures and records

Also, information from other process safety management elements may be useful (see Fig. 12-7). The auditor should understand the corporate and facility policies that specify the equipment included in the process equipment integrity program. The criteria for including equipment should be based upon how critical a piece of equipment is to ensuring process safety. For critical processes, the auditor should verify that critical safety equipment is included in the process equipment integrity program. Some examples are vessels and tanks, piping systems, emergency shutdown systems, and fixed fire protection systems.

New Equipment Design, Fabrication, and Installation

The design codes and specifications for new equipment should be in the design package, along with items such as data sheets, calculations, and equipment drawings. The auditor should confirm that the appropriate personnel address fabrication and installation recommendations from the project safety reviews. Also, it should be verified that the required tests and inspections have been performed.

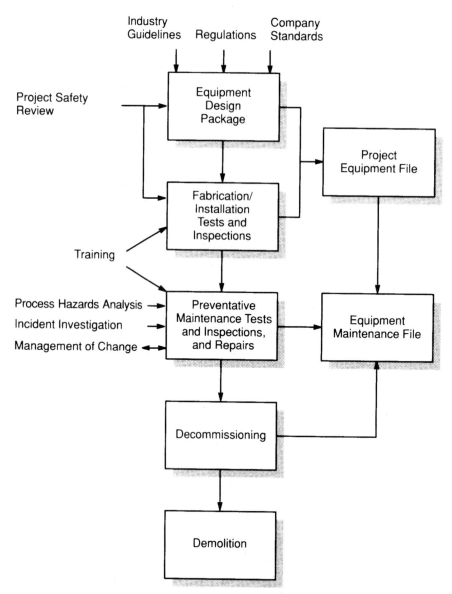

Figure 12-7. Process equipment integrity chart. (*Reprinted from Guidelines for Auditing Process Safety Management Systems, with permission. © 1993 by the American Institute of Chemical Engineers.*)

For new equipment fabrication, the auditor should ensure that the quality assurance program meets project requirements. The auditor should also verify that the following activities, when applicable, are specified before the start of fabrication:

- Welding, fabrication, and nondestructive examination requirements
- Required approvals
- Qualification of employees (e.g., welder certification)
- Quality assurance audits
- Quality assurance procedures that may be conducted during fabrication include radiographic examination of welds, hydrostatic testing, and verification of dimensions and tolerances

When manufacturers of critical equipment provide installation instructions, the auditor should make sure that there is a mechanism to ensure these are being followed. The tests and inspections carried out during installation may include soil compaction, structural steel integrity, leak testing of connections, and electrical load testing of emergency equipment. The auditor should verify that fabrication and installation records include the date of the test or inspection, the person who conducted the procedures, the results, and their acceptability. Test and inspection records should be filed to facilitate retrieval. A system must be in place to forward a copy of these records to maintenance before start-up.

Preventive Maintenance

Where critical equipment is operating in a process, predictive or preventive maintenance programs should be in place to prevent or identify defects and minor failures before they can become more serious ones. The auditor should verify that any required tests and inspections, along with their frequency and acceptable limits, are specified and that there are written procedures for conducting each test or inspection.

There should be a process to establish preventive maintenance frequencies, e.g., start with a conservative frequency and change it based on test results. The following are some examples of critical preventive maintenance:

- Test and reset pressure relief valves.
- Replace pump and compressor seals.
- Inspect vessels and tanks.
- Test and inspect emergency and fixed fire protection equipment.

The auditor should verify that there is a system to analyze the maintenance history of equipment for trends, so that methods, frequencies, or design can be changed to incorporate this experience. The auditor should then sample equipment records to verify that tests and inspections are done at specified frequencies and that results fall within acceptable limits. The auditor should also verify that there is a training program for personnel who conduct preventive maintenance.

Maintenance Procedures

Work Authorization.   Process-related maintenance activities should require a written authorization. This authorization is particularly important to ensure that maintenance and operating personnel coordinate activities to safely perform necessary maintenance and to identify potential changes to a process. It is equally important that maintenance work on equipment be documented to provide a record that can be analyzed as part of the reliability program.

The auditor should verify that process-related maintenance work requires a written description and appropriate authorization. He or she should also verify that there is a mechanism for the quality assurance of materials and spare parts used for repairs.

For critical service, or where special materials of construction are used, there must be a method to verify that these are the proper materials. The pressure and temperature ratings of equipment should be confirmed by the maintenance coordinator or contractor. Documentation of equipment repairs should be retained and retrievable.

Safe Work Practices.   Safe work practices for preparing equipment for maintenance or tie-ins are critical. Examples are hot work, confined space entry, and lockout. To control process hazards, the auditor should verify that these tagout practices exist and are conscientiously used. The auditor should: (1) identify maintenance or project work in progress that requires compliance with safe work practices, and (2) verify that these practices are being followed and that permits are complete and current.

Contractors.   A study on the use of contract labor in the U.S. petrochemical industry found that contractors perform about one-third of routine maintenance and one-half of turnaround maintenance.[8] The auditor should verify that the safety record of any potential contractor has been considered before awarding them contracts. The auditor should also ensure that a system exists for contractor orientation and that orientation includes a review of the emergency plan, hazard communication, and the facility's safe work practices. The contractor must train its employees in the procedures to work safely at the facility.

Decommissioning and Demolition.   When equipment has served its useful life, it should be dismantled, removed, and/or demolished. The auditor should verify that requirements are specified for the isolation and cleaning of equipment to be taken out of service and establish that there are special procedures for cleaning and disposal of hazardous residues or materials. An examination should be made to determine how such equipment has been handled in the recent past. Finally, the auditor should verify that any equipment that is subsequently recommissioned has gone through the appropriate safety reviews.

Process Risk Management

Process risk management is identifying, evaluating, and controlling potential hazards associated with:

- Existing operations
- Modifications
- New projects
- Acquisitions
- Toll processors
- Customer and supplier activities

The basic concepts and purposes of the components of process risk management that follow are covered in greater detail in Ref. 6.

An audit of a risk-management system begins by reviewing the elements of a facility management's program including:

- Hazard identification studies
- Risk-analysis techniques, which can rank risk-reducing alternatives
- Risk-reduction activities
- Management of residual risks
- Practices of suppliers and customers

The auditor should understand the risk-management goals, how the acceptable level of residual risk is determined, and special requirements and practices, and should locate guidelines for handling these subjects. Regulatory requirements that may influence or form the risk-management program should be known.

Hazard Identification

This involves reviewing an engineered system as thoroughly as is reasonably practical to identify deviations from the design intent that can lead to an event causing undesired consequences. As such, hazard identification is the foundation of a solid risk-management program and should be conducted at numerous stages during a process. The emphasis of the audit of process risk management should be on evaluating the system for conducting hazard identification.

The auditor should identify existing standards and guidelines for hazard identification studies and interview plant management to determine their understanding of study objectives and their role in documentation and acceptance. The auditor should then review a sample of hazard identification reports and interview key staff who produce the reports to verify that the guidelines and a number of specific considerations, such as scope study, hazard identification techniques, implementation, and study recommendations, are followed.

Scope of the Study.   The usefulness of the results of a hazard identification study depends on its initial scope. Scope is often defined by the level of detail, consequences of concern, and system boundaries. To audit the scope, the auditor has to identify the standards or guidelines that apply. The level of detail includes such considerations as looking at individual pieces of equipment versus subsystems and looking at internal versus external causes of failures. There are two general causes of failures:

1. Direct failures of equipment that process the chemicals (e.g., a gasket leak in a chlorine system)

2. External events, defined as failures unrelated to the system that can lead to a release (e.g., rupture of a lien during excavation).

Risk management also addresses undesired consequences. These include fatalities or injuries to employees or the public, property damage and business interruption, and environmental damage.

System boundaries should be defined in terms of geographical, systemic, and functional limits. The auditor must determine if the boundaries are clearly stated in selected hazard identification reports and if these boundaries are consistent with the guidelines.

Methodology Selection.   Several methods will identify the hazards of an operation. Some of the common ones reviewed in Ref. 7 are hazard and operability analysis (HAZOP), failure mode and effects analysis (FMEA), and what-if analysis.

Reference 7 provides guidelines for selecting the appropriate methodology for hazard identification. The methodology should be consistent with regulatory requirements as well as company guidelines. If there are no corporate or site guidelines, the basis for selecting the methodology should be documented. The auditor should interview those conducting such studies to determine how a methodology was selected.

Implementation Practices.   Hazard identification studies should be led by someone knowledgeable in the selected methodology. The hazard identification team must include persons who know the operational, engineering, and maintenance aspects of the process. Using this team makeup as a criterion, the auditor should verify that team leaders are qualified to conduct a valid study. The responsibility for assuring that such studies are staffed and conducted should also be clearly defined and verified by audit.

Study results should be documented. The report format can vary, depending on the methodology used and study scope and objective. The study may be reported as part of a more detailed hazard analysis report. The level of detail reported should be consistent with the study's objectives and applicable requirements. For example, some regulations require specific reporting practices. The auditor should see if this applies and review reports to ensure that they meet applicable practices. The auditor should also verify that appropriate individuals receive copies of these reports.

Hazard identification should be repeated periodically. It is the auditor's job to see if there are guidelines for the frequency of this process and then verify adherence to the guidelines.

Study Recommendations. A hazard identification study generally results in recommendations for modifications to equipment, operating procedures, or administrative controls. The auditor should determine if such recommendations are documented and if the management system provides for tracking the status of the resulting action plan. The auditor should also determine the existence of a paper trail and verify that recommendations are closed out within the specified time.

Risk Assessment of Operations

A risk assessment is a detailed look at the hazards identified in terms of their potential consequences and their likelihood of occurrence. A hazard identification study may be insufficient to support risk management decisions when the recommended modifications are extensive or costly. Further analysis in the form of a risk analysis may be needed.

The auditor should identify any corporate, site, or regulatory guidelines for risk assessment. He or she should review several risk assessment reports and interview the key staff who produce the reports to verify that the guidelines are followed, readily available, and known.

Also, the auditor should interview plant management on their understanding of the objectives of a particular study to see if the objectives are met and to determine management's role in documenting and accepting the recommendations. The auditor may find it necessary to interview corporate staff if they supplied the guidelines or risk-analysis expertise.

Objective and Scope of the Analysis. Most risk assessments are qualitative or semiquantitative. A typical and frequent objective of a risk assessment is to rank or prioritize individual release scenarios. Risk-assessment studies may be conducted on an entire processor unit or on selected hazard scenarios. A risk-assessment report should discuss:

- Objectives
- Consequences
- Equipment and operations

- Selection of methodology
- Level of detail (i.e., screening versus detailed, qualitative versus quantitative, or conservative versus best estimate)

The auditor should interview staff and review several reports to verify that such issues are addressed in the reports.

Qualitative Risk Assessment. Scoping qualitative risk studies requires engineering judgment. The auditor should verify that guidelines for scoping qualitative risk assessments address:

- The inclusion or exclusion of utilities and other sitewide services
- The treatment of secondary or domino events
- The assumptions used to develop frequency and consequence estimates
- The definitions of qualitative estimates of risk (e.g., high, medium, or low)
- The basis for risk prioritization
- The conditions under which additional analyses should be conducted

The first two items are also part of a quantitative risk assessment, but they are more likely to be overlooked in a qualitative study.

Quantitative Risk Assessment. Here, a critical aspect of scoping is the consistency of assumptions within and between studies. The auditor should verify that guidelines or standard assumptions exist for:

- The impact level of a release (serious injury or fatality) in terms of modeling criteria for overpressure, thermal radiation, or toxic exposure
- The probability of suffering a specified consequence, given exposure to a certain level of hazard
- Selection of consequence modeling packages and models, and input parameters (i.e., meteorological conditions)
- Presentation of results, for example, F-N curves
- Typical databases to determine failure rates, or use of standard or generic rates
- Treatment of uncertainty
- Treatment of population, e.g., average versus day and night

The auditor should verify the existence of and adherence to such guidelines by reviewing quantitative risk-assessment reports and interviewing staff. Corporate guidelines should include regulatory requirements.

Methodology Selection. Several methods for risk assessment are described in Ref. 9. The main difference in choice is the level of detail. This level is usually based on the materials being handled, the complexity of the process, estimated risk level, and the available resources. Corporate policies or regulatory requirements, however, may specify the methodology, level of detail, or format of the results for the risk assessment. The auditor needs to determine the existence of such guidelines and then review adherence to them. The auditor should also verify that documentation exists regarding the basis for selecting the methodology.

Implementation Practices and Reporting. The background, skills, and experience needed for conducting quantitative risk assessments are quite different from those for hazard identification. For example, the risk-assessment team may have to be experienced in fault-tree analysis and meteorology. Often these assessments are done by consultants or corporate staff. Therefore, plant personnel should review the assessment results to assure that process conditions and other assumptions reflect current operating practice.

Documentation standards or guidelines should be available, particularly for use by consultants. The reports should include objectives, scope, discussion of methodology and assumptions, results, and recommendations. The auditor should review reports to determine consistency.

Study Recommendations. Risk assessments of operations, as with hazard identification studies, often lead to recommendations for risk-reduction measures. These recommendations, their priority and implementation status, and the documentation of any actions (or nonactions) should be addressed.

Risk-Reduction Activities

The outcome of hazard identification and risk-assessment studies can be risk-reduction measures, a prioritized list of such measures, and an action plan and schedule. The auditor should determine if the facility staff has evaluated the effectiveness of the measures and reviewed any potential added risks that may result from implementing the measures. When reviewing the risk-reduction activities, the auditor should be certain to focus on the following elements:

- Corporate or regulatory risk-acceptance guidelines
- The types of risk-reducing measures considered
- Assurance that new risks are not introduced because of risk-reduction measures
- Dates and responsibilities for implementing risk-reducing measures
- Effectiveness of these measures
- The approvals of those accountable for process safety

The means to establish acceptable risk can vary. The auditor should interview appropriate decision makers to see if risk-acceptability judgments are made consistently. The program should also allow different types of risk-reducing measures—such as changes in management systems, hardware, process parameters, process design, chemicals, safety systems, monitoring systems, release mitigation devices, and administrative controls—as appropriate. The auditor should review existing guidelines to assure that all are included.

A risk-reduction plan provides an implementation schedule based on recommendations from studies and decisions of corporate and facility management. The risk-reduction plan is key to process risk management and PSM development, and it should contain clearly defined responsibilities.

The auditor should verify assignment of responsibilities for implementation and that the implementation is on schedule. If some recommendations are not implemented, the auditor should also verify that the reasons are documented.

Residual Risk Management

Risk-reduction plans may not completely eliminate risk from a particular hazard. Management of this residual risk involves ongoing review and reconsideration of process-safety-related conditions and controls. Such reviews and added risk assessment and hazard identification studies can confirm that the risk has not increased. Residual risk management requires periodic review and reanalysis of the technical studies. The corporate guidelines for residual risk management may specify how often such reanalyses must be conducted, as well as reasons why they should be started ahead of schedule. Guidelines should also provide a format for the documentation as well as the approval and communication channels for the findings.

The auditor should verify the existence of such guidelines and determine how they have been followed by reviewing older studies and following their paper trail. Existence of a paper trail for the periodic reviews, the results of the reviews, any action items, and a schedule for implementation should be verified. The auditor should probe for their existence through interviews with staff and reviews of documentation, and then he or she should verify that there is a tie-in to management of change procedures.

Technical studies for residual risk should be based on original assumptions regarding the system design parameters, parameters that may have a degree of uncertainty. For example, the reactivity of a reactant may not be known accurately. The auditor should verify that reviews revisit the areas of uncertainty to confirm that the changes in plant conditions and use of recent information have been properly evaluated for their impact on the residual risk level.

New information may come from operating experience. For example, near misses may reveal accident scenarios not considered initially. There may be significant differences between the actual failure rate and that used initially in a risk assessment. Thus the auditor should verify that the residual risk management program includes a periodic review of the assumptions used.

New or improved consequence models may become available, offering better insight into the behavior of certain releases. Also, risk acceptability guidelines may be revised because of changes in corporate philosophy. The auditor should verify that the residual risk-management process addresses these types of changes.

The auditor should also verify that the following are in the program:

- Periodic review of assumptions and methodology
- Preparation of documents reviewing assumptions and methodology
- Communication of findings of periodic reviews to appropriate persons
- Identification and completion of action items

Customer and Supplier Facilities and Practices

The auditor should see whether there is a system for defining high-risk raw materials or products. The company should assure itself that its customers and suppliers have risk-management programs for such materials and products.

The auditor should check guidelines to determine which materials are covered by this program. Selection of suppliers, customers, carriers, and shippers should be based on their risk-management practices and/or willingness to abide by the corporation's practices. Also, the guidelines should specify recommended practices for customer unloading and storage, supplier-certified product analysis, and actions facility managers should take if there are failures of the agreed-upon risk-management measures.

The auditor should interview plant management to see if contractual or other formal agreements exist between the plant and its suppliers, customers, and carriers. The auditor should verify that there are records of inspections of the facilities and procedures for these various companies. A close look should be given to transport vessels that are owned by other corporations but used by the facility.

The auditor should confirm that a program exists to review the risk-management program of the carrier to assure vessel integrity. Also, there should be an assessment of the relative risks for all contract or toll operations. The auditors should request documentation of such considerations and the measures that are in place to assure that good risk-management practices exist and are encouraged.

Incident Investigation

Incident investigation, analysis, and follow-up are essential to any process safety management system; safety cannot be properly managed without such a system to define system failures and corrective action. Because there is no industry standard for incident investigation, auditors should familiarize themselves with the philosophy and content of some sample incident investigation procedures. (Examples are provided in Ref. 10.)

There are three general categories of incidents:

1. Major accident—an incident involving multiple injuries, a fatality, or extensive property damage
2. Accident—an incident limited to a single injury or minor property damage
3. Near miss—an incident that has the potential for injury or property damage

All of these should be reported. Ideally each should be investigated to spotlight flaws in process and personal safety management programs. From a more practical standpoint, extensive investigation of all incidents falls beyond the resources of most companies. Consequently, guidelines for determining the level of investigation for an incident are essential.

The auditor should be aware that incident investigation interacts with several other management systems (see Fig. 12-8). For instance, corrective actions may:

- Change operating procedures to prevent similar incidents
- Refine process knowledge
- Provide input to process hazards analyses
- Require retraining
- Improve notification of emergency personnel and facility management

The first step in an incident investigation system audit is interviewing those responsible for coordinating the investigation system to understand the management system. Next, the auditor obtains a copy of the incident investigation system description or at least the reporting form(s). These materials are reviewed for format and content before beginning data gathering. The auditor should also establish if the system complies with facility requirements for updating and approval.

Reporting Mechanism

Definition of Incidents.   Incident investigation begins with incident reporting. Therefore, the definition of an incident that requires reporting should be reviewed to see if it is clear and concise. A definition applicable to the chemical industry is any occurrence, condition, or action that did or could have resulted in personal injury or damage to the plant, community, or the environment. While major accidents and accidents can readily be identified as incidents, near misses are difficult to define. Potential accidents or near misses are included in the above definition. Incidents are often classified to assist in deciding the extent of investigation and determining if a formal report must be made to governmental agencies or insurance companies. Some examples are fatalities, chemical releases, and equipment damage or production loss above a certain value.

The auditor should note whether such a classification procedure—with definitions—exists, and he or she should ascertain by interviews whether facility personnel understand how to classify incidents and what constitutes a reportable incident.

Reporting is a critical link in the incident investigation procedure; incidents that do not get reported cannot be investigated. Investigation of major accidents may involve special techniques and require resources that a facility might not possess. Therefore, the auditor should see if there is a plan for investigating major accidents, particularly for identifying and accessing corporate and outside resources (see Ref. 10).

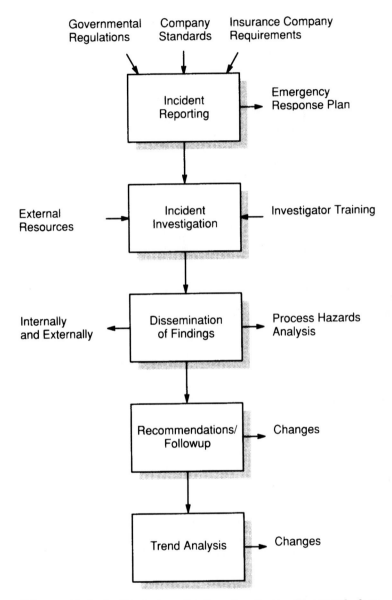

Figure 12-8. Incident investigation flowchart. (*Reprinted from Guidelines for Auditing Process Safety Management Systems, with permission. © 1993 by the American Institute of Chemical Engineers.*)

Initial Reporting.   A standard form should be used to report incidents initially, and the auditor should review the form to see if it is consistent with the written procedure. The incident reporting form should elicit the recording of facts and should provide space to list information needed to satisfy reporting requirements to governmental agencies, insurance companies, and/or corporate organizations. The auditor should review a sample of reports to verify that the incident is classified properly and that required information is reported.

Responsibilities.   The reporting chain begins with the employee. Although this may seem elementary, the auditor should verify that employees understand that they are responsible for reporting all incidents—including minor injuries and near misses—in a timely manner. Auditors should verify that incidents are reported within the time set by company policy. Also, they should verify that the system works as intended by interviewing key staff, such as the plant manager, incident investigation coordinator, and members of the emergency response team.

Investigation

Criteria for Investigation.   Although most accidents—the minor ones—are investigated by a qualified person, frequently the first-line supervisor, the auditor should see whether there is a system for determining when team investigations are required and how the teams are organized. Typically, an incident investigation should start within 48 hours. The auditor should verify that investigations are initiated within the required time and are completed expeditiously.

Investigation Team

The investigation team members should be objective and possess skills appropriate to the investigation. Therefore, the auditor should verify that there are procedures that address the use of (1) a team leader without vested interests in the investigation and (2) a system for selecting members based on needed qualifications.

The skills possessed by team members depend on the nature and extent of the incident. At least one member should have firsthand knowledge of the process. Major incidents involving extensive site damage may also call for a corporate representative, technical specialists, or consultants. The auditor should see whether guidelines link incident type or class with specific representation. For example, a corporate risk-management representative or consultant might be included for incidents involving a fatality, contractors with liability implications, or serious environmental or plant damage.

The incident system procedures should define the team's responsibilities and state to whom team members are to report. Reviewing past reports will tell the auditor whether teams meet their responsibilities. The team should be able to interview those who know about the incident and have access to additional resources (such as technical or legal expertise) as needed. The auditor should verify that the investigation team has access to these resources.

Investigation Process

The investigation team must gather facts quickly while the incident is fresh in people's minds and evidence is readily available and free of tampering. Items to consider are:

- Interviewing personnel quickly before memories fade
- Collecting physical and process evidence before it is lost
- Thinking globally in securing evidence, since it is not always possible to anticipate the direction of the analysis
- Using recording methods such as taping, photography, and radiography
- Sampling (in some cases) process fluids for subsequent analysis.

After analyzing all the facts gathered, the team determines the underlying (i.e., management system failures) causes of an incident.

An effective way to audit this area is to interview plant personnel who participate in investigations. The auditor should talk to both investigators and witnesses to obtain a balanced perspective. The auditor should also review past investigation reports to assess methodology and findings and also to verify that the findings are incorporated in the recommendations. The auditor should also verify that the investigator, or team, has experience or training in root-cause analysis.

For major accidents, governmental agencies and insurance companies may conduct their own investigations, and the legal department may become involved. The auditor should determine whether the system addresses the coordination of investigations with outside organizations.

Investigation Reporting

A written report of the investigation should be required. To evaluate compliance, the auditor should review copies of several reports spanning the audit period. This is to verify that necessary information is included in the report and that the investigation was completed within a reasonable time. The report should contain at least the date and description of the incident, factors contributing to the incident, recommendations of the investigation, and those who conducted it.

The auditor should review reports to see whether the findings and recommendations are summarized and are readily apparent from the facts. A sampling of recommendations should be reviewed to determine whether they are based on root-cause analysis. Reports should also be reviewed for conclusions and recommendations that may find general application. The auditor should discuss these with the facility staff to see if the recommendations were reviewed for applicability at other units or plants.

Dissemination of Findings

Identifying "lessons learned" is of value only if they are conveyed to individuals and organizations that might experience or be able to prevent the same incidents.

Internal Distribution.   The incident investigation system should include a report distribution requirement. During the interview with the incident investigation coordinator, the auditor should determine whether appropriate avenues, such as worker safety committees, are used to disseminate findings. The auditor should also verify that facility personnel and contractors are aware of recent incidents. This information should become a part of the process safety knowledge.

Also, there should be a policy for sharing incident reports with other plant locations. The auditor should verify that this policy is in place and is used.

External Distribution.   Leading companies in process safety often disseminate information on root causes of incidents to industry groups, particularly when similar manufacturing technology is involved. The auditor should determine whether there is a company policy regarding external dissemination of incident cause information—and whether the policy is followed.

Recommendation Implementation and Closure.   Actions recommended as a result of an incident investigation should be closed out, whether or not they are accepted. Because of the importance of this item in preventing further incidents, the auditor should examine this activity closely. The auditor should verify that there is a written procedure describing how the investigation is documented and how follow-up and closure are to be carried out. The responsibility for action and follow-up of that action should be assigned to someone with appropriate authority.

Investigation recommendations and their implementation and closure should be verified through examination of an appropriate sampling of records. The auditor should also determine whether any items with formal assignments are still outstanding. Finally, the auditor should determine whether process modifications follow the management of change system and whether investigation results are cycled to other elements that require them, such as process knowledge, equipment integrity, or human factors, as appropriate.

Incident Analysis

As already noted, learning from mistakes is the objective of incident investigation. Classifying findings by root causes can provide information about major contributors to incidents, and it assists management in placing emphasis for improving process safety. Knowing if incidents result from facility failures or deficiencies in management systems is essential for proper management of process safety. The auditor should verify that incidents have been classified according to root causes and archived so that the information can be used. Preferably, there should be a central depository for incident investigation records.

Human Factors

Human factors is defined as the application of knowledge about people to the design of both technical systems and equipment in order to provide more efficient and effective human-system interactions and enhance safety. Human factors may be poorly applied for many reasons, including the nature of the task, the design of the equipment, the capabilities and limitations of personnel, and the operating environment. This is an emerging discipline that has few standards or guidelines. Often, human factors are addressed informally.

Nevertheless, the process safety systems audit must determine how effective the process safety management system is in implementing human factors principles and practices. The audit of this area is generally limited to company standards or practices. As a result, the auditor should identify the management system deficiencies that may result in the omission of human considerations, rather than identify individual symptoms. Examples of questions asked in a human factors review appear in Table 12-4. Guidance for identifying system deficiencies follows.

Organizational Issues

A company's organization and operating environment should support and promote human factors principles and practices in day-to-day operations. A company's effectiveness in meeting these goals is measured by the nature and frequency of applying principles and the knowledge of personnel about the program. In an audit, the auditor needs to determine the presence of programs, policy statements, and feedback mechanisms and how effectively they address process-related human factors issues. The auditor must distinguish between those that simply address injuries and illnesses, and programs that deal with process design issues.

Employees should actively help implement the program to incorporate human factors, since their firsthand experience often provides early detection of human-hardware interface problems. Evidence of employee involvement includes representation of human factors or safety committees and participation in developing procedures. The auditor, for example, may review safety committee minutes or design review reports or safety inspection reports to see if employees are in fact involved and their level of participation.

The auditor should meet with those who do the training to see how well teaching materials and instruction address such issues. The auditor should determine whether the staff understands the principles of human factors and should assess their application to process safety management. The human factors expertise

Table 12-4. Examples of Items to Consider in a Human Factors Review

---

Organization and policy issues
- Have human factors engineering policies been established and communicated to employees by upper management?
- Are human factors support and expertise available within the organization?
- Is there a communication and follow-up mechanism to address human factors issues?
- Does the organization encourage employees to express human factors suggestions and concerns and allow them to contribute in the decision-making process?
- Is management willing to allocate time and resources to address human factors issues?
- Are human factors issues discussed at management meetings?
- Have formal procedures or policies been established that address the evaluation of new, modified, and/or existing processes/systems in terms of human factors principles?

Operator—process interface
- Do control and display layouts minimize the chance for operator error in terms of functional grouping, sequence of operations, and color (including use of colors appropriate for color-blind individuals) etc.?
- Are there design standards that specify proper layout?
- Is information clear, concise, and readily accessible to the operator (including left-handed individuals)?
- Are the implications on various actions and their effects on the process clear to the operator?
- Are controls accessible, easy to reach, and operable?
- Are critical controls operated in the same manner (e.g., up–down; push–pull)?
- Is there adequate space to access system elements for normal operations and maintenance?
- Have upset conditions and emergency response been considered?

Task design and job organization
- Have the operator's individual responsibilities been clearly defined?
- Have the psychological and physical demands of the job been considered for both routine and emergency operations?
- Have actions been taken to reduce the likelihood and impact of potential human errors?
- Have shift work and overtime schedules been designed to minimize operator fatigue and stress?

Workplace and working environment
- Have posture, movement, and accessibility been considered for both operations and maintenance activities?
- Have environmental conditions (e.g., noise, temperature, illumination, etc.) been considered?
- Have employees made modifications to existing systems that would indicate failure to apply human factors principles in the original design?

Training and education
- Have people been assigned to jobs based on demonstration of required skills?
- Have process-related training requirements been defined based on job requirements?
- Have training modules and methods been developed?
- Does training include hands-on exercises or simulations?
- Has the training been documented?
- Have employees received training in human factors?
- Are there training programs and support services to help employees with controlled substance use or abuse, or mental health problems?
- Have supervisors been trained in detecting the effects of substance abuse and stress on the performance or personnel?

Table 12-4. Examples of Items to Consider in a Human Factors Review (*Continued*)

Procedures

- Are procedures clear and complete, consistent in format and terminology, and compatible with the comprehension level of the user?
- Is there a system in place to ensure that procedures are periodically reviewed and updated based on process changes, the results of job-task analyses, or investigation of process incidents and near misses?
- Have operators been involved in developing and/or reviewing operating procedures?
- Have critical operating procedures been clearly identified as such?
- Do procedures properly account for other activities for which the operator may be responsible at the same time?
- Have procedures been reviewed relative to the response time available to the operator to correct a problem?

Reprinted from Guidelines for Auditing Process Safety Management Systems, with permission. © 1993 by the American Institute of Chemical Engineers.

available on site should be identified separately from that available from the corporate office, outside consultants, or governmental agencies.

Design Considerations

During each step of designing a new process or facility, the facility's policies or standards should promote consideration of human factors. To verify this, the auditor should ask engineering and management about the design standards followed for several recent projects. Also, the auditor should review any written process design procedures or ongoing reviews of process operation to look for inclusion of human factors principles. Design guidelines should address potential effects of environmental conditions and job requirements on the human-equipment interface.

Once the auditor understands how a company's or plant's design review should operate, he or she then examines operating areas to see whether the procedures function as intended. The auditor may identify instances of poor human factors design that may indicate a management system failure.

Operating Culture

Stress and fatigue often result when a system fails to consider how people and the existing organizational structure fit. The auditor should interview personnel and review design procedures to verify that psychological as well as physical demands of the system or process are considered. Problem areas include activities that workers try to avoid. Also, the effects of shift work, overtime, and other demands on employees should be considered. The auditor should verify that specified limits on overtime are followed.

The principles of human factors are integral to the operating environment, and programs should address factors that can impair performance—especially using drugs or alcohol on the job. The auditor should interview facility personnel to find out whether employee assistance programs exist. If such programs do exist, the auditor should interview employees to assess their accessibility and effectiveness.

Operating Procedures

A review of the facility's operating procedures can reveal how effectively human factors are incorporated into the operating environment. The auditor should ensure that these procedures are clear and complete, consistent, and comprehensible. Using interviews and documentation, the auditor can determine whether operators participate in writing or reviewing operating procedures.

In addition, the auditor should determine whether there is a formal means of periodically reviewing and updating procedures from a human factors perspective. The updating is based on process changes, the results of job safety analyses, employee suggestions, and incidents influenced by human factors. The auditor should interview operators to verify that the operating procedures are complete and well matched to actual performance. The auditor should also determine that procedures help operators identify and respond to upsets. For critical procedures, the following questions should be addressed:

- Do operators have adequate time to respond?
- Is there adequate verification that the job has been done properly (i.e., by cross checking by others or through instrumentation or controls)?
- Has parallel review of maintenance procedures been conducted?

Environmental Conditions

The industrial environment may inhibit the operator's ability to perform. Ameliorating severe environmental conditions must be part of an effective human factors program.

Lighting

With improper lighting, objects may be seen incorrectly or not at all. In addition to eyestrain, process incidents may result. The audit should include an examination of critical visual tasks and a determination that the lighting allows these to be done safely. The auditor should determine the lighting standards that are used and whether lighting surveys are done. Interviews with operators can help illuminate problems with illumination.

Other Factors

Excessive noise can interfere with communication, cause psychological stress, and alter workers' performance. Temperature extremes can also affect a worker's performance. Hot conditions may reduce productivity because of fatigue or discomfort or may even discourage work. Cold weather outdoor work can lead to decreased productivity and potential errors because gloves sacrifice dexterity. The auditor should use interviews and documentation to ensure that such factors are considered in new facilities and processes and during process modifications.

Process Control Issues

Process control design issues have become increasingly important with the growing use of automation and computer control. Operators must interact effectively with control systems to ensure that the process runs safely and efficiently.

Display Design and Layout.   Displays should be located and spaced by function, relative importance, and relationship to associated controls. Factors such as viewing distance, the number of displays, legibility, labeling, lighting, and glare should be considered when selecting displays. Their configuration should facilitate monitoring, comparison, and sequencing, so that normal, out-of-normal, and emergency situations are readily apparent.

The auditor should interview those who design and install the displays to determine what consideration has been given human factors issues. In addition, the auditor should determine whether the operators feel that the displays are clear, properly located, and provide information properly.

Alarms.   Critical alarms must be differentiated from those that merely provide information, and operators must understand which alarms are critical and why. The system must prioritize alarms and determine which alarm sounds first when there is a multiple alert. The auditor should conduct interviews to find out how decisions are made regarding alarm system design. The auditor should determine whether the system design identifies critical process safety parameters, alarm criticality, and possible suppression of noncritical alarms during multiple alarms.

Match between Process and Terminal Displays.   For computer displays to communicate effectively, graphics and symbols must match process flows and system components. The auditor should interview operators on the effectiveness of computer displays in presenting process information. The auditor should also verify that there is a system to update displays when a change in processes design or operation occurs.

Monitoring Multiple Screens.   The ability to monitor multiple screens is essential to tracking process parameters. Screen position should facilitate viewing without requiring the operator to move about unnecessarily. By interviewing operators, the auditor can determine whether these conditions are met. System designers must consider these requirements during design and computer program development. This should be confirmed by the auditor in interviews.

Training and Performance

Process safety management requires that persons in an organization understand their roles and have both the knowledge and skills to perform them. As such, training is critical to a process safety management system.

The role of the auditor is to assess the existence and functioning of a system for establishing:

- The type of employee who needs training
- The type of training required
- The frequency of training
- Documenting training completed

Those in charge of the training program should be responsible for measuring the effectiveness of training. The auditor should not evaluate effectiveness. He or she should only examine system quality and verify that effectiveness is being checked.

Needs Assessment

There are two types of process safety training:

- General training—covering topics that everyone at a facility needs to know
- Specific training—covering topics important to particular employees

To establish a training program, an assessment must be made to determine who should be trained, in what, and how frequently. The auditor can begin to define the management system for training and to understand the needs analysis by asking questions, such as:

- How are process safety training objectives set?
- At what level of the organization?
- How frequently is this assessment made, and by whom?
- At what level of the organization does this training take place?
- What subjects are included?

- After development of the training materials, who has the responsibility and authority for validating their completeness and accuracy?
- Does validation actually take place?
- Is the training program documented in a syllabus that is adequate? Updated? Used?

Once overall process safety training needs are determined and implementation mechanisms are established, specific needs are set for a site, unit, department, function, or individual. The auditor should verify who determines these needs, the methodology used, and how frequently this assessment is updated.

Most process safety training includes new and transferred employees. The auditor should review the training requirements for such workers and for those who fill in when someone is ill or on vacation. Contractors, too, must be trained before working on site. Training includes notice of special hazards. The auditor should confirm that such training is done. The system that the company uses to ensure that contract employees are trained by the contractor should also be reviewed.

Program Content and Presentation

The design of a training program should be based on a needs assessment. However, to be effective, training must be integrated to ensure consistency among subjects and plant functions. The auditor should confirm that there is a guideline for deciding who presents each course. The instructors' qualifications for courses they have taught should be examined. The auditor should also verify that training modules address identified needs. Course documentation should permit the auditor to understand the essence of the training program.

General Process Safety Training. This should cover plant safety rules; emergency alarm signals such as alerts, evacuation, and all clear; and smoking and no smoking areas.

Process-Specific Training. Training subjects covered here deal with basic process chemistry, design, and critical parameters that affect the reactions and productivity. This training should be given to personnel involved with the process, including operators, maintenance personnel, and contractors.

Task-Specific Training for Operators. Such training generally starts with basic principles and then becomes specific to the process. For instance, general training could address common unit operations found on the site. Once such general information is taught, trainees will have a basis for specific unit training.

Specific unit training starts with details of units, such as how a specific unit fits in the overall process. Next, normal operation is covered, including standard operating procedures (SOPs). SOPs include:

- Established safe operating limits
- Consequences of operations outside these limits
- Start-up procedures
- Normal operation and shutdown
- Emergency procedures
- Equipment troubleshooting

Training Maintenance and Contractor Personnel.   Training maintenance personnel in process areas should include the general process safety training and relevant aspects of process-specific training. The curriculum then considers standard maintenance operations, such as working in specific hazardous locations and the needed tools and techniques. Training in preventive maintenance should also be included. Facility staff should provide contractors with process-hazard and site-specific information so they can train their employees. The auditor should verify that there is a system to ensure that contractors are trained.

Training for the Operations and Maintenance Interface.   Operations generally prepares the equipment for maintenance. Therefore, both groups should be trained in work authorization procedures and safe work practices. For example, operations needs to know how to prepare equipment for maintenance and testing and how to verify that it is safe before start-up.

Process Safety Management Training.   Also, some persons need training in management of change procedures. Process safety training should include techniques for conducting process hazards analysis and risk assessment. Training in accident and incident investigation is also necessary for some individuals.

Training Frequency

These programs should be held often enough to maintain skills. The auditor should confirm that training frequency requirements are specified and are being met.

Training Records

The auditor should verify that training records include a description of the training course, the date it was presented, the instructor, and the attendees. Attendance records should be compared with employee rosters or training schedules to confirm that all employees are trained within a specified period. The system must provide makeup sessions. Training records may be kept in a central location or in the individual departments. Where practical, the records should be kept by the individual.

Training Program Effectiveness

There are short- and long-term indicators of effectiveness. Short-term indicators are written and oral tests and field demonstrations. Long-term indicators are random spot-check tests, incident reports, log sheets, and log books. The auditor should interview employees who have had different types of training to obtain their view.

Emergency Response Planning

Emergency response planning can mitigate the effects or control of process upsets, fires, explosions, chemical releases, and other sudden, unplanned events that might result in damage or loss. The objectives of emergency response planning are to try to prevent or limit losses in the following areas:

- Acute health effects to workers, emergency responders, and the public
- Environmental damage
- Property, equipment, or product damage
- Production loss
- Loss of goodwill and public trust
- Third-party liabilities

The auditor's role is to assess a facility's emergency response plan to determine whether there is a system in place to address the site's hazards, develop adequate plans, implement the plans, train plant personnel, and document all of the above.

Needs Analysis

The auditor should verify that the system for emergency response planning is founded in the potential hazards identified by hazards analyses and risk assessments. Historical incident data are another source for developing response plans. Factors influencing the identification of emergencies and subsequent development of plans include the nature of the process hazards, potential consequences, proximity to public and environmentally sensitive areas, and internal and external resources such as staff, emergency response agencies, and equipment. The auditor should verify that plans address these issues and that there is a system to review and update emergency response plans when facility modifications change the nature of the hazards.

Emergency response planning is affected by a host of regulations. An auditor should verify that there are means to identify and address those that apply.

Emergency Response Plan Content

Most facilities have a variety of emergency response plans and procedures for spills, fires, chemical release, and the like. The auditor should identify the plans of the facility. At a minimum, the emergency response plan should be examined to ensure that it addresses safe control of processes in emergency conditions and provides training for employees and contractors. Minimum provisions include:

- Alarm and notification
- Emergency evacuation or shelter
- Spill containment and control
- Loss of critical power and utilities
- First aid medical care
- Response procedures to fires, explosions, and chemical releases

The auditor should be sure that the basic elements of emergency response planning are in place, as outlined in Fig. 12-9. In reviewing emergency response planning documents, the auditor should see if the management system can respond to a major emergency and whether planning details, such as strategy, resources, and organization, facilitate training and implementation of the plan. A complete emergency response plan should provide mechanisms for accomplishing each of the emergency response planning elements.

Auditing Emergency Response Planning

In preparing for the audit, the auditor should identify those responsible for emergency response planning. Typical documentation, which should be available to the auditor, includes:

- Facility description, including organization and staffing
- Emergency response plans
- Plot plans
- Policy regarding emergency response planning
- Description of emergency systems and equipment
- Emergency organization plan
- Description of process hazards
- Description of external resources and support organizations
- Regulations applicable to the facility

These items may be included in a comprehensive emergency response plan document or in separate documents. An audit of this element should include the following steps:

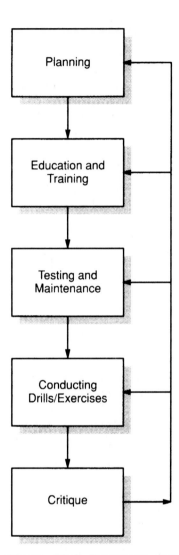

Figure 12-9. Emergency response planning elements. (*Reprinted from Guidelines for Auditing Process Safety Management Systems, with permission. © 1993 by the American Institute of Chemical Engineers.*)

- Review of documentation of the facility organization and staffing as described earlier
- Review of emergency action plans and procedures (including external planning efforts)
- Interviewing department managers to understand the management systems that are in place

- Sampling records for compliance with procedures and completeness
- Interviewing emergency response personnel, other employees, and contractors to verify that training requirements are met
- Sampling emergency systems and equipment specified in the emergency response plans
- Review of critiques on conducted drills and exercises

The auditor should verify that the facility has an incident command system to manage emergency responses. Based on the information gathered above, the auditor should then examine emergency plans to see whether each of the following areas is addressed and that there are systems to specifically teach each:

- Integrated communications
- Comprehensive resource management
- Predesignated command facilities
- Consolidated action plans
- Manageable span of control
- Modular organization

Another set of emergency response plan considerations is enumerated in Table 12-5. An audit of these components should include identifying components in place; determining a sampling strategy for each group of components; and verifying through sampling that the components in the plan are in place, maintained, and functioning. The auditor should verify that the required tests and inspections for maintenance of emergency response equipment are conducted and documented.

Training and Records

Training requirements should be documented, and the auditor should sample training records to verify that training is done according to those requirements.

Emergency Response Plan Effectiveness

A facility with a well-executed process safety management system will not have the number or size of incidents that would keep emergency responders skilled in their assigned duties. An emergency plan, to be effective, must be tested periodically by exercises and drills. The auditor should verify that exercise schedules are met, that the exercise results are documented, and that changes are made when appropriate.

Table 12-5. Typical Emergency Response Plan Considerations

A. Company policy and plan objectives

B. Facility planning basis

C. Emergency response organization—structure and duties

D. Detection, alarm, and notification procedures

E. Emergency communication systems

F. On-site evacuation and security

G. Emergency facility shutdown procedures

H. Medical emergency procedures

I. On-site emergency response teams

J. Personal protection of response teams

K. Fire response procedures

L. Spill containment and cleanup procedures

M. Environmental and spill monitoring

N. Public relations in emergencies

O. Application of plan to natural hazards

P. Off-site postincident recovery

Q. Off-site sources of assistance

R. Resource listings—supplies and supplemental services

S. Hazardous material data sources

Reprinted from Guidelines for Auditing Process Safety Management Systems, with permission. © 1993 by the American Institute of Chemical Engineers.

References

1. Center for Chemical Process Safety, "Guidelines for Auditing Process Safety Management Systems," American Institute of Chemical Engineers, New York, 1993.

2. Center for Chemical Process Safety, "Plant Guidelines for Technical Management of Chemical Process Safety," American Institute of Chemical Engineers, New York, 1992.

3. Anthony, Robert N., "Management Control Systems," Richard D. Irwin, Inc., Homewood, Ill., 1984.

4. Dermer, Jerry, "Management Planning and Control Systems," Richard D. Irwin, Inc., Homewood, Ill., 1977.

5. Eilon, Samuel, "Management Control," Macmillan and Company, Ltd., London, 1971.

6. Center for Chemical Process Safety, "Guidelines for Technical Management of Chemical Process Safety," American Institute of Chemical Engineers, New York, 1989.

7. Center for Chemical Process Safety, "Guidelines for Hazard Evaluation Procedures:

Second Edition with Worked Examples," American Institute of Chemical Engineers, New York, 1989.

8. John Gray Institute (JGI), Managing Workplace Safety and Health: The Case of Contract Labor in the U.S. Petrochemical Industry, July 1991.

9. Center for Chemical Process Safety, "Guidelines for Chemical Process Quantitative Risk Assessment," American Institute of Chemical Engineers, New York, 1989.

10. Center for Chemical Process Safety, "Guidelines for Investigating Chemical Process Incidents," American Institute of Chemical Engineers, New York, 1992.

PART 6

Industry-Specific Approaches to EHS Auditing

13

Allied-Signal's Health, Safety, and Environmental Audit Program

Ralph L. Rhodes

Director, Corporate Health, Safety, and
Environmental Audit, Allied-Signal

The Allied-Signal health, safety, and environmental audit program was launched in 1978, following a series of costly and embarrassing environmental incidents at facilities of the then Allied Chemical Company. A special committee of the board of directors, which had led a corporationwide review of health, safety, and environment (HS&E) programs, concluded that while programs and practices were basically sound, additional formality and management oversight would provide greater assurance of continuing compliance with regulatory requirements and the corporation's internal policies and procedures.

The HS&E audit program was one of several new programs developed in response to the review findings. The program was modeled on operational audit practices that had been practiced in industry since the early 1940s. The writings of Larry Sawyer, a widely respected internal audit program innovator and interpreter of sound audit principles, were particularly influential in the design of the program. Over the first 5 years of the program, procedures were developed and refined. Acceptance by line management (which had been understandably reserved initially) began to grow. The number of audits grew from 3 in 1978 to 37 in 1983. Early in this period Arthur D. Little, Inc. was retained to assist in program development and to serve a third-party auditor role in the program, reporting independently to the board of directors.

Since 1978, both the corporation and the HS&E audit program have changed greatly. The corporation evolved from a chemical company with $3 billion in annual sales to a diversified manufacturer of aerospace and automotive products and producer of fibers and other engineered materials with $12 billion in annual sales. The HS&E audit program is now a state-of-the-art program that occupies an important and respected position in the corporation's health, safety, and environmental management program.

Background

The corporation operates 240 manufacturing facilities with 100,000 employees worldwide. It is organized into three sectors: Automotive, Aerospace, and Engineered Materials. Principal products include automotive and truck brake systems; safety restraints; aircraft engines, components, and avionics; fibers for industrial and home use; and specialty chemicals.

The Corporate Health, Safety, and Environmental Sciences Department is headed by a vice president who reports to the chief operating officer. The department, organized as shown in Fig. 13-1, is responsible for overseeing and coordinating the corporation's health, safety, and environmental programs.

Each sector is supported by environmental, health, and safety staff (structured like the corporate staff but without an auditing function). Each sector also has a safety function and a medical services function. The sector staff is responsible for

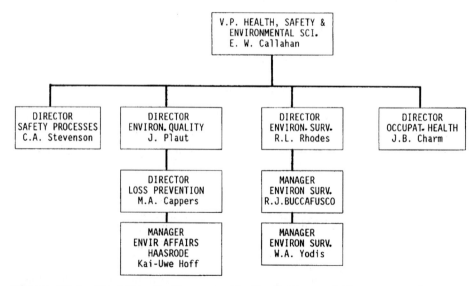

Figure 13-1. Allied-Signal's Corporate Health, Safety, and Environmental Science Department organization.

day-to-day health, safety, and environmental compliance in their sector. Each group is headed by a director or vice president who reports at a high level within the sector organization. A dotted-line reporting relationship exists with corporate counterparts. HS&E personnel assigned to specific facilities report to facility management but have a dotted-line reporting relationship to sector health, safety, and environmental counterparts.

About 400 full-time HS&E professionals are employed corporatewide. Large locations have HS&E organizations structured like sector organizations. At small locations, HS&E functions are performed by personnel on a part-time basis with support from business-unit HS&E professionals.

Health, safety, and environmental programs are guided by a formal health, safety, and environmental policy issued by the chief executive and chief operating officers (Fig. 13-2). The policy provides for programs (1) to ensure that applicable laws and regulations are known and obeyed and (2) for the adoption of the corporation's own standards where laws and regulations may not be adequately protective or where laws do not exist. Detailed guidance is provided in written corporate guidelines for each discipline (e.g., pollution control, safety, and health). In addition, sectors, divisions, and individual facilities maintain written operating procedures to supplement and elaborate on the corporate guidelines and addresses health, safety, and environmental concerns specific to the individual businesses.

Program Objectives

The objective of the Allied-Signal health, safety, and environmental audit program is to provide independent verification to management and the board of directors that:

- The corporation's operations are in compliance with the law and with corporate policies and procedures
- Systems are in place to ensure continued compliance

While the program's primary customers are the senior management of the corporation, reporting elements of the programs are designed to communicate information gathered in the course of audits to all involved management to assure them that their health, safety, and environmental responsibilities are being met.

Organization and Staffing

The original design of the audit program was based on several key objectives:

- Independence of the audit program from the day-to-day management of corporate and sector environmental programs, yet organizationally located

Health, Safety and Environmental Policy

It is the worldwide policy of Allied-Signal Inc. to design, manufacture and distribute all its products and to handle and dispose of all materials without creating unacceptable health, safety or environmental risks. The corporation will:

- Establish and maintain programs to assure that laws and regulations applicable to its products and operations are known and obeyed;

- Adopt its own standards where laws or regulations may not exist or be adequately protective;

- Conserve resources and energy, minimize the use of hazardous materials and reduce wastes; and

- Stop the manufacture or distribution of any product or cease any operation if the health, safety or environmental risks or costs are unacceptable.

To carry out this policy, the corporation will:

1. Identify and control any health, safety or environmental hazards related to its operations and products;

2. Safeguard employees, customers and the public from injuries or health hazards, protect the corporation's assets and continuity of operations, and protect the environment by conducting programs for safety and loss prevention, product safety and integrity, occupational health, and pollution prevention and control, and by formally reviewing the effectiveness of such programs;

3. Conduct and support scientific research on the health, safety and environmental effects of materials and products handled and sold by the corporation; and

4. Share promptly with employees, the public, suppliers, customers, government agencies, the scientific community and others significant health, safety or environmental hazards of its products and operations.

Every employee is expected to adhere to the spirit as well as the letter of this policy. Managers have a special obligation to keep informed about health, safety and environmental risks and standards, so that they can operate safe and environmentally sound facilities, produce quality products and advise higher management promptly of any adverse situation which comes to their attention.

Alan Belzer

Alan Belzer
President
and Chief Operating Officer

Larry Bossidy

Larry Bossidy
Chairman of the Board
and Chief Executive Officer   *Revised April 1992*

Figure 13-2. Allied-Signal's corporate HSE policy.

where communication and resolution of problems and differences of opinion between auditors and program managers would be most efficient.

- Minimizing the full-time staffing commitment to the HS&E audit program, yet having a readily available supply of competent, objective team members, continuity in the conduct of audits, and long-term accountability for the program.

Several design options considered were:

- Relying on external auditors (consultants)
- Establishing an independent internal group housed within the corporate Audit Department or the corporate Health, Safety, and Environmental Sciences Department
- Using task forces made up of persons drawn from throughout the corporation

Each option had advantages and disadvantages. For example:

- An external auditor would not require the addition of any full-time employees and would have a high degree of independence, yet would probably involve higher total costs. Additionally, coordination and communication with an external group was likely to be more difficult than with an internal group.
- An independent group within the corporate Audit Department would have a high degree of independence from health, safety, and environmental program management, but would require staffing with technically qualified HS&E personnel.
- A separate, independent group of full-time auditors within the corporate Health, Safety, and Environmental Sciences Department would have good opportunity for communication with health, safety, and environmental management, but its independence might be questioned.
- A task force would have broad participation and high flexibility in terms of level of effort and relatively low cost, but it would carry a potential for loss of continuity as task force membership changed and for disruption of the regular functions of task force members.

In order to achieve the best mix of the qualities of the options above, a composite approach was adopted and remains in use today. The health, safety, and environmental audit program was established within the corporate Health, Safety, and Environmental Sciences Department and is now staffed by four full-time professionals. To ensure continuity and accountability, the team leader for each audit is one of those four professionals. The remainder of the audit team (which varies from two to four people depending on the review scope and size of facility) is composed of:

- Corporate and sector health, safety, and environmental professionals familiar with the review subject but not directly involved in the programs being reviewed

- A representative of Arthur D. Little (which provides the advantages of an external auditor)

Audit team members drawn from within the corporation are members of a specially selected and trained cadre of environmental auditors who participate in from two to six audits each year in addition to carrying out regular health, safety, or environmental functions in a plant or business organization.

Arthur D. Little personnel participate as audit team members and, as a second responsibility, observe the conduct of the audit from the perspective of a third-party auditor. These observations are periodically reported to the board of directors by the Arthur D. Little case leader.

Audit Scope and Frequency

All of Allied-Signal's facilities worldwide fall within the scope of the audit program. Operations assessed as having lower health, safety, and environmental risk receive less attention than those assessed as high-risk operations; however, all facilities are subject to audit. The functional scope of the program includes:

- Air pollution control
- Water pollution control and spill prevention
- Solid and hazardous waste disposal
- Occupational health and medical services
- Safety and loss prevention
- Product safety/product integrity

Most audits are limited to from one to three of the functional elements listed above (e.g., air pollution control only) in order to maximize the amount of in-depth review conducted in the 1-week period allocated for each audit. Thus, while the audit program covers the full range of health, safety, and environmental programs, the scope of a specific audit is often narrow.

Frequent audit of all six functional elements at 240 locations would be a formidable task. Instead, an audit sample is selected each year that represents all major business units, geographic areas in which the corporation operates, and functional elements. Since the health, safety, and environmental programs at facilities worldwide are for the most part mature and are subject to ongoing oversight by sector and business unit HS&E professionals, they tend to be relatively consistent in quality within a given country, business unit, and functional element. A sample that incorporates these characteristics has been found to be satisfactory to meet program objectives.

Corporate environmental staff, the director of audit, and Arthur D. Little annually develop a selection pool of facilities to be audited that includes facilities from each of the operating companies and major business areas. Facilities are then selected randomly using a computer program through a process that reflects their assessed environmental risk. Approximately 50 audits are selected. Audit locations and scope are chosen to represent a cross section of the corporation's business interests and health, safety, and environmental concerns where there is significant potential risk.

At the beginning of each year, the schedule of locations to receive audits during the year is sent to corporate and sector health, safety, and environmental staff. One month prior to the audit, the team leader notifies the facility manager of the functional scope of the audit.

Audit Methodology

Health, safety, and environmental audits employ a number of tools and techniques such as:

- Formal internal control questionnaires
- Formal audit protocols
- Informal interviews with facility personnel
- Physical observations
- Documentation review
- Verification

A written audit protocol has been developed for each functional element. The protocol guides the auditor methodically to an understanding of the management system audited and, ultimately, to a conclusion as to compliance with applicable criteria. Auditors carefully document their work to provide a formal record of the completion of all audit steps and a record of all deficiencies in the systems audited.

The basic phases in Allied-Signal's audit process (illustrated in Fig. 13-3) include:

Phase I: Preparation

Among the preaudit activities conducted by the audit team leader are the confirmation of audit dates and organization of the audit team based on the functional scope of the audit. One month in advance of the audit, the team leader notifies the facility manager in writing of the specific audit dates and scope. Corporate files are screened to obtain audit information on the facility and its processes (e.g., process flow diagrams, plant layout diagrams, policies and procedures, operating manuals, and permits). Regulations applicable to the facility

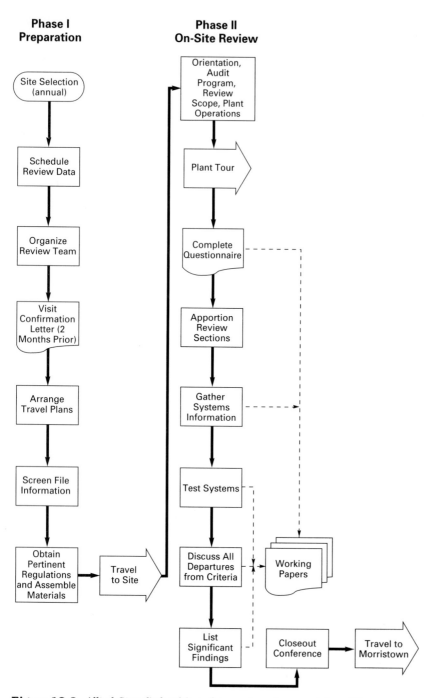

Figure 13-3. Allied-Signal's health, safety, and environmental auditing process, Phases I and II..

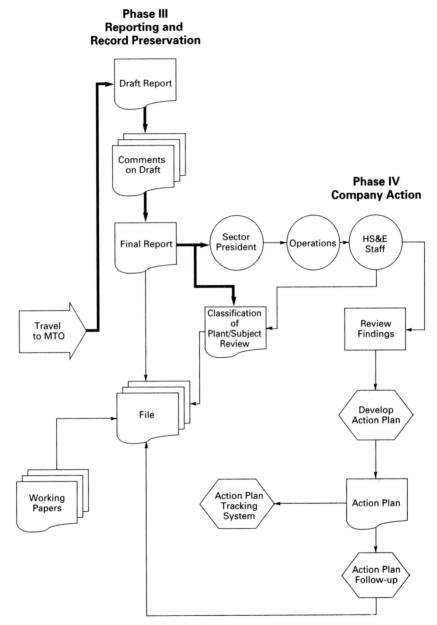

Figure 13-3. (*Continued*) Phases III and IV.

are also obtained. Team members are provided with selected information to help them prepare for their role in the audit.

Phase II: On-Site Audit

The on-site audit commences with a meeting of the audit team, the facility manager, and appropriate facility personnel. During this meeting, the team leader discusses the objectives and scope of the audit. This is followed by a facility presentation providing an overview of products, processes, facility organization, and operations of significance within the scope of the audit. The audit team then tours the facility, with a member of the facility environmental or production staff, to gain a general understanding of facility characteristics.

After the tour, the audit team and appropriate facility environmental staff meet to complete the internal controls questionnaire (see Fig. 13-4 for a typical page). This questionnaire, administered by the audit team leader, aids the auditors in developing an initial understanding of facility operations, processes, personnel responsibilities, and health, safety, and environmental management controls.

Working from the audit protocol (see Fig. 13-5 for a typical page), with major sections divided among the audit team members, each auditor gathers system information and performs relevant tests. In the course of the audit, the auditor must use sampling techniques and exercise professional judgment in selecting the type and size of samples to be used to verify that the key controls in the system under review are in place and working. No testing may be done until the system is well understood and a carefully reasoned plan of testing is worked out. The understanding may come from interviews with facility staff, review of facility operating procedures and systems, and other sources.

Testing of the systems in place can take a variety of forms. For example, verification testing for water pollution control can include:

- Visual observation of physical features related to permitted emissions and discharge monitoring or regulation performance.

- Comparison of strip charts, basic laboratory data, and discharge monitoring reports to applicable criteria.

- Review of correspondence related to regulatory issues

- Determination that composite samplers, effluent flow measuring devices, and in-place monitoring devices are properly maintained and calibrated

Each auditor carefully documents all testing plans and the results of each test. Observations and information collected are shared with fellow audit team members and plant personnel involved in the audit. Time is set aside at the end of each day for a meeting where information gathered is shared and plans are made for the next day of the audit. The audit team is instructed to continuously feed back any impressions being formed about the facility's compliance with

<u>Yes</u> <u>No</u> <u>N/A</u>

13. Is there a calibration and preventive
maintenance program for all monitoring
and pollution control equipment? ____ ____ ____

14. Does the location maintain records of
all monitoring activities including all
original strip charts and calibration and
maintenance records for at least 3 years? ____ ____ ____

15. Does the location have separate storm sewers? ____ ____ ____

16. Do storm sewers receive:

   a. Process wastewater ____ ____ ____
   b. Sanitary waste ____ ____ ____
   c. Contaminated surface water ____ ____ ____

17. a. Has any of the plant property been
used for process operation, storage or
disposal resulting in contamination by
hazardous chemicals which could leach
into groundwater or surface waters? ____ ____ ____

   b. Is there any known or suspected
groundwater contamination under or
in the vicinity of the plant site? ____ ____ ____

18. Is incoming water periodically
analyzed for pollutants controlled in
plant permits? ____ ____ ____

19. Are there any surface water flows from
adjoining properties that could influence
the quality of plant discharge(s)? ____ ____ ____

20. Does the discharge from the location
contain any detectable amounts of
Aldrin/Dieldrin/D.D.T/D.D.D./D.D.E./
Endrin/Toxaphene/Benzidene/PCBs? ____ ____ ____

21. a. Have any excursions beyond the limits
occurred for any permit parameter
within the review period? ____ ____ ____

   b. Have the appropriate agencies been
notified of all excursions in a
timely fashion and the report
documented? ____ ____ ____

22. Are all excursions and incidents promptly
reported into the Excursion/Incident
reporting system? ____ ____ ____

23. a. Are laboratory analytical procedures
for the analysis of pollutants in
compliance with the guidelines and
test procedures as outlined in 40 CFR,
part 136? ____ ____ ____

   b. Have alternate test procedures been
approved by appropriate authorities? ____ ____ ____

24. Has the location complied with the
provisions of state and local pollution
control regulations; i.e., sewer
ordinances, on-site disposal regulations,
etc.? ____ ____ ____

25. Have operators of waste treatment
facilities received appropriate
certificates? ____ ____ ____

Figure 13-4. Internal controls questionnaire, a typical page.

B. Detailed Understanding and Testing of Facility Programs

Availablity of Spill Prevention Facilities

5. Utilizing the process flow diagrams, plant layout maps, the SPCC plans and direct observation; identify and locate all oil and chemical storage and handling facilities.

    a. Record the location and capacity of all oil and chemical storage and process tanks and drum storage areas. Determine whether, based on storage capacity, an oil SPCC Plan is required (40 CFR 112.1(d)(2)).

    b. Record the location of transfer equipment, piping, process equipment, and waste handling equipment containing oil or chemicals.

    c. Record the location of all spill monitoring and control equipment.

    d. Note any locations where visible evidence of past or present spillage or leakage exists.

5.a.	5.b.	5.c.	5.d.

Spill Prevention Plans

6. Review the location's oil SPCC Plan for the following:

    a. Consideration of all applicable portions of 40 CFR 112.7.

    b. Certification of the plan by a registered professional engineer (40 CFR 112.3(d) and 112.5(c)).

    c. Consideration of all oils and oil-containing substances.

7. Review the location's chemical SPCC plan for substances identified as hazardous under 40 CFR 116. Determine if the response plan is adequate for the circumstances.

6.a.	6.b.	6.c.	7

8. Review the requirements for the location to have a Facility Response Plan under Section 311(j)(5) of the Clean Water Act. If the facility is required to prepare a plan, review the plan and verify that it contains the following elements:

    a. Identity of the person responsible for implementing the plan.

    b. A discussion of emergency scenarios with worst case discharge, potential threats and prioritized actions for each scenario.

    c. Identity of measures to remove and mitigate a "worst case" discharge.

    d. Description of requisite training, equipment testing, unannounced drills and response actions.

    e. EPA and/or Coast Guard approval.

8.

Adequacy of Spill Prevention Facilities

9. Review supporting documentation: i.e., plans, specifications, engineering studies, testing records, etc., for bulk storage and process tanks and drum storage areas.

    a. Determine that material and construction of the tanks and drums are compatible with the material stored and conditions of storage such as pressure and temperature

    b. Confirm that storage and process tank and drum storage area secondary containment volume is at least 10% greater than the volume of the largest tank contained.

Figure 13-5. Audit protocol, a typical page.

established criteria to the facility personnel involved in the audit. This continuous feedback is intended to:

- Eliminate misconceptions and false trails for the team member who may have misunderstood what he or she was originally told
- Encourage the team members to organize their thoughts
- Give facility personnel an opportunity to participate in the audit process

At the conclusion of the audit all significant findings are listed by each audit team member and are organized by the team leader on a summary sheet for discussion with facility management. The on-site audit concludes with a closeout meeting between the audit team and facility management. Each finding on the summary sheet is thoroughly discussed to assure that the finding statement is both factual and properly stated.

Phase III: Reporting and Record Preservation

Report Format.   Built on the oral reports presented during the audit, a written report is prepared on each audit. The purpose of the report is to provide information to all interested management on the more significant findings of the audit. An exception report, it includes only matters that do not conform to established criteria. The report is based on findings listed on the audit findings summary form. Findings related to regulatory standards are qualified with a statement that they have not received a detailed legal review.

The report follows a standard format that consists of four parts. Section I contains the background (who, what, where, and why) information. The next two sections include all significant instances of noncompliance with:

- Regulatory standards (federal, state/provincial, and local)
- Allied-Signal policies and procedures (corporate, sector, or facility)

The final section includes any significant deficiencies in the facility control systems that would make continued compliance with the law or company policy questionable (such as record retention, documentation, and clear assignment of environmental responsibilities).

The length of the report depends on the number of findings; typically it is three to five pages long.

Report Distribution.   The written audit report is issued in draft form one or two days after the audit by the team leader. Copies are provided to involved line and HS&E personnel at both the business unit and corporate levels, the facility manager, and the audit team. Comments on this draft report are required within

2 weeks of its issuance. When comments necessitate significant revision of the first draft, a second draft of the report may be prepared and circulated for review.

The final written report is issued to the sector president approximately one month after the audit, with copies to the law department, HS&E vice president, corporate environmental specialists, business area management, facility manager, and the audit team. The final report is accompanied by a request that the business unit respond in writing with an action plan for correction of the deficiencies noted.

Performance Classification.   When the final report is drafted, the audit team leader reviews the findings in the report and the working papers as input to the evaluation of performance for each subject audited at the facility. The auditors follow a formal procedure to classify each subject reviewed into one of four categories of performance:

A—Meets or exceeds requirements

B—Substantially meets requirements

C—Generally meets requirements with certain exceptions

D—Requires substantial improvement

The rating is developed in accordance with a formal procedure that involves rating component parts of each discipline audited. Each functional area is rated for severity and prevalence of deficiencies found. After evaluating component parts, an algorithm is used to determine the overall performance factor for the subject of the audit. The performance classification is included in the final report as a summary indicator of the status of the operations audited. The Corporate HS&E audit department uses the information as a basis for trend analysis.

Records Retention.   The HS&E Audit Department has a formal records retention policy to ensure that all records relating to audits are retained for a period of time that is consistent with their utility in the program and keeps the records volume at a manageable level. In general, audit working papers are retained for 3 years and audit reports and action plans are retained for 10 years [50 years where subject to Resource Conservation and Recovery Act (RCRA) or Comprehensive Environmental Response, Compensation and Liability Act (CERCLA) records-retention requirements].

Other Reporting.   In addition to the formal, written report on individual audits described earlier, the Corporate Responsibility Committee of the board of directors receives twice yearly reports on audit program activities. The director of HS&E audit reports on program status and any significant issues arising from audits conducted since the previous report to the committee. A representative of

Arthur D. Little provides an independent report on program management and issues identified in the course of audits.

Phase IV: Corrective Action

The job of the audit team ends with the submission and management's understanding of the report. The audit process, however, continues until those responsible for correcting any deficiencies noted have prepared an action plan for correcting the deficiencies and ultimately complete the action plan steps.

The action plan is developed by facility personnel and approved by line management and the sector health, safety, and environmental director. A copy of the action plan is provided to the HS&E audit director and other managers concerned. The HS&E audit director reviews the action plan to confirm that the final report has been understood and that the response is consistent with the findings of the report.

Action plans are typically received within two or three months of the issuance of the final written report. The plan reports on corrective actions already taken and describes those that are planned. Operating management assumes primary responsibility for follow-up and monitoring of the corrective actions. Action plan steps to be completed are entered into a corporatewide action-plan tracking system to aid in management oversight. The HS&E audit group performs follow-up reviews to confirm the completion of approximately 40 percent of these action plans. These follow-up reviews have indicated that most of the findings are corrected as planned.

Assuring Quality

Several steps are taken to ensure audit quality:

1. A member of the HS&E audit staff participates in each audit as team leader. The team leader's role includes making sure there is good communication among the team members and that each team member understands and carries out his or her role consistent with audit program procedures.

2. Allied's audit protocols provide a structured framework that guides the auditors through a series of steps designed to create an understanding of the system under review, conduct appropriate tests to confirm that the system is working, and determine specific deficiencies.

3. Audit working papers document all audit activity. The credibility of the audit depends on how well each auditor documents what he/she has done and the conclusions reached. Each team member must prepare working papers that document the information gathered in completing their portion of the audit protocol. At the end of each audit, the team leader carefully reviews, initials,

and dates each page of the working papers. The working papers are then reviewed by at least one other member of the audit staff. The working papers serve as support for the audit report and a base for evaluating the audit and the performance of each team member.

4. Arthur D. Little provides an additional quality control check. Representatives of the consulting firm participate in a number of audits sufficient to understand and report on the program quality. All audit reports are reviewed by the consultant to ensure accurate and consistent audit reporting. Working papers for audits that do not have direct Arthur D. Little participation are reviewed by A.D. Little staff. Detailed feedback is provided.

Program Benefits

The HS&E audit program provides a number of benefits. Among them are the following:

- For top management and the board of directors, the program provides independent verification that operations are in compliance with applicable requirements of environmental law and the corporation's environmental policy.

- For environmental management, the program serves as an important source of information on the status of operations and on both individual deficiencies and patterns of deficiencies. The information, along with information from other sources, is used to plan actions to correct deficiencies and provide for improved assurance of continued compliance.

- For line management, the program is a stimulus to become more familiar with the detailed implications of environmental requirements. The program identifies problems with operations that require corrective action, or, more frequently, it confirms that environmental requirements are being met.

Conclusion

The HS&E audit program is fully consistent with generally accepted audit practices. It has been designed to be consistent with the structure, needs, and culture of Allied-Signal. More than 600 audits over the past 15 years have more than adequately demonstrated the utility and reliability of the program as an element in the corporation's health, safety, and environmental management system.

14

Eastman Kodak Company's Health, Safety, and Environmental Assessment Program

Alfred E. Fields
Eastman Kodak Company

Eastman Kodak Company has sales of $20 billion, employs 130,000 people worldwide, has manufacturing or other major operations in over 50 countries, and consists of three organizations: Chemicals, Imaging, and Health.

Imaging, the largest of the three, manufactures and sells photographic films, papers, chemicals, cameras, photocopiers, printers, and other imaging products for industry, the government, and consumers.

Chemicals is represented by the Eastman Chemical Company, which manufactures chemicals, fibers, and plastics.

Health includes Sterling Winthrop, Inc., a pharmaceuticals subsidiary, Clinical Diagnostics, which makes medical diagnostic products, Health Sciences, which manufactures industrial and medical x-ray films, and L&F Products, a manufacturer of household products.

The development of an efficient and workable assessment program for a corporation with such a diverse product mix and with so many facilities and properties worldwide presented an ambitious challenge. The success of the company's corporate health, safety, and environmental (HSE) assessment program is attributable to four key factors:

- From its inception, the program has had strong involvement and support from top-level management.
- A systematic quality model was used in developing the overall corporate HSE management program (of which the assessment program is part). That same process is used on an ongoing basis to ensure continuous improvement in the program.
- Assessments for health, safety, and environmental issues are integrated into one comprehensive assessment program that looks at all three areas during the same site visit.
- Multidisciplined teams of highly qualified and well-trained experts make the integrated approach possible.

Strategic Planning

The mission of the corporate HSE assessment program, which was initiated in 1988, is to evaluate the compliance of Eastman Kodak Company organizations worldwide with corporate health, safety, and environmental policies and standards, with sound HSE management practices, and with the laws and regulations of national, regional, and local governing entities. To carry out this mission, the company developed a plan with the following major elements:

1. Use the quality leadership process to develop and implement the program.
2. Work with corporate organizations to develop a uniform, comprehensive set of performance standards for use in HSE management and assessments so that facilities worldwide will be evaluated in a consistent and equitable manner.
3. Determine the location and function of every Kodak facility and property worldwide.
4. Prepare manuals for use by the teams making the HSE assessments.
5. Identify potential assessment team members throughout the world, get commitment to use them from their management, and train them.
6. Determine the priority and frequency with which Kodak facilities should be assessed and establish the types of assessments required for each facility.
7. Establish attorney-client relationships for the assessment processes to ensure the confidentiality of the findings.
8. Develop a management-approved, effective system for reporting findings.
9. Develop a system for monitoring the progress of corrective action plans.

10. Maintain the quality of the assessment system by having outside consultants review the program periodically.

To ensure that Kodak's HSE performance remains a high priority throughout the organization, the company decided that the director of the assessment program should report functionally to the chairman of the Management Committee on Environmental Responsibility (MCER), who is the chairman, president, and chief executive officer (CEO) of the corporation, and operationally to the vice president and director, corporate health, safety, and environment. This dual reporting system keeps top management fully involved in the process and maintains high visibility for HSE issues.

The Quality Leadership Process

Kodak used the quality leadership process model (Fig. 14-1) in developing its corporate HSE management program. The corporate HSE assessment program is the verification step in that cycle. Therefore, the assessment program is set up so that—in addition to reporting on the compliance status of each facility—the findings can be used in the review step for the corporation's compliance plans, and modifications can be made as necessary.

In addition to reporting on compliance with all applicable standards, policies, and regulations, the assessment evaluates whether facilities have effective HSE management systems in place. These management systems are essential in organizations where continuous improvement is expected.

Corporate Health, Safety, and Environmental Standards

Corporate health, safety, and environmental standards are established by MCER, based on recommendations from the Health, Safety, and Environment Coordinating Committee (HSECC).

MCER—which is composed of senior Eastman Kodak Company managers and is chaired by the chairman, president and CEO—establishes health, safety, and environmental policies and performance standards for the corporation.

HSECC is chaired by the vice president and director of the corporate Health, Safety, and Environmental organization, and includes staff representatives from Corporate and the Imaging, Chemicals, and Health business groups. HSECC provides recommendations on HSE policies, performance standards, and issues.

A list of the corporate health, safety, and environment performance standards is shown in Fig. 14-2. These comprehensive standards, along with the corporate policies, are used in HSE management and assessments and allow facilities worldwide to be evaluated in a consistent and equitable manner.

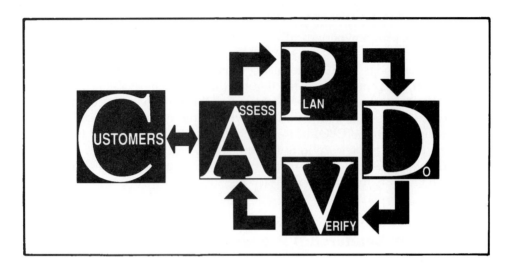

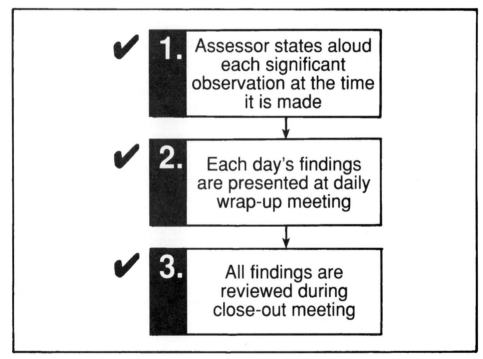

Figure 14-1. The QLP Cycle.

Medical standards
Medical surveillance examinations
Medical records
Emergency medical care
Fitness to work
Preventive medical services
Medical facilities and staff
Medical services liaison activities

Environmental standards
Air emission control
Release reporting
Waste minimization
Groundwater protection
Wastewater management
Waste management

Safety standards
Emergency preparedness and community involvement
Facility safety
Chemical process and equipment safety
Personal protective equipment
Storage, handling, and distribution of materials
Contractor safety
Equipment safety
Fire protection

Health standards
Product responsibility
Exposure assessment and control
Occupational health record systems
Ergonomics
Health, safety, and environmental education and training
Chemical management control
Company exposure limits

Figure 14-2. Corporate health, safety, and environment performance standards.

Site Locations and Prioritization

With the help of the corporate Real Estate Organization, a database was developed to track transactions and provide updated information on sites owned and/or operated by the company. Initially, a questionnaire was sent to each site so that preliminary risk assessments could be made. As a result of these assessments, sites were prioritized into a logical ranking for site visits. Risk assessments of new facilities are made using information from preacquisition audits.

Assessment Scheduling

The risk assessments drawn from the HSE questionnaires identify the facilities that should be assessed first. When a facility is scheduled for assessment, it is often advantageous to assess other sites in the geographic area at the same time in order to take advantage of shared travel expenses. Assessing various types of facilities also helps management get a better overall perspective on the range of HSE activity levels across the corporation.

The frequency of assessment is also an important issue to consider. MCER decided that the major manufacturing, research, and distribution sites will be assessed every three years. This time frame allows for corrective actions to be incorporated into the business planning cycle and for major remedial projects to occur before the next assessment. Other facilities, such as photofinishing operations, small distribution centers, and warehouses, are assessed less frequently— every three to five years.

Once the sites are identified and approved by MCER for assessment, the following issues (among others) must be considered in finalizing the assessment schedule:

- *Weather.* Snow and rain can obscure damage to the environment and make site inspections difficult.

- *Holidays.* Holidays are different throughout the world. In some areas (parts of France, for example), most vacations are taken during one month.

- *Shutdowns.* Many plants have scheduled maintenance shutdowns. Since the team needs to observe the equipment and personnel under normal operating conditions, assessments should not be scheduled during shutdowns.

- *Immunizations.* The assessment schedule is reviewed by the Medical Department, which determines the types of immunizations and travel advice the assessment team members should have. Some vaccines have to be given over a period of several months to be more than 90 percent effective.

- *Visas.* Visas are necessary for some countries. Some visas may be valid for less than one year and could expire if they are obtained too early or if the assessment is postponed.

- *Security.* U.S. Department of State travel advisories are reviewed to determine if unacceptable risks exist in the country where the assessment site is located.

Assessment Team Members

The team of experts selected for a given assessment depends on the operations of the facility. The pool of assessors comes from line and staff organizations around the world; their management allows them to participate in one to three assessments per year. This is beneficial to both the assessors and their management because assessments are a great learning opportunity. Team members become familiar with the operations of the company, critique management systems, and establish intracompany communications with their peers. Team members do not assess their own facilities, however.

Kodak takes particular pride in the diversity among the members of its assessment teams. Members come from both U.S. and international operations. The pool of team members includes minorities, with African-Americans, Native Americans, Indians, Hispanics, and Japanese-Americans represented. Many of the team members are women.

Team members are sent to the Arthur D. Little, Inc. course Mastering Environmental, Health, and Safety Auditing Skills and Techniques. There they learn interviewing and observational techniques and the overall assessment process.

A team may range in size from three to 13 members, depending on the size of the facility and the number of topics that must be addressed. Typically, a team will include members with expertise in such areas as:

- Industrial safety
- Industrial hygiene
- Occupational medicine
- Environment
- Biosafety
- Process safety
- Legal/regulatory issues

Team members from other disciplines are added as required. Cross-trained team members are used at small sites to minimize the number of assessors at the facility.

Each team includes a consultant (not a Kodak employee) who brings additional technical expertise and who helps to continually evaluate and improve the assessment process itself. The consultant assists team members during the assessment, helps develop findings, and reviews other team members' findings to assure that they are correctly documented. When possible, the consultant is from the country in which the facility is located. In non-English-speaking countries, the consultant also acts as a translator.

Team leaders coordinate the assessment at the site. At minimum, they are unit directors and are selected because they have good communication skills and leadership qualities, and they participate in Kodak's in-house team leader training program. The team leader's duties are to:

- Establish and maintain communications with site management
- Coordinate logistics at the site
- Help new team members get started
- Coach team members on assessment techniques
- Chair the opening meeting and daily wrap-up meetings
- Present the team's findings at the closeout meeting
- Act as an assessor at small sites and help other assessors finish on time

An integrated, comprehensive HSE assessment under one team leader has a synergistic effect. If health and safety assessments were separate from the environmental assessment, one group might observe things that are the responsibility of the other, which often results in lost findings. But by encouraging the health people to look for safety and environmental issues, for instance, the team can develop a much more complete picture of the health, safety, and environmental issues at the site.

In the course of one assessment, for example, the environmental person was investigating emission sources on the roof of a building. Nearby, two employees opened a cover on an exhaust system to change a filter and were knocked back by the fumes. The environmental assessor immediately called the industrial hygienist and medical person to investigate. If this had been a purely environmental assessment, the opportunity to investigate the problem might have been lost.

The comprehensive approach of the corporate assessment program also serves as a model for sites as they work to improve their HSE management processes.

Protocol Manuals

Arthur D. Little, Inc. worked with Kodak to prepare two manuals for assessors. The first, which is used for assessments in the United States, contains auditing procedures, questions, and checklists, and references U.S. regulations. (Figure 14-3 lists the topics covered in that manual.) The second manual, which is more generic and omits U.S. regulations, is used to assess facilities in other countries. These standardized assessment protocols provide consistent evaluation criteria, no matter where the facility is or who is on the assessment team.

Regulations

One of the most difficult tasks is obtaining up-to-date regulatory information. This is especially true in non-English-speaking countries, where language barri-

Environment
  Air pollution
  Emergency planning
  Polychlorinated biphenyl (PCB) management
  Spill control and emergency planning
  Underground storage tanks
  Waste—solid and hazardous
  Water pollution control

Safety
  Employee safety
  Loss prevention and emergency response

Process safety
  Industrial hygiene
  Radiation safety
  Drinking water
  Ergonomics
  Industrial hygiene
  Laboratory health and safety

Biosafety

Medical
  Occupational medicine

Other
  Product safety/chemical control (Toxic Substances Control Act [TSCA],
  Federal Food and Drug Administration [FDA], Federal Insecticide, Fungi-
  cide, and Rodenticide Act [FIFRA])

Figure 14-3. Topics in assessors' manual for the United States.

ers may impede the search for information. Site HSE personnel can be helpful in obtaining regulations, but the team often draws on independent sources to verify that they have all the applicable regulations. Consultants and local law firms typically provide Eastman Kodak Company with regulatory information for international assessments. Other information sources used include:

- RegScan, by Regulation Scanning, Williamsport, Pa. This database, containing United States federal HSE regulations, runs on laptop computers. It is updated monthly.

- Bureau of National Affairs, Inc. (BNA), Washington, D.C. Hard copies of state regulations are available from BNA.

- Enflex Info, ERM Computer Services, Exton, Pa. This system contains federal, state, and some international regulations on a compact disc (CD).

The Assessment

Preassessment Visit

Larger facilities often receive an informal preassessment visit from the director of the corporate HSE assessment program. This is an excellent opportunity for site management to learn more about the assessment process and what the assessment team members will need during their visit, such as:

- Who will they need to interview?
- What areas will they need to see?
- What activities or operations will they need to observe?
- What records or other documentation will they need to review?
- How much time will their activities require?

Preassessment Questionnaire

Each facility receives a detailed preassessment questionnaire prior to the assessment. The questionnaire requests specific information that the assessment team will need in order to prepare for the site visit. It outlines the records and other documentation that may be requested during the assessment, including input on applicable regional and local laws and regulations.

Assessment team members review these materials before going to the site so that their time at the facility can be used most efficiently.

The Site Visit

The site visit usually takes from two to five days. The visit begins with an informational meeting between the assessment team and the facility managers, supervisors, and site health, safety, environmental, and medical staff members. A 30-minute videotape is used to help the site personnel understand more about the assessment process and what the team will do during the visit. The video—which was customized for Kodak and produced by Environmental Video Products of San Francisco—also acts as a refresher for the assessors.

The opening meeting is followed by a group tour of the facility. Next, individual team members—using the established protocols as a guide—conduct in-depth interviews and make observations and investigations, taking extensive notes. Each team member is usually accompanied by a site escort who can assist in interpreting site operations. Interviews can be very beneficial to the site people because they provide an excellent opportunity to discuss current operations and suggestions for improvement.

At the end of each day, the assessment team meets with site management to review the day's findings. These daily wrap-up meetings allow site management to ask questions, clarify any information that might have been misinterpreted, and provide additional input where the information is incomplete.

The team's findings are entered on-site into a laptop computer equipped with Microsoft™ Word for Windows® software. Clearly legible overheads summarizing the findings are then produced on a Kodak DICONIX™ printer.

On the final day of the assessment, the team leader presents the team's findings to management at a closeout review meeting. Once again, site management has an excellent opportunity to ask questions, verify findings, and provide input into the final report.

Verifying the Findings

One of the more challenging and time-consuming aspects of HSE assessments occurs when site personnel dispute (sometimes at great length) the assessors' findings during the closeout meeting. Kodak's assessment incorporates a three-step verification process that has proven very helpful in minimizing disputes relating to a given finding. Figure 14-4 outlines those steps.

No Surprises

Site management, then, is fully aware of all the assessors' findings long before the assessment report is released; the report should contain no surprises at all. The three-step process provides ample opportunity to discuss each finding, but because the findings are thoroughly and clearly verified, there is very little dispute over the facts uncovered during the assessment.

Checkpoint 1

Each assessor is accompanied by a site escort during the assessment visit. The assessor mentions each significant observation aloud at the time it is noted (e.g., Some line workers are not wearing steel-toed shoes). The escort is fully aware of everything the assessor notes.

Checkpoint 2

Each of the day's findings is presented at the daily wrap-up meeting. Site personnel have an opportunity to ask questions and to clarify the findings at that time. The site escort can verify everything the assessor noted.

Checkpoint 3

The findings are reviewed a third time at the closeout meeting at the end of the assessment visit. Since no new findings are brought up at this time, the meeting generally runs quite smoothly.

Figure 14-4. A three-step process to verify the findings.

The Assessment Report

With the approval of the legal department, a draft report is left at the site after the assessment so that remediation activities can begin immediately. The final report, which is sent to the organization's legal department within four weeks of the assessment visit, does not make recommendations for corrective action. Its function is to establish the current status of the facility with respect to the assessment criteria. Typically 30 to 100 pages long, the final report is confidential and protected by attorney-client privilege. The information it contains is available only on a need-to-know basis.

The assessment report must be useful to both the MCER and to the assessed facility. A guidebook provides tips on writing reports in a clear and concise manner, using factual, consistent, and neutral statements that are legally acceptable. Below each finding, the appropriate regulation, corporate policy or standard, or professional judgment statement is cited in parentheses. That information is for reference purposes only and does not imply violation of the standard, law, policy, or regulation. A sample report outline is shown in Fig. 14-5.

Postassessment Meeting

Six weeks after the site receives the formal report, a postassessment meeting is held—on-site or by teleconference—to discuss the major assessment findings, plans for corrective action, and progress made. The meeting includes the following people:

- Organization manager
- Site manager
- Health, safety, and environmental affairs director for the organization

---

1. Title page
2. Reviewer signature page
3. Table of contents
4. List of assessment team members
5. Introduction
6. Executive summary
7. Positive findings section
8. Findings
9. Appendix (professional judgment findings)

Figure 14-5. Sample report outline.

- Director of environmental, health, and safety legal staff
- Director of corporate health, safety, and environment
- Director of the corporate health, safety, and environmental assessment program

Attorney-Client Privilege

The assessment program, which is operated under attorney-client privilege to ensure confidentiality of the findings, functions as an agent of the legal department to obtain information on the state of compliance of each site. The team leader sends the report to the legal department, which distributes it to the MCER and to the facilities. All copies are stamped, and distribution is strictly controlled to comply with the conditions necessary to establish the attorney-client privilege.

Since the attorney-client privilege does not exist in all countries, it is important to determine what degree of confidentiality can be obtained for the assessment report. Penalties for failure to comply with health, safety, and environmental laws can be very severe in some countries (Australia, for example), and the assessment report could establish past or present noncompliance.

Corrective Action Plans

The key to any effective assessment program is an equally effective program for appropriate corrective action and follow-up. Once a negative assessment finding has been established, it is essential that timely response and follow-up occur.

Failure to address assessment findings will ultimately place the corporation in the untenable position of being in possession of knowledge of possible noncompliance with no plans for correction. Therefore, when the site managers receive the report, they must work effectively with their health, safety, and environmental affairs directors and HSE staffs to develop corrective action plans. It is the responsibility of site management to ensure that corrective action takes place. The health, safety, and environmental affairs directors of the organizations routinely submit progress reports to their management.

The corporate HSE assessment program director keeps track of the number of outstanding corrective-action items for the corporation and publishes a quarterly report for the organizations' legal departments and the Health, Safety, and Environmental Coordinating Committee. These reports are reviewed periodically by the MCER.

Program Evaluation

A questionnaire is sent to site managers within three weeks of an assessment visit for their evaluation of the team's performance while at the site. The completed questionnaire is sent directly to the vice president and director of the corporate Health, Safety, and Environment Organization.

The Corporate HSE Assessment Advisory Committee provides guidance on the structure and implementation of the corporate HSE assessment program. This committee—composed of legal and HSE representatives from corporate staffs and the business groups—is led by the vice president and director, corporate health, safety, and environment.

The overall corporate health, safety, and environmental assessment program has been evaluated by McLaren-Hart, a subsidiary of Sandoz, Inc. The firm interviewed plant managers and HSE staff members, reviewed records and reports, and observed two assessments, one of which was outside the United States. They also evaluated the program against the U.S. Environmental Protection Agency (EPA) Auditing Policy Statement and against generally accepted practices in comparable companies. While the evaluation team offered some suggestions for improvement, their rating of Kodak's assessment program was excellent.

The corporate HSE assessment program uses all these sources of input for self-evaluation:

- Feedback from site managers.
- Feedback from assessment team members.
- Periodic third-party assessments.
- Evaluations by the consultants on the assessment team.
- Feedback from HSECC and MCER.
- The Corporate HSE Assessment Advisory Committee.

The quality leadership process takes all these evaluations into account to ensure the continuous improvement of the corporate HSE assessment program.

Learning from the Kodak Experience

As companies work toward developing and improving their HSE management and assessment programs, they will do well to pay close attention to the key factors for success:

- Support from and accountability to top management
- Using a systematic quality methodology throughout
- Integrating the health, safety, and environmental areas into one comprehensive team with one central focus: efficient, effective assessments
- Continuous evaluation and review of the programs by both internal and external experts

15

Manufacturing: Environmental Auditing at 3M

Allen Dressler

Quality Assurance Manager, 3M Environmental Engineering and Pollution Control

Daniel Schmid

Auditing Supervisor, 3M Environmental Engineering and Pollution Control

3M is a transnational corporation with a long history of environmental leadership. Its corporate environmental policy, pollution prevention programs, sustainable development activities, and customer service philosophy have led to the development of a total quality environmental management (TQEM) process, and the environmental audit program is an essential component of that process. The following describes the background, approach, and audit procedures performed by the Environmental Compliance Audit staff. Health and safety audits are performed by other corporate functions.

Environmental Compliance Auditing

The environmental compliance auditing process had become a valuable instrument used by 3M management in measuring environmental performance and ensuring that the goals of continuous improvement are maintained. Since 3M began its formal environmental auditing program in 1981, the process has continued to evolve. The objectives of the environmental compliance audit function are to:

- Measure and ensure that procedures, practices, and programs comply with environmental regulations and 3M corporate policies
- Identify potential environmental concerns
- Keep top management informed on compliance issues, and
- Assist 3M operations in managing beyond compliance

To better understand the 3M environmental auditing program, it should be studied in the context of total quality environmental management at 3M.

Global Total Quality Environmental Management

The goal of 3M is to become a sustainable-development corporation—one whose products, facilities, and operations have minimal environmental impact, so that its business can continue to grow without adverse effect on the earth and its inhabitants. The organizing principle of this effort is total quality environmental management. It is a comprehensive philosophy encompassing 3M's proactive environmental policy and corporate goals, as well as the processes and tools needed to realize them. TQEM is implemented corporationwide by the Environmental Engineering and Pollution Control (EE&PC) Department. Within EE&PC, the environmental compliance auditing function operates as part of the corporate quality effort.

The 3M commitment to TQEM began in 1975 with the adoption of the corporate environmental policy. The environmental compliance audit program was an outgrowth of 3M's commitment to this policy.

3M Environmental Policy

1. Solve own environmental problems
2. Prevent pollution at the source
3. Conserve natural resources
4. Develop environmentally sound practices
5. Comply with all regulations
6. Assist governmental organizations

Part of Q90s Corporate Quality Program

TQEM at 3M is evolving as part of the larger Q90s corporate quality process, modeled after the Malcolm Baldrige National Quality Awards. It is a comprehensive management system that emphasizes continuous improvement in areas that result in greater customer satisfaction. At 3M, customer satisfaction in environmental terms means more than total compliance with governmental regulations. It means adherence to strong corporate environmental policies and the principles of sustainable development.

3P—Pollution Prevention Pays

In 1975, 3M formally began an official companywide project known as Pollution Prevention Pays, or 3P. The 3P project was conceived to protect the environment while providing economic benefits to the company. It is based on the belief that the best way to prevent pollution is not to generate it in the first place. 3P is a voluntary program designed to encourage all 3M employees to develop innovative technologies for pollution prevention. It represents the first organized application of the concept of pollution prevention throughout one company, worldwide. Since 3P began, 3M employees have carried out more than 3000 projects, preventing over 600,000 tons of pollution and achieving a savings for the company of more than $600 million (cumulative first year savings).

Year 2000 Goals (Challenge '95)

In addition, 3M has set ambitious goals for continuing the improvement of its global environmental performance. As part of the Challenge '95 initiative, 3M will, by 1995:

- Reduce waste 35 percent over a 1990 base
- Reduce energy use 20 percent over a 1990 base

And by the year 2000, 3M expects to achieve:

- A 50 percent reduction in waste generated
- A 90 percent reduction in all releases, worldwide

International Chamber of Commerce Business Charter for Sustainable Development

Sustainable development is a concept that recognizes that economic development and environmental protection can and must coexist. It involves meeting the needs of the present without compromising the ability of future generations to meet their own needs. As an original signatory to the International Chamber of Commerce Business Charter for Sustainable Development, which was adopted November 27, 1990, 3M actively pursues the achievement of the 16 principles outlined in the charter.

CMA Responsible Care Initiative

The Responsible Care® program was initiated in 1989 by the Chemical Manufacturers Association (CMA), of which 3M Specialty Chemicals Division is a member. Member facilities are routinely audited against the Responsible Care principles and Codes of Management Practices. This auditing process is in addition to the environmental compliance auditing function performed by EE&PC.

3M Corporate Policies

Standards of Practices for Operating Units

To ensure 3M operating units achieve sustainable development goals and to drive them beyond the demands of regulatory compliance, EE&PC develops, publishes, and updates environmental statements. The intention is to emphasize and clarify environmental practices and procedures. Audits gauge 3M facility compliance with these statements. There are three types of environmental statements.

1. An *environmental policy* is a directive that originates from a corporate-level committee. It must be implemented at all targeted 3M facilities. Policies generally establish a new approach, a change in philosophy, or a far-reaching goal.

2. An *environmental standard* is a description of an existing regulation or requirement, further explaining the requirement or reemphasizing its importance.

3. *Environmental guidelines* suggest helpful day-to-day operating procedures.

Environmental Management Plans

An environmental management plan is a document containing the organizational structure, responsibilities, procedures, and resources in place at operating unit facilities to deal with all applicable environmental issues. Individual facility plans are created as part of the overall operating unit plans. Each plan is developed by the corporate Environmental Engineering Services (EES) facility contact in conjunction with facility management. Emphasis is placed on the prevention of adverse environmental effects, rather than resolving problems after they occur. The goal of a completed environmental management plan is to address the following:

1. Corporate environmental policies and objectives

2. Review of permit requirements

3. Responsibilities and time lines for the resolution of issues

4. Current and anticipated environmental regulations and impacts

5. Environmental review of current and planned activities

6. Waste minimization activities

7. Environmental releases

8. Methods for interaction with EE&PC management systems

9. Emergency response planning

10. Establishment of facility environmental management teams

Waste Stream Profile Program

The waste stream profile program identifies, evaluates, and provides instructions on the disposition of all waste streams at 3M. The program was developed to ensure full corporationwide compliance with all applicable regulations and company policies. It includes a centralized database, which is reviewed by the auditor as part of the preaudit process.

Environmental Compliance Audit Program

All 3M facilities in the United States participate in the environmental compliance audit program. Air and water quality, solid and hazardous waste, spill prevention, storage tank management, and environmental management systems are the primary issues that these audits address. The Environmental Compliance Audit staff also audits recycling and treatment and disposal facilities, while site assessments of corporate acquisitions and divestitures are performed by the Environmental Engineering Services Remediations Group.

All environmental compliance audits are conducted under the direction of the Office of General Counsel. Counsel advises appropriate facility management of any potential corrective actions or other legal implications. All reports are issued under the attorney-client privilege.

Facility audit schedules are individually determined on the basis of the size and nature of the operation and the environmental activities at the facility. The frequency of audits can range from one to four years.

Goals and Objectives

The goals and objectives of the audit program include:

1. Providing independent verification that 3M operations are in compliance with applicable internal and governmental requirements

2. Ensuring that management plans are in place for continued compliance

3. Improving overall environmental performance at the facility

4. Assisting facility management

5. Increasing environmental awareness

6. Improving risk management

7. Protecting from potential liabilities

8. Optimizing environmental resources

9. Ensuring that facilities and process units are operated and maintained properly

10. Providing informal compliance education to facility staff

11. Identifying 3P opportunities

Air Issues.   Using the regulations and corporate environmental policies as a foundation, the auditor reviews each air-emission source in the facility to ensure that the appropriate permits have been received and/or applied for. Permit requirements are reviewed for reporting, record keeping, and operation and maintenance requirements. After consulting with the 3M senior environmental counsel, the auditor may call the state, regional, or local enforcement office for clarification regarding applicability of requirements or exemptions. Attention is given to process modifications and whether these changes and/or planned changes will affect the compliance status of the process. Air issues are documented using detailed audit checklists. Checks on compliance with 3M corporate policies (air emission reduction program, ozone depleting chemicals, early reductions, etc.) are also included in the review.

Hazardous Waste Issues.   During the facility walk-through, all containers and tanks are reviewed for appropriate labeling, marking, container type, closures, and other items (see list at end of this chapter). All paperwork (e.g., weekly storage-area inspection records) and procedures required by federal, state, and local hazardous waste regulations are reviewed. Corporate waste-stream profiles are reviewed for accuracy. The audit examines waste management practices including the use of 3M-approved recycling, treatment, and disposal facilities. Facility compliance with corporate programs, such as 3P, Challenge '95, and 3M corporate policies, is also included in this audit. These reviews are documented in detailed audit checklists.

Tank Issues—Spill Prevention, Containment, and Countermeasures.
All aboveground and underground tanks are inventoried. Containment areas and loading and unloading areas are inspected for conformance with 3M and regulatory requirements: Tank-management and spill-response procedures and practices are evaluated. These practices are compared to those written in the facility spill prevention, containment, and countermeasures (SPCC) plan. The SPCC plan is then reviewed in detail for accuracy and completeness along with other related SPCC and tank records. Compliance with 3M corporate policies for aboveground and underground tanks is also included in this review.

Wastewater Issues.   All wastewater discharge sources, including both contact and noncontact cooling-water usage, are examined for management practices. All permit discharge limits, reporting, record keeping, and other requirements are reviewed. Storm water permitting status and any groundwater issues are also assessed. These issues are documented in the detailed audit checklists.

Preaudit Preparation

The effectiveness of a facility audit depends on the preaudit preparation by the audit team members. The goal is to understand all of the issues before the site visit. Each audit is tailored to the specific needs of a facility. Auditors team up with facility personnel to help ensure a thorough audit. Preplanning includes:

1. Notifying the operating unit management.
2. Contacting facility management and scheduling the audit.
3. Selecting the audit team. The audit team consists of up to three members:
   a. A full-time auditor, the team leader who coordinates and oversees the audit process
   b. The appropriate Environmental Engineering Services (EES) contact
   c A specialist who has expertise in an identified area of concern (with selected participation by the 3M Internal Audit Department)
4. Reviewing federal, state, and local regulations.
5. Reviewing the facility environmental management plan with the EES contact in order to update knowledge of facility issues and their status.
6. Reviewing EE&PC facility files and corporate computer databases to become familiar with the issues at the facility. The facility files are part of the total compliance program designed to maintain and control critical records. It is a centralized filing system maintained by full-time staff. The files contain up-to-date records on permits, process and facility monitoring, emergency plans, and other regulatory activities.
7. Understanding the requirements of current federal, state, and local regulations. The *Federal Register* and other sources are used for federal regulations coverage. Due to the number of states in which 3M operates facilities, a variety of sources are used to identify state-specific regulations
8. Involving 3M internal auditing on selected audits.
9. Consulting as appropriate with 3M senior environmental counsel.

On-Site Audit

The Environmental Compliance Audit Group has developed the following audit process:

Initial Meeting—Orientation.   The audit team meets with personnel at the facility to review the purpose and scope of the environmental audit program. Together they develop a strategy for proceeding with the audit. The team:

1. Reviews objectives of 3M environmental audit program
2. Discusses potential agency concerns
3. Reviews past environmental audit findings and compliance history from time of last audit
4. Discusses audit follow-up and post audit survey
5. Schedules date and time for wrap-up meeting

The Audit.   The audit can last from one to five days, depending on the size and complexity of the facility. It will generally contain the following elements:

1. *Facility walk-through.*   The audit team inspects all indoor and outdoor locations of the facility.
2. *Facility paper audit.*   During the site visit, the audit team reviews pertinent documents and operating procedures with facility personnel to verify that all systems are in place and functioning as intended.
3. *Environmental Management Systems Review.*   Throughout the audit, facility personnel are questioned to determine the effectiveness of the environmental management system. A high-level system is essential if continuous compliance is to be assured.
4. *Inspection checklists.*   Checklists are used as a guide to help the auditors identify areas of concern. Areas covered by checklists include air, water and wastewater, solid and hazardous waste, and SPCC and tanks. (See the checklists at the end of the chapter for a sampling of issues addressed.)

Draft Report.   3M environmental compliance auditors are equipped with portable computers, enabling them to generate a draft audit report before leaving the site. The draft report details the findings and includes recommendations for improvement. This report is the basis for the exit meeting discussion and

---

Environmental Management Systems

1. Trained and experienced personnel
2. Clearly defined responsibilities
3. Systems for project review/approval
4. Internal verification and communication
5. Protective measures
6. Policies, programs, procedures
7. Record-keeping system

issue resolution planning. The draft report is reviewed by the senior environmental counsel.

Exit Meeting and Final Action Plan. At the conclusion of the audit, the draft report is reviewed at the site by facility and EE&PC personnel. Together, they clarify the draft report, then develop a final action plan. The audit team leader, the facility manager, and the environmental contact agree on target dates for each audit finding and assign a coordinator who will be responsible for resolving each issue.

Train for Self-Inspection. The audit group also develops facility self-inspection checklists that site personnel can use to continually monitor their operations.

Postaudit Follow-up

1. *Final audit report.* The audit team drafts the final report incorporating the information gained from the exit meeting. The report is reviewed and issued by the Office of General Counsel, under whose direction the audit was conducted. The audit report is a confidential and attorney-client–privileged document and is targeted to be issued within two weeks of completion of the audit. The report is addressed to the facility manager who coordinates the follow-up and keeps the audit team leader informed of progress. It is also copied to the next two higher levels of operating unit management.

2. *Tracking audit findings.* Every 30 days the audit computer tracking system (ACT) issues confidential status reports on all open audit findings. The reports go to the EES facility contact, the facility manager, and the senior environmental counsel until all audit findings are resolved. If audit findings are not resolved in a reasonable period, additional levels of management and senior environmental counsel become involved in resolving the issues.

3. *Postaudit facility survey.* The EE&PC staff vice president issues a survey to the facility manager two months after an audit has been conducted. The survey responses help determine the effectiveness of the audit program and identify any problem areas.

4. *General counsel review and follow-up.* The senior environmental counsel reviews all audit findings and provides legal advice and direction throughout the follow-up process.

5. *Audit wrap-up.* A confirmation letter is sent to the operating unit management indicating that all audit findings have been satisfactorily resolved.

Audit Computer Tracking Program

The audit computer tracking (ACT) program is a secured computer-based information system used to manage audit information and to assist in the audit follow-up process to ensure that all audit findings are resolved in a timely manner. In addition to expediting audit follow-up, the program is used to help manage the overall audit program by automatically generating a variety of reports and documents.

At the time of the audit, each audit report and each audit finding is assigned a unique number (e.g., 225-1, or audit no. 225, item no. 1), which is entered into the ACT system.

The information acquired by auditors from this database or from another centralized EE&PC database includes:

- Facility-specific information (site location, division name, etc.)
- Names and addresses of the facility manager, the next two higher levels of operating unit management, EES manager and the facility environmental contact, the audit team leader, and team members
- Start and end dates of audit and the audit report date
- The audit number, audit status, etc.

For each audit there is specific information on each audit issue, including:

- Audit number and item number
- Type of item (hazardous waste, wastewater, etc.)
- Description of item
- Remedial action plan
- Target date or completion date
- The person responsible for resolving the issue

All data entry is through menu-driven programs with built-in checks and balances to prevent data entry errors. Access to ACT is strictly limited to audit team leaders and the environmental counsel on a 24-hour-a-day, seven-day-a-week basis.

Audit Reports

The following are examples of the variety of reports and documents for use by the auditors to manage the various aspects of the audit program.

Weekly

1. *Audit reports not completed.* A report listing audits that have been conducted where the final reports have not been issued is generated for audit team leaders.

Monthly

1. *Audit preparation letter.* A letter by the audit team leader is generated and sent to the facility manager two months prior to audit.

2. *Audit findings status letter.* A letter updating the status of the audit findings, including an attachment listing open findings, target dates, etc., is generated by the audit team leader for the facility manager with copies to EES manager/EES contact and the senior environmental counsel.

3. *Audit survey letter.* A letter to obtain feedback on the performance of the audit is generated by the EE&PC staff vice president for the facility manager, two months after the audit.

4. *Second request audit survey letter.* If no response is received to the original survey, a reminder letter is generated by the EE&PC staff vice president for the facility manager.

5. *Audit wrap-up letter.* This letter from the EE&PC staff vice president for the facility's operating unit management confirms that all audit items have been resolved.

6. *Audit summary accounting progress (ASAP) report.* A management report is generated to assess audit effectiveness. It includes the number of audits, time to complete each audit, time to complete audit reports, categories of audit findings (if any), and measurements of responses to audit status letters.

7. *Audits completed report.* This document lists all facility audits that have resolved all issues during the month.

8. *Audit schedule report.* This report is reviewed by audit team leaders, EE&PC management, senior environmental counsel, and internal auditing. It shows which audits have been completed and those remaining scheduled for the calendar year.

Quarterly

1. *Audit survey new results.* A report is generated from the audit survey responses and is reviewed by the EE&PC staff vice president and quality assurance manager.

2. *Audit summary accounting progress (ASAP) report for all audits.* This management report covers the performance records of all facility audits since the beginning of the program. It is reviewed by EE&PC management, senior environmental counsel, and internal auditing.

Annually

1. *Audit announcement scheduling letter.* This letter is generated by the auditing supervisor for the facility's operating unit management.

2. *Active 3M sites not audited.* This report is used by the auditing supervisor for scheduling the next year's audits. It contains justifications for not including a facility in the audit schedule (e.g., office only).

Recycling, Treatment, and Disposal Facilities Audit Program

The purpose of this program is to minimize 3M's potential legal liability from entering into agreements with waste handling facilities. Prior to a 3M facility engaging a third-party recycling, treatment, and disposal facility (RTDF), it must be reviewed by EE&PC and the senior environmental counsel and approved for use. If approved during the initial review, the RTDF is then audited on a routine basis (one to three years, depending on need) to ensure continuing compliance with 3M requirements. These facilities must meet the following requirements including:

- Compliance with all applicable environmental regulations and 3M standards
- Demonstration of financial strength
- Retention of capacity to meet 3M needs
- Satisfaction of 3M requirements (contracts, service, costs, etc.)

Under the direction of the senior environmental counsel, other 3M departments are involved in this process, including Resource Recovery and Packaging Engineering (drum reconditioners only). A listing of approved facilities is maintained by EE&PC.

Benefits of the 3M Environmental Audit Program

In addition to protecting the company and measuring the effectiveness of environmental management programs, a well-designed and implemented compliance auditing program will help to create awareness of environmental issues among facility staff. Audit can also help to facilitate improved integration of environmental management systems into daily operations. They also serve as a key element of a corporate information management system designed to keep top management informed of environmental activities throughout the company.

Despite a dramatic increase in the volume and severity of environmental regulations, the number of findings per audit have been reduced—a direct result of a strong audit and follow-up program.

After each audit, the department sends a postaudit survey to the facility management. The results have shown that an increasing value is being placed on the audit function along with increasing customer satisfaction with both the process and the results.

3M's Environmental Record

Even with a tremendously positive environmental culture and tradition, formally institutionalized in 1975, full compliance with the ever-increasing multiplicity of environmental requirements can be a daunting challenge for any organization. The Environmental Compliance Audit Program has played an essential

---

3M Environmental Post-Audit Survey

1. Was the audit preparation letter helpful?
2. Did the auditors conduct the audit in a professional manner?
3. Did the EE&PC audit team cover all areas of environmental concern?
4. Did the wrap-up meeting clearly define audit findings and follow-up process?
5. Was the audit report clearly written, concise and fair?
6. Was the audit report issued in a timely manner?
7. Is the audit follow-up process working?
8. Is the environmental audit process a good tool to assess compliance status?
9. How frequently should your audits be schedule?
10. How can the environmental auditing program be improved?

---

role in helping 3M enhance its respected environmental record. The February 1991 *Quality Digest* highlighted 3M's environmental programs for quality. The November 5, 1990, issue of *Industry Week* recommended 3M as the company to benchmark for environmental management programs. A survey by *Fortune Magazine* (February 8, 1993) rated 3M as number one in community and environmental responsibility. Franklin Research and Development Corporation, a Boston-based firm that rates companies on their social responsibility, gave 3M its highest environmental issue rating (*New York Times*, February 3, 1991). To date, many major multinational companies have selected 3M EE&PC for benchmarking, focusing on its environmental auditing program.

3M Environmental Audit Checklists

The following checklists were prepared for use by 3M internal auditing functions. They have been included here only as examples of 3M procedures, not as recommendations. Each industry must develop checklists appropriate for its needs. These checklists are changed as regulations and corporate policies change.

3M Environmental Inspection—Air

Permit Applications

- Locate and identify all potential sources and note likely sources not permitted
- Compare with permit file, and review compliance status with state and local regulations
- Confirm that all pollution control equipment is properly permitted and registered

- Check permit expiration dates and fees paid
- Review emissions rates and proposed increases

Permit Conditions

- Review permit conditions and regulatory requirements
- Review calibration and instrumentation procedures and records
- Review monitoring data for compliance
- Review internal records of corrective action for exceedances
- Review reporting record keeping and any compliance schedules
- Review procedures in place to verify continuing compliance with permit requirements
- Determine compliance via inquiry, observation, and agency inspection records; review annual/biennial air emissions reporting

New Stationary Source Standards (If Applicable)

- Review new source and specific facility standards, noting changes or construction since 1975
- Review Environmental Protection Agency (EPA) construction notification records and quarterly Continuous Emission Monitoring's (CEMs) reporting requirements
- Review record keeping of start-up, operating conditions, and monitoring reports
- Verify performance tests conducted within 180 days after start-up
- Confirm that sampling methods are approved by the agency and that agency has been notified of test dates prior to testing
- Confirm opacity standards are met
- Review maintenance and inspection procedures and records
- Review specific monitoring requirements and confirm equipment is available and functioning

Hazardous Air Pollutants (If Applicable)

- Determine if National Emission Standards for Hazardous Air Pollutants (NESHAP) limits are being met for any NESHAP sources at the site: asbestos, beryllium, benzene, fugitive dust emissions, inorganic arsenic, mercury, radionuclides, radon 22, vinyl chloride
- Review documents related to emission of hazardous air pollutants and check: timeliness with which information was filed, completeness of information, agency approvals received
- Review procedures established for handling of asbestos

Miscellaneous

- If an air pollution alert and emergency plan is required, review plan.
- Is sulfur content of fuel oil and coal in conformance with prescribed regulations?
- Do odorous emissions exist? Are complaints/impacts reviewed, etc., as required?
- Does all air pollution control equipment have written operating and maintenance procedures?
- Is the facility in compliance with the 3M corporate ozone-depleting chemicals policy?
- Is the facility in compliance with the 3M corporate asbestos policy and asbestos removal requirements?
- Review facility compliance with 3M corporate air emissions reduction program?
- Confirm compliance with standards for fuel dispensing operations (greater than 550 gallons).
- Verify vapor control system or submerged fill lines, compliance in labeling, and purchase records.
- Ensure that waste minimization activities and projects are recorded (3P projects, Challenge '95, Programs for Profit, etc.).

3M Environmental Inspection— Solid and Hazardous Waste

Waste Determination

- Confirm generator status:
  - TSDF—treatment, storage, and disposal facility
  - LQG—large-quantity hazardous waste generator
  - SQG—small-quantity hazardous waste generator
  - CEG—conditionally exempt small quantity generator
  - Verify Environmental Protection Agency (EPA) identification (ID) number
- Waste stream profiles are completed, submitted, and correct
- Waste stream profile ID numbers are used on labels
- Waste stream profile reference sheets are used
- 3M waste manual is on site

Pretransportation Requirements

- 3M, EPA, and Department of Transportation (DOT) labels are complete
- Proper DOT containers are used

- DOT placards are available
- Approved hazardous waste transporter is used

Site Security

- Site has 24-hour surveillance, fence, or barrier
- Site has controlled access and entry

Satellite Accumulation Areas—55-Gallon Hazardous Waste Limit

- Containers are in good condition
- Wastes are compatible with containers
- Containers are stored closed except when waste is added or removed
- Labels are complete (minus date)
- Containers are under the direct control of the operator
- Full containers are dated and moved to an approved storage area within three days of being filled

Ninety-Day Hazardous Waste Storage

- Correct hazardous waste labels are used
- Accumulation dates are on labels (satellite containers when full)
- Containers are in good condition
- Containers are compatible with wastes
- Containers are stored closed
- Containers are protected from damage
- Funnel covers are closed
- Accumulation areas contained
- Incompatible wastes are separated
- LQG—less than 90-day storage
- SQG—less than 180-day storage (or 270 days if TSDF is more than 200 miles away)
- Appropriate drum gaskets are used
- Correct bungs and gaskets are used
- Weekly inspections of 90-day hazardous waste areas documented
- Flammable and reactive wastes stored more than 50 feet from property line
- Empty drums have less than 1 inch of residue and no free liquids
- Flammable wastes protected from direct sunlight (state specific)

■ No hazardous wastes in nonregulated drums

Hazardous Waste Tanks

■ Correct hazardous waste labeling

■ Tank and hazardous wastes are compatible

■ No incompatible hazardous wastes in tank

■ Continuous feed system cutoff controls exist and are operable

■ Daily, documented inspections of hazardous waste tank level, controls, and monitoring equipment

■ Weekly documented inspections for corrosion and leaks

■ Verify name of person maintaining tank logs

■ Uncovered hazardous waste tanks have at least 2 feet of freeboard or emission controls

■ Tank integrity tested and professional engineer (PE) reviewed report on-site

■ Verify review date

Personnel Training

■ Verify employee name and job title

■ Verify written job description

■ Verify description of training

■ New employee training

■ Annual training review

■ Training records on file

■ Verify person maintaining Resource Conservation and Recovery Act (RCRA) training records

■ Verify location of RCRA training records

Preparedness and Prevention

■ Internal communications and alarm system

■ Phone or radio for outside help

■ Adequate fire and spill control equipment

■ Adequate water and/or foam for fires

■ Emergency equipment maintained and accessible

■ Adequate aisle space

■ Police, fire, hospital familiar with facility and types of wastes

■ Emergency phone numbers posted

- Primary and secondary authority designated
- Written agreements with emergency response authorities and contractors

Contingency Plan and Emergency Procedures Plan Contents (TSDF)

- Documented emergency response actions
- Documented arrangements with emergency response teams
- Emergency equipment list with use and capacity descriptions
- Current evacuation plan
- Plan approved by EE&PC and distributed to police, fire, hospital, and other local emergency responders
- Current list of emergency coordinator and alternates with work and home phone numbers
- Written authority for emergency coordinator to use all resources necessary to respond to a release
- Notification and reporting procedures and activities reporting during and after emergency response
- Decontamination procedures for cleanup equipment exist and emergency equipment is maintained
- Closure plan maintained
- Postclosure plan maintained (if required)
- Financial assurance plan filed and a current copy on-site
- Hazardous waste operations (HAZWOPER) plan current

Contingency Plan and Emergency Procedures Plan Contents (LQG)

- Documented emergency response actions
- Documented arrangements with emergency response teams
- Emergency equipment list with use and capacity descriptions
- Current evacuation plan
- Plan approved by EE&PC and distributed to police, fire, hospital and other local emergency responders
- Current list of emergency coordinator and alternates with work and home phone numbers
- Written authority for emergency coordinator to use all resources necessary to respond to a release
- Notification and reporting procedures and activities reporting during and after emergency response

- Decontamination procedures for cleanup equipment exist and emergency equipment is maintained

Small-Quantity Generator (SQG) Requirements

- Facility has an emergency coordinator assigned
- Posted near phone: names and phone numbers (work and home) of emergency coordinator and alternates
- Phone number of local fire department
- Fire and spill control equipment location
- Current evacuation plan posted

Record Keeping

- Manifests are completed and correct
- Manifest copies are distributed and filed
- Manifest copies are returned from TSDF within 45 days
- Manifests are retained longer than three years
- Land ban notification is used
- Land ban copies are retained longer than five years
- Waste characterization testing records are maintained
- State and/or county waste stream registrations are current
- Biennial or annual reports due March 1, 1990 ('92, '94) were submitted
- Exception reports are filed if required
- Hazardous wastes to reclaimers is manifested
- No hazardous wastes are received from off site
- Waste minimization activities and projects are recorded (3P projects, Challenge '95, Programs for Profit, etc.)
- Pollution prevention plans on file and annual progress reports submitted to the agency
- SARA Title III 313 forms available on site for agency inspection (state-specific)

Miscellaneous

- Contracts obtained with waste haulers/hazardous material disposal facilities
- Contracts obtained with hazardous material recyclers
- Corporate compactor policy is reviewed and enforced
- Approved drum reconditioners are used

3M Environmental Inspection— SPCC/Tanks

Site Spill Prevention, Containment, and Countermeasures Plan is Required if There:

- Is more than 1320 gallons of oil (or oil product) stored aboveground or in a single tank of 660 gallons (55-gallon drums included)
- Is more than 42,000 gallons of oil (or oil product) stored underground

If answer is yes to either above question then site needs an SPCC plan.

SPCC Plan Requirements

- Verify SPCC coordinator
- Verify first alternate
- Plan contains chemical tank storage inventory
- Addresses tank inspection and testing and site security
- Procedures exist for dealing with spill prevention, control, and cleanup
- Procedures exist for liquid transfer operations
- Procedures exist for emergency notification and reporting
- Plan is current and accurate with site, equipment, and procedures changes (amendments made within six months of changes; routine plan review/evaluation every three years)
- Plan contains employee training records and training outline
- Plan certified by a registered professional engineer (registered in the state in which the facility resides) and signed by the facility manager

Spill Control Procedures

- Site or storage area fenced and gates are locked
- Secondary containment is provided and is adequate
- Dike drain valves are normally closed and are locked
- Spill response equipment and materials are available, adequate, and accessible

Transfer Operations

- Loading and unloading areas are contained
- Documented transfer procedures are followed

SPCC Training

- New employee training is given and documented

- Annual review is given and documented

Tank Inspection Program

- Documented daily inspections are made of all aboveground tanks
- Underground tanks are in compliance with 3M corporate policy
- Review underground tank inspection and/or replacement programs

3M Environmental Inspection—Water/Wastewater

Process Review

- Is site subject to categorical pretreatment standards?
- Is site classified as a significant industrial user?
- Review sources and discharge points of noncontact cooling water
- Review site discharge water associated with industrial activity
- Determine if all wastewater pollution control equipment has written operating and maintenance procedures

Review Existing Permits

- Verify National Pollution Discharge Elimination System/State Pollution Discharge Elimination System (NPDES/SPDES) permit number
- Verify Publicly Owned Treatment Works (POTW) permit number
- Review existing site permits
- Permits current/accurate with equipment, site, process changes

Conformance with General Effluent Standards

- Review pretreatment standards, if applicable
- Baseline monitoring report submitted, if applicable
- Review monitoring records for compliance
- Procedures exist/followed to ensure that unauthorized material is not put down sewers
- Treatment operators trained and licensed
- Wastewater treatment equipment adequate and maintained
- Review calibration and maintenance records for water pollution control to verify equipment conformance with regulatory requirements
- Representative sample point locations
- Sample frequency agrees with requirements

- Proper sample containers used
- Proper sample preservation techniques used
- Proper sample holding times observed
- Proper analytical and quality assurance procedures used
- Lab analytical procedures conform to prescribed text procedures or best management practices
- Sampling and analytical testing is reviewed by EE&PC laboratory quality control personnel
- Files maintained for sampling, analysis, equipment maintenance, and reports
- Required reports submitted to regulatory authority

Storm Water

- Site subject to NPDES permit for storm water
- Verify storm water permit number
- Site in compliance with individual storm water permit
- Site in compliance with group storm water permit
- Site in compliance with general storm water permit
- Review storm water pollution prevention plan

Safe Drinking Water Act

- Site obtains drinking water on site
- Site subject to national primary drinking water standards
- Site collects required samples
- Site meets national primary drinking water maximum contaminant levels
- Site submits required reports to the state
- Required reports kept on site
- Site analyzes samples for secondary drinking water parameters
- Site meets secondary drinking water maximum contaminant levels

Groundwater Monitoring Requirements

- Wells capped and locked when unattended
- Groundwater wells located on site
- Groundwater samples collected at required frequency
- Monitoring data available and reported, if applicable

16

Health, Safety, and Environmental Auditing at Polaroid Corporation

Harry Fatkin
Director of Health, Safety, and Environmental Affairs, Polaroid Corporation

Polaroid's health, safety, and environmental audit program has evolved over 15 years, and during this period we have developed and used a series of audit designs. One important lesson we have learned has been the value in taking whatever time is needed for thoughtful management discussions aimed at clarifying the purpose of an audit.

Circumstances change rapidly in the health, safety, and environmental (HSE) field, and the purpose of an audit may need to change to reflect new circumstances. If the purpose changes, the design and structure of the audit may need to change. What might have been a useful audit approach in the past may not be as effective today. A company's internal HSE program may be different, or the attitude of government may have changed in ways that would influence a company's approach to auditing. Conversely, an approach used earlier—and perhaps abandoned—may turn out to be the most effective way to address contemporary issues.

In our 15-year audit history, we have experienced such changes and learned the need to periodically reexamine, and clarify, the audit purpose. This reexamination includes looking at:

- Organizational structure
- Key activities for health, safety, and environmental management
- Roles and responsibilities of individuals and organizations

Polaroid's current audit program will be described in the context of three questions that should be asked by anyone when beginning—or modifying—a health, safety, and environmental audit program:

1. What is the purpose of the audit?
2. What audit design best satisfies the audit purpose?
3. What is the structure of the audit process?

What Is the Purpose of the Audit?

At Polaroid the primary purpose of the audit is to assist line managers in meeting the objectives that they have set and in getting the results that they want. This may seem obvious, but some corporate audit programs are structured with a very different purpose. One purpose might be to gather independent data, another might be for the purpose of reporting to a chief executive officer (CEO) or to a board of directors. Our purpose, to assist line managers, does not preclude other uses of audit learnings, but it does clearly position the audit as a tool for the line organization and for continuous improvement. Continuous improvement is one of the total quality ownership principles of the company.

The line organization structure at Polaroid has, as a key focal point, the plant manager. In manufacturing, the plant manager is accountable for a clearly defined organizational unit and for performance in seven areas:

- Cost
- Quality
- Schedule
- Inventory
- Diversity
- Safety
- Environment

Audit findings are directed to the plant manager to assist in meeting HSE performance goals.

The schematic (Fig. 16-1) outlines key activities related to health, safety, and environmental management. It was developed and used in our management discussions aimed at clarifying the purpose of our audit. These key activities are ongoing and can be thought of as elements in a *plan-do-check-act* cycle (Fig. 16-2) for continuous improvement under total quality programs.

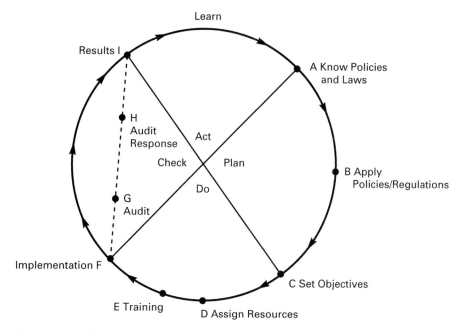

Figure 16-1. Key activities related to HSE management.

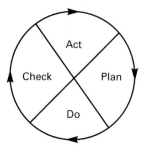

Figure 16-2. Plan-do-check-act cycle.

The plan-do-check-act cycle for key HSE activities starts at (A)—Know policies and laws, ends at (I)—Results, and leads to continuous improvement when an organization both learns from that cycle and acts on what it has learned before it begins the next one.

Polaroid has many plants, each of which is individually moving through this cycle, and different parts of the company are at various stages of learning on the whole range of health, safety, and environmental subjects. There are hundreds of such subjects that can apply to any given plant, e.g., law, company policy, and public policy issues.

A. Know policies/laws	Maintain awareness of laws, policies, "best practices," and Polaroid experiences.
B. Apply policies/regulations	Determine which part of which law (or Polaroid policy) applies to which part of divisional operation and specify local requirements.
C. Set objectives	Informed by (B), determine divisional objectives.
D. Assign resources	Assign human and financial resources to meet objectives (C).
E. Trainings	Transmit information and guidance necessary to satisfy requirements of (B).
F. Implementation	Take necessary actions to fulfill local requirements (B) and meet divisional objectives (C).
G. Audit	Assess implementation of (B).
H. Audit response	Address issues raised in audit.
I. Results	The bottom line (in addition to audit data, performance metrics can be developed to foretell results).

Figure 16-3. Description of key activities related to HSE management.

For any one subject, however, the cycle is the same. It begins with knowledge (A) about the subject. As the line organization works its way through the steps of the cycle, management actions are required—(B) through (F)—to arrive at acceptable results (I), relative to the subject (A). Brief descriptions of the activities at each of these steps is listed in Fig. 16-3. These activities, with their descriptions, were used at several points in our discussion of audit purpose. One very valuable use was to sort out important roles and responsibilities.

For instance, a plant manager typically has responsibility for all organizational resources necessary to produce a finished product, or a significant product component. These resources include functions ranging from production and quality to human resources and finance. Some individuals in these plant functions have dual reporting relationships through a matrix organization to corporate groups. Those responsible for HSE activities in a plant are in this category. They report through the plant manager and also have a functional relationship to the Corporate Office of Health, Safety, and Environmental Affairs.

For each of the activities, the organizational relationship was evaluated. Knowledge (A) was identified as a primary responsibility of the corporate office. Individuals in that office are expected to be fully aware of new requirements and of regulatory or public policy trends. An individual plant is not expected to have sufficient resources to cover the broad range of health, safety, and environmental subjects. This does not imply ignorance at the plant level because knowledge (A) is a joint responsibility, but the corporate organization takes the lead. The corporate office has the responsibility of communicating this knowledge to a plant and the plant has the responsibility to apply this knowledge to local circumstances.

Application (B) of knowledge also is a joint responsibility with the corporate office, but here the plant takes the lead. Those who work in the plant know the processes and plant activities that are impacted by HSE requirements. With corporate and plant HSE professionals working together, broad knowledge about a subject gets translated into the depth needed to satisfy local needs. Organizational roles and responsibilities for each of the activities were similarly clarified before the auditing program was finalized.

The plant, provided with this knowledge (A) and informed of its applicability (B) to plant activities, is responsible for setting objectives (C). Although an individual plant manager sets objectives (C) for a plant, this work is not done in isolation. Individual plant objectives are aligned through all the levels of the line organization from plant manager through manufacturing director to vice president.

Responsibility for results (I) rests with this line organization, and the audit is an "assist" to all levels. The concept of an audit as an assist is depicted by the way the audit is positioned in Fig. 16-1. Acceptable results are not dependent on auditing. What is needed for acceptable results (I) is effective implementation (F) of programs put in place to meet objectives (C).

The audit is an assist and, in fact, line managers do not need auditing to inform them of results. Line managers who set objectives have many ways of knowing results. They know results when:

- Accidents occur or individuals are injured
- An environmental spill or release takes place
- Environmental samples are taken and analyzed
- A government inspection is completed

In other words, life goes on without auditing, and a manager knows how things are going.

Acceptable results are one thing, but unacceptable results may do human or financial damage. Damage may be found in the pain of injury, the cost of environmental cleanup, or the violation notice of noncompliance. So, once programs are implemented (F), the organization will experience results (I). These results, however, may be acceptable or unacceptable.

An effective line organization wants to prevent unacceptable results, not experience them, and audits are a tool to help line managers do this. That is the reason for the positioning of the audit as a dotted cord in Fig. 16-1. The audit assists line managers by testing the likelihood that the organization will experience acceptable results.

The audit (G) and response to the audit (H) are directly tied to the continuous improvement cycle as an opportunity to learn. The organization learns after the fact from results and before the fact from the audit. The audit is an assist to line managers both in meeting the objectives that they have set and in getting the results that they want. That is because audits (G) provide data from which the line organization learns. Acting on this learning leads to greater knowledge (A) to be applied (B) as the line organization begins another cycle of continuous improvement.

What Audit Design Best Satisfies the Audit Purpose?

To satisfy our audit purpose, we have actually developed three audit designs:

- The awareness audit
- The compliance audit
- The management system audit

On occasion we have combined these designs into a single audit. The design chosen for a given audit, however, depends on the subject being audited and the results sought by that audit.

The Awareness Audit

The result of an awareness audit is the discovery of the level of knowledge within an organization about the audit subject. Recall that knowledge (of corporate policies or laws and regulations) is a starting point in our continuous improvement cycle. An awareness audit is often the audit design selected to deal with a new subject area.

At the plant level, getting ready for an audit and experiencing it is, in itself, a learning opportunity. If a new law has been enacted or new regulations are being implemented, commissioning an awareness audit will serve not only to discover the organization's level of knowledge about the law or regulation but also to create a vehicle to raise that level of knowledge.

Awareness audits can target a specific subject, such as a policy or a law, or they can be used to cover a broad range of subjects. In the latter case, Polaroid used awareness audits to accelerate the dissemination of a revised set of companywide safety policies. These policy revisions were extensive, covering a broad range of subjects from electrical to mechanical to chemical safety.

The policies had been written, approved, announced, and published. The audit was done to discover the level of awareness of this revised series of requirements at all Polaroid plants. The audit was scheduled and publicized well in advance, and plants increased their awareness of the revised policies as they prepared for the awareness audit.

The audit findings informed both the plant and the company as a whole of the level of awareness of these revised policies. As stated earlier, we sometimes combine audit designs, and for this audit we included elements of our compliance audit design.

Compliance Audit

The objective of a compliance audit is to examine and document the compliance status of a facility with respect to a set of standards, whether these standards are laws, regulations, or company policies.

A typical compliance audit is targeted at a go/no-go assessment of compliance and, in fact, is often modeled on government inspections that focus on regulatory command-and-control requirements. This audit design is most commonly used in the environmental area where compliance with command-and-control regulations is a key plant objective. The compliance audit design enables audited plants to discover vulnerabilities and to improve performance against these criteria.

We have, on occasion, attempted more sophisticated assessments of compliance with gradations in scoring replacing the go/no-go criteria. The previously discussed safety policy awareness audit is one that used this approach in a combined awareness/compliance audit. Compliance, in that case, related to Polaroid's safety policies and not to governmental requirements.

The primary response of that audit was to assess the awareness of revised companywide safety policies, but we decided to include scoring as a way to also determine the level of implementation, or compliance. The score was on a scale of zero to four, using the method described in Fig. 16-4.

---

4   EXCELLENT.
There is clear evidence that this policy's objectives are met. This instruction is substantially implemented.

3   VERY GOOD.
There is widespread evidence that this policy is understood and that its objectives are being substantially met. Responsibilities under this instruction are understood and accepted, and there is frequent evidence of implementation of the procedures.

2   GOOD.
There is frequent evidence of accomplishment of the objectives of this policy. Responsibilities under this instruction are accepted and there is usually evidence of implementation of the procedures.

1   FAIR.
There is some evidence of accomplishing the objectives of this policy. There is infrequent evidence of implementation of the procedures.

0   POOR.
There is no clear evidence that the objectives of this policy are being seriously addressed. There is no apparent awareness of this instruction.

Figure 16-4. Safety policy audit scoring method.

In addition to numbers (0, 1, 2, 3, 4), the scoring method allowed for evaluative judgments (poor, fair, good, very good, excellent). The definition of each of these five levels of performance also included words (no, some, frequent, widespread, and clear) related to the level of audit observations.

Scoring invariably has a subjective element, and most individuals do not like to be scored. While scoring can be contentious, it does command attention. But scores need to do more than command attention, they need to be understood. The reason for including numbers, judgments, and observations at each performance level is to enrich the conversation and increase the level of understanding about performance. The conversation is more important than the score, since the purpose of the audit is to assist continuous improvement.

In the safety policy audit, each plant was given a score for each of the revised safety policies—33 in all. An overall score, averaging each of the individual policy scores, was calculated for each plant. This overall score was an assessment of the level of implementation of the new policies. The score for each individual safety policy enabled a plant manager to identify those safety areas that needed the most improvement.

Since this was an audit of 15 major plants, the scoring also had value companywide. Based on average scores in the company for a particular safety policy, general weaknesses related to implementation of that policy could be identified. Where scores were low, plans were developed to better communicate (knowledge), or to increase awareness (training), related to those policies.

Management System Audit

The management system audit has the broadest focus in that it strives to examine the managerial, human, engineering, and equipment systems in place for compliance or for health, safety, and environmental protection. One of our audits evaluated performance against a management system model that was originally developed at Polaroid for safety prevention but which is now also used for prevention of environmental spills or releases.

This model, known as "The Preventer," describes a management system built on six safety performance areas:

- Knowledge
- Ability
- Motivation
- Design
- Maintenance
- Actions of others

The flowchart, Fig. 16-5, indicates how these six performance areas are linked in a management system model that encourages implementation of preventive measures for safety—or environmental—incidents.

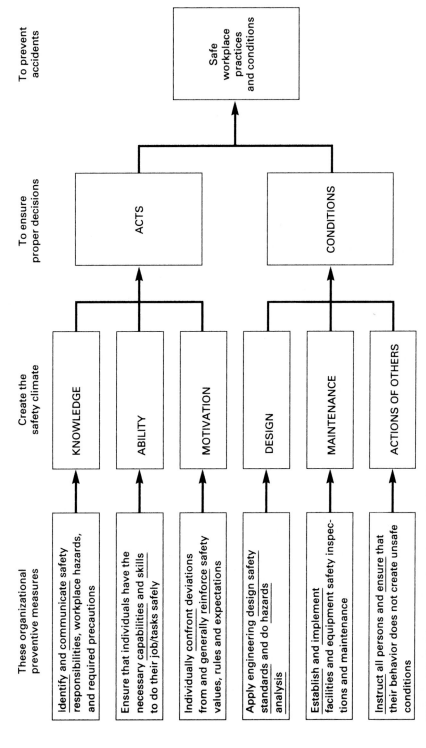

Figure 16-5. Management system audit flowchart.

A variation of this model—the mirror image of Fig. 16-5—is routinely used to investigate accidents and environmental releases. The objective of such an investigation is to determine the cause of an unacceptable result, and the plant manager uses the model to identify those deficiencies in the six performance areas that contributed to that result.

While the mirror image is used to analyze unacceptable results, The Preventer model strives to prevent unacceptable results, and the management system audit based on The Preventer assists line managers by examining the effectiveness of their organization in each of these six areas.

The Preventer audit included an evaluation of each of the six performance areas using a scoring system similar to the one described earlier. Figure 16-6 describes the scoring method for knowledge, one of the six performance areas.

Each of the six areas was assessed during the audit and assigned a score according to a series of definitions that included a number (0, 1, 2, 3, 4), a judgment (poor, fair, good, very good, and excellent) and observations (very little, some, usually, widespread, and clear and comprehensive). Scoring is not a common feature of our audits, but when we use it, we keep the methodology as sim-

---

4   The organization is EXCELLENT at identifying and communicating personal safety responsibilities, workplace hazards, and required precautions. When sampled, there was CLEAR and COMPREHENSIVE evidence that obligations were understood, accepted, and implemented.

3   The organization is VERY GOOD at identifying and communicating personal safety responsibilities, workplace hazards, and required precautions. When sampled, there was WIDESPREAD evidence that obligations were understood, accepted, and implemented.

2   The organization is GOOD at identifying and communicating personal safety responsibilities, workplace hazards, and required precautions. When sampled, there was USUALLY evidence that obligations were understood, accepted, and implemented.

1   The organization is FAIR at identifying and communicating personal safety responsibilities, workplace hazards, and required precautions. When sampled, there was SOME evidence that obligations were understood, accepted, and implemented.

0   The organization is POOR at identifying and communicating personal safety responsibilities, workplace hazards, and required precautions. When sampled, there was VERY LITTLE evidence that obligations were understood, accepted, and implemented.

Figure 16-6. Example of Preventer scoring: Knowledge.

Score

Knowledge   ———
Ability   ———
Motivation   ———
Design   ———
Maintenance   ———
Actions of Others   ———
                             ———
Total Score   ———

Scoring Key:

4   EXCELLENT. There was clear and comprehensive evidence that required preventive measures were understood, responsibilities accepted, and procedures implemented.

3   VERY GOOD. There was widespread evidence that the required preventive measures were understood, responsibilities accepted, and procedures implemented.

2   GOOD. There was usually evidence that the required preventive measures were understood, responsibilities accepted, and procedures implemented.

1   FAIR. There is some evidence that the required preventive measures were understood, responsibilities accepted, and procedures implemented.

0   POOR. There was very little evidence that the required preventive measures were understood, responsibilities accepted, and procedures implemented.

Figure 16-7. The Preventer: critical areas of safety performance.

ilar as possible. This helps the auditor develop skills and assists in communication to audited plants.

In addition to individual scores on each of the six performance areas, an average total score was calculated for each plant and reported as shown in Fig. 16-7. Average scores were also reported companywide. Plants could see how their performance compared to that of other plants, and, since scoring was consistent on all audits, it pointed the way to plants from which they might learn ways to improve. Companywide comparative scores also helped Polaroid identify internal best practices in each of these six areas.

Audit Process and Structure

Polaroid's audit process and structure, in its entirety, includes the following elements:

- Audit quality group
- Audit subjects

- Plant selection
- Audit teams
- Audit findings
- Audit response
- Shared learnings

We have attempted to assemble these elements in ways that also support our audit purpose of assisting line management. For example, while we do make some use of outside third-party consultants as auditors, we largely depend on Polaroid environmental specialists to conduct audits. Many of these individuals work within plants and learn as part of the audit process.

Audit subjects are also selected with plant participation to help ensure that audit resources are focused on subjects of value to plants.

Overall management of the audit process is the responsibility of the audit quality group. This group consists of members of the Corporate Office of Health, Safety, and Environmental Affairs as well as professionals from other parts of the company who report into the plant manager line structure. With such diverse membership, audit support to the line organization is present at many levels. Corporate staff are particularly sensitive to regulatory and public policy trends that might impact the company as a whole. Members of the plant organizations are the most knowledgeable about those local needs that are to be served by the audit. The chairperson of the audit quality group balances these needs to ensure that the audit process satisfies the purpose of Polaroid's audit.

The audit quality group approves the ongoing 12-month audit calendar, which identifies corporate audit subjects and selects plants to be audited. The audit design is also identified by this group. Most audits are announced in advance, although, on occasion, unannounced audits are performed. Over the course of a typical 12-month cycle, four audit subjects are covered company-wide.

This group, either collectively or by delegation to individual members or sub-committees, reviews and approves written audit findings, which are sent to plants. The group also monitors plant responses to audit findings. The group periodically publishes data on the timeliness of the audit process, both findings and responses, and provides executive reports to senior management.

Health, safety, and environmental professionals from throughout the company participate in selection of audit subjects. Individuals from most plants meet together at least monthly to discuss health, safety, and environmental issues. Since most of Polaroid's manufacturing facilities in the United States are geographically clustered in eastern Massachusetts, frequent face-to-face meetings are practical. One agenda item at these meetings is advice to the audit quality group on future audit subjects.

Not all plants are audited on all subjects, so plant selection for an audit is based on the relevance of the subject to the plant, as well as the frequency of auditing of that plant. The goal is for most plants in the United States to be audited at least annually. Some plants, because of their size or complexity, will experience multiple audits over a 12-month period.

Typically, audit teams consist of internal specialists led by the team member who is most expert in the audit subject. When consultants are used as team leaders to augment internal resources or to inject outside expertise, internal specialists also participate. Participation in an audit team is one way for internal specialists to learn and to grow professionally. Before embarking on an audit, all team members are trained, by the team leader, on subject matter as well as on auditing techniques or methods such as scoring.

The size of the team depends on the audit design that is selected and on the complexity of the subject matter. A typical team is composed of three or four individuals. In some cases, such as those audits that narrowly target very specific compliance subjects, one person may conduct all audits.

The structure of audit teams is intended to gain a high degree of involvement and participation from individuals throughout the company. For any given audit subject there may be 10 or more people serving on various audit teams. Each team includes at least one individual who participates in most of the audits for that subject. A member from the plant being audited participates on the audit team for the purpose of gaining immediate insights on plant performance. In this way, audits also serve as a kind of peer review.

At the conclusion of an audit, the audit team conducts an exit interview with plant management, informing them verbally of significant audit findings. Since each audit team includes a member from the audited plant, first-hand knowledge of findings is available to and resides in the plant.

Written audit findings are prepared after the audit and reviewed by the audit quality group for content, consistency, and quality, and the findings may request an audit response by the plant. In addition to audit content, the audit quality group monitors and reports on the timely submission of audit findings and on the timeliness and completeness of requested audit responses.

The shared learnings meeting is the vehicle that ties together the continuous improvement purpose of the audit on a companywide basis. The audit structure is a sampling of different plants on different subjects over a 12-month cycle.

The structure intentionally involves a diverse group of individuals and plants in the audit process. However, an individual plant may experience only one audit on one subject during a 12-month cycle.

At the shared learnings meeting, the audit quality group convenes management and others from throughout the company to review significant generic learnings from all audits. In this way, plants that were not audited for a particular subject will gain the benefit of learning from those plants that were. With this knowledge, they can examine their own operations and develop an appropriate internal response for their plant.

Through these shared learnings meetings, the company at large is able to take advantage of audit learnings. Even though all plants are not audited for all subjects in a 12-month cycle, by participating in the shared learnings meeting, each plant has the benefit of audit findings for each subject.

Conclusion and Next Steps

It may be a universal truth, given the human condition, that people do not like to be audited. Polaroid's program has encountered its share of difficulties in this regard, despite the fact that the purpose of the audit is to assist line management. It is difficult to change the traditional perception of an auditor from that of an inspector, who is looking to find fault, to that of a doctor, who is helping to diagnose a problem so that it can be corrected. Our program strives to have the auditor perceived as a partner with the line organization in a continuous improvement activity.

A metaphor that has been useful in describing this role of an auditor is that of a sparring partner. Even the best of the professional boxing world's champions rely on skilled sparring partners to help them get ready for a major fight. The sparring partner helps the champion anticipate what might happen in the boxing ring and prepare for the real thing. At Polaroid, auditors are sparring partners for plant managers. They help plant managers anticipate and prepare for realities. For instance, if the subject is compliance, the auditor, as sparring partner, does the things a government inspector might do to test the plant's effectiveness in meeting its compliance objectives.

One of the reasons we have involved plant HSE professionals heavily in the audit process is to further this notion of partnership. Members of plant staff who participate in the selection of audit subjects and in choosing the awareness, compliance, or management system design, do so to reflect the needs of their plants. Plant staff, as members of audit teams, discover audit findings on the spot and are immediately able to help their plant organization respond.

Global partnership in this regard is an emerging area for continuous improvement. Plants outside the United States have participated only in those audits related to corporate policies, such as safety and The Preventer. Compliance audits typically relate to United States government regulations and are limited to facilities in the United States. These audits, because of command-and-control regulatory demands, have consumed most audit resources in recent years.

As health, safety, and environmental practices become harmonized among European countries, Polaroid environmental professionals in Europe have formed a working group that will evolve into an audit quality group for that region. Global harmonization of environmental issues is an emerging worldwide public policy trend. International developments, particularly within the European Union, that support creation of corporate environmental management systems are likely to be the basis for harmonizing all of our audits worldwide.

The European notion of eco-audits as an integral part of a corporate environmental management system is consistent with Polaroid's continuous improvement audit cycle. The next phase of worldwide auditing within our Polaroid audit structure will likely combine an awareness audit of environmental management systems with a management system audit of some of the system elements themselves.

European environmental management system elements include:

- Program objectives
- Standards of performance
- Performance indicators
- Measurement methods
- Roles and responsibilities
- Planning and resources

These fit readily into our structure and, in fact, are related to the key HSE activities described at the beginning of this chapter. One of our next steps will be to reexamine those activities as well as all other aspects of our program and again ask, What is the purpose of our audit? as we begin another cycle of worldwide continuous improvement.

17

Environmental Management Audits at the Tennessee Valley Authority

A Key to Quality Environmental Performance

John R. Thurman, James W. Bobo, Madonna E. Martin, and David C. Fuller

Tennessee Valley Authority

Introduction

Regulatory and Enforcement Developments in the United States

On February 3, 1992, in West Palm Beach, Florida, United States Sugar Corporation (USS) was sentenced to pay a record fine in a guilty plea to an eight-count felony charging violation of the Resources Conservation and Recovery Act (RCRA).[1] USS was fined $3,750,000, the largest fine ever imposed for RCRA-related criminal violations. The chief executive officer of USS was present at the sentencing and made a public statement on behalf of the corporation admitting guilt and promising future compliance with all Environmental Protection Agency (EPA) regulations and environmental laws. The conviction resulted from USS's illegal transportation of hazardous waste without a manifest, transportation of hazardous waste to a nonpermitted facility, and illegal disposal of

445

hazardous waste, which included thousands of gallons of spent solvents. This was just one of many recent cases where corporations and their officers and managers have been found guilty of environmental crimes and have received large fines and even jail sentences.

In fiscal year (FY) 1993, EPA reported that it brought a record number—more than 2100—of administrative, civil, judicial, and criminal enforcement actions.[2] The actions included 140 criminal cases, 1614 administrative penalties, 338 civil judicial cases, and 18 enforcements of existing consent decrees.

Specifically, EPA reported that during FY 1993:

- It levies an estimated $133.5 million in criminal fines and civil penalties, including a 66 percent increase over the previous year in RCRA-related sanctions

- The record 140 criminal cases represent a 31 percent increase over the previous year's record of 107 cases

- Criminal charges were brought against 161 defendants, resulting in 135 convictions and 943 months of prison sentences

- EPA emphasized that not only is it assessing large penalties to recapture the money saved by environmental violators, but also it is requiring these parties to change the way they do business

In addition to EPA's increased enforcement activity, there has been a change in whom the actions are taken against. Until the late 1980s, state and federal environmental enforcement activity was primarily aimed at companies rather than individual officers, managers, and employees. Now, the enforcement focus has changed from companies to individuals. The government learned that companies were accepting environmental fines as a cost of doing business, and overall compliance trends were not improving. Recent legislation and regulatory changes make it easier for the government to criminally prosecute individuals. Figure 17-1 shows how the enforcement environment has changed in the last few years.[2] It can now be said that Companies don't go to jail, people do.

As we move further into the 1990s, more and more companies are realizing that environmental compliance must be viewed as an integral part of doing business. Accordingly, responsible companies have placed a high priority on developing environmental management systems to ensure effective, long-term compliance, and they are incorporating environmental requirements into business operations.

Expanding Global Interest in Environmental Management Systems

The environmental management movement has expanded internationally. In conjunction with the 1992 international environmental summit held in Rio de Janeiro, the United Nations requested the International Organization for Standardization (ISO) to consider the merits of having an environmental manage-

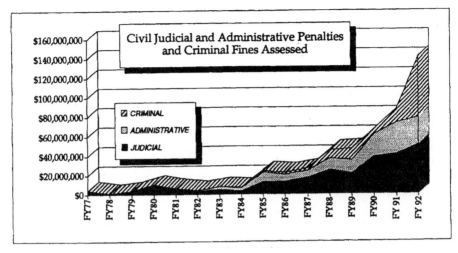

Figure 17-1. The changing enforcement environment.

ment standard. The ISO had previously established, in August of 1991, an ad hoc Strategic Advisory Group on the Environment (SAGE).[3] One of the charges ISO issued to SAGE was to recommend an overall strategic plan for environmental performance and management standardization. Subsequently, SAGE established a subgroup for environmental management auditing. In the near future, there will be ISO certification programs for environmental management systems and environmental management auditing. Companies may gain a market advantage by having ISO-certified environmental management systems and environmental management audit programs.

The United Kingdom has developed an environmental management standard.[4] Both the United States and Canada have drafted environmental management system standards that are now under review. The U.S., Canada, and numerous other international standards development groups are also working on environmental management audit standards.

Elements of an Environmental Management System

To achieve effective, long-term compliance and assume a leadership role in environmental protection, a company must have an effective environmental management system. Such a system provides the company with the organizational structure, procedures, processes, and resources for achieving and maintaining quality environmental management.

It is interesting to note that the U.S. Department of Justice (DOJ) and the U.S. Sentencing Commission (USSC) encourage implementation of an "effective environmental compliance program."[5,6] For example, USSC guidelines list the following elements that judges are to consider as mitigation factors for sentencing:

- Line management attention to compliance
- Integration of environmental policies, standards, and procedures
- Implementation of auditing, monitoring, regulatory, and tracking systems
- Establishment of regulatory training
- Institution of incentives for compliance
- Institution of disciplinary procedures
- Continuing evaluation and improvement
- Additional innovative approaches

While the above elements are not referred to by the USSC as components of an environmental management system, their similarity to management system elements is readily visible. Significantly, environmental auditing is common to all of the various environmental management system development efforts as well as to the DOJ and USSC guidelines. It is difficult, if not impossible, for companies to meet due-diligence requirements without having an environmental management system that features an environmental audit function.

The environmental management system comprises several interrelated components, and a number of these basic elements are common to each of the international environmental management system development efforts:

- A formal, written corporate environmental policy
- A pollution prevention program featuring reduction, reuse, and recycling
- Integration of environmental standards, objectives, and meaningful performance measurements into business planning and budgeting
- Clearly defined roles and responsibilities for all personnel
- A formal training program that includes all personnel
- A system that holds both managers and employees accountable for their compliance duties and promotes environmental excellence
- An environmental record-keeping system
- Routine self-assessments that provide management feedback on compliance status
- Periodic audits that check compliance status and verify the effectiveness of the management system and controls
- Timely corrective action on deficiencies identified in self-assessments, audits, and regulatory inspections
- Communication, both inside and outside the company, of compliance information and environmental policy, goals, and objectives
- Periodic risk assessments that are used for planning and budgeting

The complexity of a facility's environmental management system is determined by the level of environmental risk inherent to the facility. (Environmental risk can be defined as a factor or factors that adversely affect operations and/or the environment.) The level of risk is a product of the probability of occurrence and the consequence of its occurrence. Some examples of typical risk factors that companies must use to determine how to best design and implement an environmental management system include:

- Facility size, design complexity, and age
- Compliance history
- Regulatory climate
- Potential financial liability
- Shareholder, public, and other group expectations

Environmental Management Auditing

Defining Environmental Management Audits

As the 1980s began, environmental auditing was in its infancy and hardly known or used. By the mid-1980s, a number of companies and several federal agencies were conducting environmental compliance audits. Also, many engineering consulting firms were getting into the business of conducting audits. In 1986, EPA published an environmental auditing policy statement in the Federal Register.[7] In this policy statement, EPA defined environmental auditing as "any systematic, periodic, and objective review by a firm (or regulatory entity) of facility operations and practices related to meeting applicable requirements." EPA's policy outlined the elements of an effective auditing program:

- Independence
- Top management support
- Adequate resources and staffing
- Explicit audit program objectives, scope, resources, and frequency
- A process which collects, analyzes, interprets, and documents information sufficient to achieve audit objectives
- Procedures
- Quality assurance

These elements were very similar to the standard for internal auditing published by the Institute of Internal Auditors.[8]

The EPA policy statement also encouraged companies to go beyond typical compliance audits and evaluate the extent to which "systems and procedures" provide for:

- Development of organizational environmental policies
- Training and motivating personnel
- Communicating environmental information
- Providing for trained and qualified personnel
- Incorporating environmental protection into operating procedures
- Applying best management practices and good housekeeping
- Initiating preventive and corrective maintenance systems
- Exploiting source reduction, reuse, and recycle potential
- Auditing risk and uncertainties

EPA stated that environmental management audits have the potential to provide agencies and corporations with more useful and valuable information than compliance audits. By the 1990s, as previously mentioned, international standard development groups and a number of corporations were advocating EPA's position and promoting management audits as the key to achieving and maintaining effective, long-term compliance.

Differences between Compliance Audits and Management Audits

Most environmental audit programs are initially established to determine the status of environmental compliance with regulations and company procedures. They are not designed to evaluate management controls and to look for probable causes of deficiencies. These compliance audits identify specific nonconformances and report this information to management. As such, they tend to focus on identifying symptoms rather than probable causes of problems. Consequently, compliance audits are limited in their usefulness to management because they do not provide sufficient information regarding causes of problems, and they provide little basis for effective, long-term compliance because they are not focused on identifying risks. Accordingly, the problems will in all likelihood recur. This can be compared to treating a sore shoulder by injecting cortisone. The pain goes away for a while, but it is bound to return. The cause of the problem was not addressed, and proper corrective action was not taken.

There is an increasing recognition throughout industry that auditors as professionals have a responsibility to inform management of environmental risks and the effectiveness of the management system and controls to manage those risks. The scope of the typical compliance audit simply is not sufficient to accomplish this.

In contrast, environmental management auditing can be defined as a systematic evaluation to determine if the environmental management system is adequate to control risks and help the corporation meet regulatory requirements and its environmental policy, procedures, and goals.[9] In management audits, the auditor probes for the causes of the problem such as lack of procedures, insufficient training of personnel, or lack of funding. Deficiencies are reported to management with the audit team's best assessment as to the probable cause and an appropriate recommendation for timely, corrective action. The objective of the management audit is to work closely with the client (the auditee) to:

- Evaluate the effectiveness of the environmental management system to control risks
- Identify risks and develop effective corrective action plans
- Determine compliance status with regulations and company policies and procedures
- Help the client gain a better understanding of compliance requirements

Figure 17-2 outlines the major differences between compliance and management audits.

The term "environmental management audit" may connote the lack of a compliance evaluation during the audit. On the contrary, an environmental management audit includes a significant amount of effort in determining the status of compliance with federal, state, and local environmental regulations and company policies and procedures. Detailed compliance information is the evidence the auditors gather to determine the effectiveness of the management system and controls. Another way to define the environmental management audit might be to call it a "compliance-plus audit."

Compliance	Management
Emphasis is on noncompliance with regulatory requirements	Emphasis is on evaluating the environmental management system and identifying risks. Looks for noncompliance with regulatory and company requirements and effectiveness of management controls. Nonconformances are the evidence or indicators used to evaluate the environmental management system.
Reports specific nonconformances	Reports specific nonconformances but also examines probable causes.
Recommendations for corrective actions are narrowly focused	Provides expanded recommendations including management controls needed to prevent recurrence of the nonconformance.

Figure 17-2. Compliance versus management audits.

Skills Needed to Conduct Management Audits

To conduct quality management audits, the audit team needs more than basic audit experience and knowledge of regulatory requirements. The team must be knowledgeable and proficient in the following areas:

- Facility operations
- Management systems and controls
- Information gathering—especially interviewing, observation, and sampling
- Environmental regulations
- Company policies and procedures
- Written and oral communication
- Interpersonal skills

During a management audit, auditors will normally be interviewing a large number of persons, including high-level managers such as facility managers and company vice presidents, as well as lower-level staff. Therefore, auditors must have substantial audit experience and excellent interpersonal and communication skills. It is critical that auditors conduct themselves professionally and maintain credibility with managers and employees at all levels of the organization. It is one thing to simply report instances of nonconformance and make straightforward recommendations as is normally done in a compliance audit. It is quite another thing to attempt to explain to a facility manager—and gain agreement—that the facility should have an environmental management system that may require significant additional resources. Auditors must be able to identify and talk with managers about real risks to the business operation and be able to explain how the management system and controls can help minimize these risks and protect the facility, company, and management and employees.

Management audits typically require more time to plan and conduct than compliance audits. Careful crafting of interview questions, audit plans, audit guides, and checklists is time-consuming, and management auditors must be seasoned and experienced. These attributes are critical for successfully conducting management audits. It is not realistic to expect new, inexperienced auditors to conduct management audits.

Conducting the Environmental Management Audit

The Audit Process

Management audits should be conducted in accordance with modern, accepted internal audit standards such as the EPA policy statement guidelines and the

Environmental Auditing Roundtable standards.[7,10] An environmental management audit, like any internal audit, consists of three phases:

- Pre-site-visit activities
- On-site activities
- Post-site-visit activities

The objective of the pre-site-visit phase is to provide the audit team with sufficient information about the facility to allow team members to arrive onsite prepared to focus their efforts in order to effectively evaluate the facility's environmental management system and compliance status. The following are standard steps associated with this phase:

- Select facility
- Define scope
- Schedule visit
- Select team
- Gather and review facility background material
- Develop audit plan
- Develop audit guides/checklists
- Plan and coordinate
- Arrange entrance conference

The objective of the on-site visit phase is to verify information obtained in the previsit phase, reconcile discrepancies between regulatory requirements and evidence of performance, gather additional evidence, and evaluate cause-and-effect relationships. The following are the elements of the on-site visit phase:

- Understand the facility and its environmental management system
- Gather evidence through interviews, observation, sampling, and records
- Evaluate evidence against criteria
- Report results to management

The objective of the post-site-visit phase is to communicate audit results to the proper levels of management. The following are the associated elements of the post-site-visit phase:

- Prepare draft report
- Conduct audit report peer review

- Prepare final report
- Distribute final report
- Track and close audit items

Use of Audit Guides

The audit plan for a management audit is developed to ensure evidence is gathered in such a way that potentially harmful risks are identified. The audit plan establishes the audit objective, scope, coordination, and methods to be used. It is the basic outline for the audit. Audit guides and checklists are then developed for each regulatory area, e.g., hazardous wastes, underground storage tanks, and oil management. These tools provide the auditor with a road map to follow during the audit that will lead to a thorough assessment of adequacy of controls in each regulatory area and a reasonable determination of the facility's compliance status and of the facility's overall environmental management system.

Audit guides direct the auditor to gather information for each regulatory area included in the audit scope to determine if adequate controls are in place. For example, the audit guide for polychlorinated biphenyls (PCBs) would require the auditor to evaluate the following:

- Are staff PCB roles and responsibilities clearly defined?
- Are PCB risk assessments completed?
- Are staff trained and qualified in PCB responsibilities?
- Are PCB records in order?
- Are PCB procedures in place and adequate?
- Is there a PCB self-assessment program?

The audit guide is a plan that forces the auditor to examine in a logical, objective, and methodical manner whether adequate controls are in place to protect the facility from risk and also to ensure an accurate determination of the facility's compliance status. A detailed compliance checklist is used to determine the compliance status in each regulatory area. Figure 17-3 shows an example audit guide for PCBs.

Verifying the Environmental Management System

During the audit, the management auditor gathers compliance information (the evidence) and evaluates the level of risk to which the facility is exposed. The auditor determines whether the management system and controls are adequate to manage risk and works with the client to develop corrective action plans for deficiencies. Elements of the environmental management system that are evaluated are listed on page 462.

			Auditor
Facility: ————————	Auditor: ————————	Date: ———————	

Audit step	Reference*	Instruction	Auditor note page
I. *Presite visit*			
A. *General applicability*	N/A		
1. *Determine if facility has oil-filled electrical equipment and/or associated tanks.*		This information is obtained by the lead auditor through telephone interview.	
2. *Determine if oil at facility contains PCBs at detectable levels.*		This information is obtained by the lead auditor through telephone interview. Also, check previous audit reports and file materials such as compliance procedures, oil spill control plans, and correspondence.	
B. *Regulatory review*	N/A	Review regulations (be sure to review regulatory definitions) for each applicable audit step.	
C. *Files review*	N/A	Review files to see if there is PCB-related correspondence and review facility PCB procedure.	
D. *Develop personal checklist*		Develop checklist as needed for use at site.	
II. *Site visit*			
A. *Site overview*	N/A	After entrance meeting, tour the facility with knowledgeable person and gain an overview of the types and kinds of oil-filled equipment.	
B. *Site documents*	N/A	Gather site documents.	
C. *Management controls*			
1. *Risk Assessment and Planning*		Interview environmental manager and other staff as appropriate.	
a. *Determine if facility has completed a PCB risk assessment.*	N/A	Interview environmental manager and other staff as appropriate.	
b. *Determine if oil-filled equipment has been tested.*	N/A	Ask environmental manager if lab or kit test was used.	

Figure 17-3. Example of an audit guide for a management audit—PCB audit guide.

	Audit step	Reference*	Instruction	Auditor note page
c.	Determine if facility has PCB procedure and, if so, whether it is adequate.	N/A	Interview environmental manager and ask if facility has PCB procedure. Review procedure to see if it is up to date and covers areas pertinent to the facility. Do they contain basic control elements?	
2.	*Roles and responsibilities*	N/A		
a.	Determine key facility PCB roles: ■ Equipment sampling and inventory ■ Written annual document log and annual records ■ Maintenance of PCB equipment ■ Quarterly and annual inspection ■ PCB storage ■ Leak and spill cleanup ■ Manifest preparation and signing ■ Marking ■ Records		Interview environmental manager and other staff as appropriate.	
b.	Determine if responsibilities of staff involved in PCB compliance are written.	N/A	Interview environmental manager and other staff including Human Resources Officer, and ask to see documentation. Review procedures to ensure clear lines of responsibility are outlined.	
3.	*Training*			
a.	Determine if facility has a PCB training program for above staff.	N/A	Interview environmental manager and other staff as appropriate.	
b.	Determine if PCB training program is documented.	N/A	Interview environmental manager and other staff as appropriate and examine training records.	
c.	Determine if training is adequate.	N/A	Review records and interview environmental manager and other staff as appropriate. Determine if training covers elements needed for facility. Review procedures to see if training program is outlined.	

Figure 17-3. (*Continued*)

Audit step	Reference*	Instruction	Auditor note page
4. *Communication*			
a. Determine if facility manager has communicated importance of PCB compliance to employees.	N/A	Coordinate with lead auditor, who interviews plant manager. Also, interview environmental manager and related staff.	
b. Determine how new PCB requirements are communicated to employees.	N/A	Interview environmental manager and other staff as appropriate. Ask for evidence as to how most recent PCB change was communicated. Check to see if procedure includes latest regulation changes.	
c. Determine if employees and management are aware of PCB risks and liabilities.	N/A	Interview environmental manager and employees. Coordinate with lead auditor's interview of facility manager.	
d. Determine if roles and responsibilities have been communicated to employees.	N/A	Interview environmental manager and others as needed.	
5. *Record keeping*			
a. Determine if facility has a formal PCB record-keeping system.	N/A	Examine PCB records (obtained in Step II.B) for completeness, retrievability, organization.	
b. Determine if record-keeping system is documented in writing.	N/A	Examine records procedure if available. Interview person responsible for record keeping.	
6. *Self-assessment*			
a. Determine if facility has internal assessment functions.	N/A	Interview environmental manager and other staff as appropriate.	
b. If so, determine if it is documented and adequate.	N/A	Evaluate systems for: ■ Thoroughness ■ Follow-up ■ Accountability	

Facility: ————————— Auditor: ————— Date: —————

Figure 17-3. (*Continued*)

Facility: —————————   Auditor: ————————   Date: ——————

Audit step	Reference*	Instruction	Auditor note page
D. *Compliance status*			
1. *General Records*			
a. Determine if facility needs a written annual document log.	761.180	Interview environmental manager and other staff as appropriate, review inspection log, and determine what PCBs are in storage.	
b. Evaluate written annual document log.	761.180	Check logs for previous two calendar years. Check mass balance by reviewing manifests and PCB inspection logs, and compare with written annual document logs.	
c. Evaluate annual records.	761.180	Obtain and review copies of manifests and certificates of disposal for past two years.	
d. Determine if manifests are properly completed.	761.207 .208 .209 .210 .211 .215	Check sample of manifests for past two or three calendar years. Compare information on manifests with written annual document logs. Also, note evidence of spills and leaks by keying in on wastes shipments and refer them to step C.6.	
e. Determine that certificate of disposal (CD) requirements are met.	761.215 .218	Review CDs for past two calendar years.	
f. If facility has PCB transformers, determine if there are inspection and maintenance records.	761.30 (a)(1)(xii) (xii)(xiv) (xix)	Interview environmental manager or others as appropriate and obtain inspection and maintenance logs.	
g. Determine if inspection and maintenance log meets requirements.	761.30 (a)(1)(xii) (xii) (xiv)	Review a sample of transformers in the inspection log. Note evidence of leaks and spills. Cross-reference these with manifest and spill cleanup records.	
h. Determine if leak/spill cleanups have been properly recorded in inspection logs.	761.30 (x)	Check log entries that note leaks/spills and make sure required information has been recorded.	

Figure 17-3. (*Continued*)

Facility: ———————————   Auditor: ——————   Date: ——————

Audit step	Reference*	Instruction	Auditor note page
i. Verify that spill cleanup records are adequate.	761.125 .130 .135	Cross reference evidence of spills found in inspection logs and manifests with documented records of spill cleanup. Check documentation for required information.	
2. *Notification* *a.* If facility has PCB transformers, determine if fire department has been notified in writing.	761.30 (a)(1)(vi)	Ask facility staff for a letter (or memo if facility has its own fire brigade) from facility to fire department that would respond to a fire; review letter.	
b. If transformers are in or near commercial buildings, determine if building owner has been notified.	761.30 (a)(1)(vii)	Check to see if PCB transformers are within 30 meters of commercial buildings.	
c. If storing PCBs for more than 30 days, has EPA been notified?	761.65 (b)	Interview environmental manager and determine facility storage policy. Check manifests and storage area logs for dates.	
3. *Reporting* *a.* Verify PCB spill reporting requirements have been met.	761.125 (a)(1)	Check spill records and determine amount and concentration of PCBs spilled and if proper reporting was made to NRC, EPA, and state.	
b. Verify that EPA was notified if there has been a PCB transformer fire.	761.30 (a)(1)(xi)	Interview environmental manager or other staff and review records.	
4. *Storage* *a.* Determine if PCB equipment, liquids, or wastes are stored at facility.	761.65	Interview facility environmental manager or others and ask if items are stored for disposal or reuse. Be sure to point out that EPA inspectors would need to be convinced about reuse; ask if PCB liquids are stored for disposal or reuse.	
b. Determine maximum length of time PCB items have been stored for disposal since previous audit.	761.65	Interview environmental manager and others to determine if 30-day or 9-month option is used. Examine manifests; compare removed from service for disposal dates with shipping dates.	

Figure 17-3. (*Continued*)

Facility: ————————   Auditor: ————————   Date: ————————

Audit step	Reference*	Instruction	Auditor note page
c. If 9-month storage option is used, determine if facility has notified EPA.	761.65	See II.D.2.c. above.	
5. *Disposal* *a.* Determine if facility is involved in PCB disposal activities.	761.60	Interview environmental manager and others and discuss if PCBs are burned in facility boiler or handled in any other way than shipped to a TSDF.	
b. Verify disposal requirements are met.		Compare facility practices to requirements.	
6. *Leak and spill cleanup* *a.* Determine if staff understand leak and spill cleanup requirements.	761.125 (a)(1)(x)	Interview environmental manager, employees responsible for inspection and cleanup (including operators who conduct shift inspections), and determine if they have: ▪ Been trained ▪ Understand the difference between a leak and a spill ▪ Know the requirements and where to find them	
b. Verify that spills have been properly cleaned up and documented.		Compare spill cleanup records with information from interviews, manifests, transformer logs.	
7. *Marking* *a.* Determine if staff understand label requirements for equipment, containers, storage areas, doors, and entryways.	761.40	Interview environmental manager and employees who conduct inspections.	
8. *Combustible material* *a.* Determine if staff know combustible materials storage requirements.	761.30 (a)(1) (viii)	Interview environmental manager and employees who conduct inspections.	

Figure 17-3. (*Continued*)

	Audit step	Reference*	Instruction	Auditor note page
9.	*Site walk-through* a. Conduct inspections.	N/A	Tour facility with employees who conduct transformer inspections: ■ Select a sampling of PCB transformers ■ Walk-through transformer and switchyards ■ Check OCBs and capacitors as well as transformers. ■ Check PCB waste storage area ■ Be sure to check for containment; labels on equipment, combustible materials, leaks, spills, labels on doors and entryways to storage areas, and PCB transformers. Containment must be complete—caulking not permitted.	
D.	*Overall status* 1. Evaluate adequacy of management controls.	N/A	Note in margin of auditor notes how each nonconformance is handled—i.e., finding, observation, or verbal.	
	2. Determine effectiveness of compliance program.	N/A	Review above information and use compliance status categories form to rate PCB program.	
	3. Evaluate significance of nonconformances noted.	N/A	Determine what items should be included in the report and prepare written drafts of finding and observations.	
	4. Assign overall rating by using Appendix.	N/A	Summarize all observations and findings for exit meeting presentation.	

Facility: ——————— Auditor: ————— Date: —————

Figure 17-3. (*Continued*)

1. Top management support

2. Organizational roles and responsibilities

3. Planning and budgeting

4. Risk assessment

5. Pollution prevention programs

6. Environmental compliance procedures

7. Regulatory review

8. Record keeping and reporting

9. Training

10. Communication

11. Performance evaluation

The following describes each element and how the audit team verifies its effectiveness. Remember that the complexity of the facility's environmental management system will be determined by the level of environmental risks relevant to the facility operation.

Top Management Support.   The key to the effectiveness and success of the environmental management system is the strength of the facility manager's commitment and belief in environmental compliance. Does the manager believe that compliance is an important and necessary part of conducting business? The best way for the manager to demonstrate such a commitment is through his or her actions. It must be more than "management-orchestrated symbolism." The manager should develop a formal environmental policy statement, and this should be communicated to all employees. The policy should describe the manager's vision for compliance and inform employees about such things as performance standards and accountability. The written policy statement should be carefully considered and prepared by the manager, not assigned to lower-level staff. The policy by itself, however, is not enough. The manager must frequently communicate belief in and commitment to the policy both in writing (e.g., company newsletters) and orally. The manager must also demonstrate this commitment through actions. For example, the manager could form an environmental committee that conducts periodic plant environmental walk-throughs and includes environmental compliance staff in management meetings.

During the audit, the auditor verifies the status of this system element by interviewing the facility manager, the environmental engineer, and other staff assigned environmental responsibilities. If a policy exists, staff are interviewed to determine if they are aware of it. Also, the manager and staff are interviewed to determine how employees are held accountable for their actions.

Figure 17-4 shows examples of criteria that pertain to top management support and a possible approach the auditor might take.

Criteria	*Auditor verification*
A. *Policies and goals*	
1. Top management support of environmental excellence.	Ask facility manager if an environmental policy statement has been written and issued.
2. Policy must be from an authority high enough to communicate importance.	Determine the person and title of individual who issued the policy.
3. The policy must be specific enough to allow the company to meet its environmental goals.	Review the policy to see if it contains specific goals and objectives.
4. The policy must be clearly understood throughout the facility.	Ask the facility manager how the policy was distributed.
	Interview selected personnel throughout the facility to determine their understanding of the manager's environmental policy.
B. *Administrative decisions*	
1. Top management demonstrates support for environmental programs through administrative decisions.	Determine if environmental personnel attend staff meetings.
	Evaluate management's support of environmental budget requests.
	Interview environmental personnel to determine if they have adequate resources.
2. Support is demonstrated by creating the right culture for compliance.	Ask the facility manager how he/she communicates goals and expectations to employees.
C. *Knowledge of regulations*	
1. Managers should have a general understanding and appreciation for environmental requirements.	Interview senior managers to determine the extent of their understanding of the regulations.
2. Managers should have sufficient environmental training to enable them to make informed decisions on environmental matters.	Interview senior managers to determine their environmental training and background.

Figure 17-4. Top management support.

Criteria	Auditor verification
D. *Environmental performance measurements*	
1. To assess the effectiveness of the EMS, managers must develop environmental performance measurements.	Determine if the facility has developed environmental performance measures. Verify that the environmental performance measurements track written goals and objectives.
2. Top management's commitment is demonstrated through routine reporting on the status of environmental performance.	Ask the manager how he keeps employees apprised on their environmental performance. How frequently is this communicated? Review status reports.
3. Personnel responsible for environmental management must be held accountable for their performances.	Interview both environmental and nonenvironmental staff to determine if accountability exists for their performance. Determine if there is a system to reward excellent performance.

Figure 17-4. (*Continued*)

Organizational Roles and Responsibilities. If compliance is to be achieved on an ongoing, consistent basis, environmental staffing levels must be sufficient to meet established goals and objectives. In addition, all persons who have compliance responsibilities, including managers, must know and understand their roles and responsibilities. All employee environmental responsibilities should be defined in writing and included in position descriptions and/or in procedures. Compliance responsibilities should ideally be incorporated into operational procedures.

One of the most difficult tasks for the auditor is to assess the adequacy of environmental staffing levels. Some first symptoms of inadequate staffing are:

- Frequent noncompliances
- Excessive overtime
- Not following procedures

The auditor should interview facility management to determine how the organization determines staffing levels and whether requests for additional staff have been made and denied.

During the audit the auditors interview managers and employees with environmental responsibilities. All regulatory area responsibilities (PCBs, solid and hazardous waste, oil management, etc.) are addressed. Employees are asked to describe what their responsibilities are, and their position descriptions are reviewed. Processes and procedures may also designate duties and will need to be reviewed. Managers are asked to describe the duties of their employees, and employees are interviewed to determine if there is a common understanding of who is supposed to do what. Figure 17-5 shows typical criteria and auditor verification steps.

Planning and Budgeting. The business planning process should take into account the potential impact of proposed environmental regulations and legislation, results of audits, regulatory inspections, risk assessments, self-assessments, and industry standards and practices. It should also include corporate and organizational environmental goals and objectives.

To determine whether environmental considerations are factored into operational planning and budgeting, the audit team must review the organization's business plan and interview management and staff. Review of performance

Criteria	*Auditor verification*
A. *Environmental staffing levels*	
1. Environmental staff levels are sufficient to meet environmental goals and objectives.	Determine how the organization assesses environmental staffing needs. Ask if additional environmental staff is needed; if so, have they been approved or denied, and why.
2. Staff must be adequate to assure compliance with regulations, policies, and procedures.	In the course of the audit, look for symptoms of inadequate staffing: ■ Frequent noncompliances ■ Excessive overtime ■ Procedures not followed
B. *Clearly defined and communicated responsibilities*	
1. To hold personnel accountable, environmental responsibilities must be clearly defined, communicated, and understood by all personnel.	Review procedures to identify environmental responsibilities. Review job descriptions for environmental responsibilities. Interview staff to verify that job descriptions and procedures are accurate.
2. Environmental personnel are held accountable through systematic performance standards and appraisals.	Review performance standards for selected managers to determine if they contain environmental goals. Interview personnel to determine if appraisals accurately match performance.

Figure 17-5. Roles and responsibilities.

Criteria	Auditor verification
A. *Environmental planning*	
. Environmental considera- tion is factored into the overall planning process.	Interview budget staff to determine if envi- ronmental planning is part of the strategic planning process.
2. Environmental planning should include both short- and long-term factors.	Interview environmental staff to determine if the facility has a system to identify future environmental projects or regulations that could impact the facility.
B. *Establishing priorities*	
The facility should have a sys- tem to establish priorities that include not only products but environmental considerations as well.	Examine funding priorities to determine if they reflect environmental goals and objectives.
	Investigate environmental projects that have been delayed or canceled.
	Determine if risk assessments are used to set priorities.
C. *Environmental budgeting*	
Sufficient capital and mainte- nance funds must be available for environmental goals.	Determine if environmental concerns are taken into consideration in budget decisions.
	Examine rejected budget requests.
	During the audit, look for symptoms of lack of funding for environmental concerns such as: ▪ Nonconformances with the probable cause being lack of funding. ▪ Poorly maintained pollution control equip- ment.

Figure 17-6. Planning and budgeting.

goals and the status of budget requests for capital, operational, and maintenance projects are important areas of inquiry. Figure 17-6 shows how the auditor can verify how well the planning and budgeting process addresses environmental matters.

Risk Assessment.   Periodic risk assessments can identify conditions where the potential for significant impact to the operation and/or the environment may occur. Risk assessments can provide valuable information to the organiza- tion for the planning and budgeting process and keeping the facility in compli- ance. Risk assessments, if properly done, can help guide the company toward a more efficient expenditure of funds. The old cliche, Pay me now or pay me later, is appropriate in this case.

To evaluate this component, an auditor interviews the facility manager to determine whether the facility conducts risk assessments. Generally, if a man- ager has such a process in place, it features a management environmental review committee that conducts formal, periodic meetings and inspections in coordination with environmental staff. Other staff members are also interviewed to verify how risk assessments are conducted. The approach to evaluating the risk assessment element is shown in Fig. 17-7.

Pollution Prevention Programs.   An important part of an effective environmental management system is a pollution prevention program that features the three R's—reduce, recycle, and reuse. Corporations with leading-edge environmental programs feature aggressive pollution prevention programs. Pollution prevention can significantly reduce risks and usually will result in the company saving money. The challenge for the organization is to pursue pollution prevention aggressively in all regulatory areas and to remain aware of the latest industry advances in this area.

Criteria	Auditor verification
A. *Risk assessment program*	
1. Formal systematic reviews should be conducted to identify risks.	Determine if the facility has a formal risk assessment program.
	Review program elements to ensure the program is complete.
	Interview environmental engineer to determine: ■ How assessments are conducted ■ Who conducts the assessments ■ How often assessments are completed
2. The facility should have mechanisms to mitigate identified risks.	Interview environmental and senior management to determine how results for risk assessments are factored into the budget process.
	Review results of past risk assessments to determine if actions have been taken to mitigate risks.
B. *Environmental reviews*	
1. Environmental reviews should be performed on all new programs or activities that could impact the environment.	Evaluate the environmental review process (for federal facilities this is the NEPA review process) to ensure its accuracy and completeness.
	Determine what criteria are used for assessing the impact of a project.
2. A mechanism should be in place to track environmental commitments made in environmental reviews.	Review past environmental reviews to verify actions are taken on environmental commitments.

Figure 17-7. Risk assessment.

The audit team interviews plant management to determine whether there is a mandated program and, if so, how it operates. The pollution prevention plan is reviewed and each regulatory area (e.g., PCBs, solid and hazardous waste, and water use) is reviewed to determine what is being done and to identify opportunities for improvement. It is essential for the auditor to be aware of the latest developments in pollution prevention so that the team can sell the benefits to management and obtain their buy-in. Figure 17-8 shows how this element can be evaluated.

Environmental Compliance Procedures.   The auditor must also determine whether the facility has environmental compliance procedures and, if so, whether they are adequate, up to date, and being followed. Auditors also interview staff to determine whether they are familiar with procedures and also check to see whether there is a process in place to keep procedures up to date. Figure 17-9 shows a possible approach.

Regulatory Review.   The facility should have access to a system that keeps it up to date with all regulatory changes. An auditor can check the efficiency of the facility's system by asking staff about a recent regulatory change that affects the facility. Figure 17-10 shows an approach for evaluating this element.

Record Keeping and Reporting.   A system should be in place to ensure that all required records are available and retrievable. The system should ensure that all required reports are prepared properly and submitted on time. An auditor checks for the evidence of a system by interviewing staff and conducting a detailed review of required records and reports. Figure 17-11 offers an approach.

Training.   Management and employees must not only understand their roles and responsibilities, but they must also be trained and qualified to carry them out. The best way for an organization to achieve this is to have a formal training program that not only identifies and provides for regulatory-agency-required training, but also includes awareness training of management staff as well as employees. On-the-job training for employees is a valuable training tool that is often overlooked. Hands-on training is often more effective than classroom training. The training program should be documented and include a system that ensures all employees receive the required training.

An auditor should ask the facility manager, other managers, and employees to describe how training is completed. Records should be inspected to verify whether mandated training is completed. One important element of the formal training program is the development of a matrix listing all training and individual positions that need the training. Figure 17-12 shows a suggested auditor approach for evaluating the training element.

Communication.   Management should ensure that environmental policy, goals, objectives, and performance standards are clearly and effectively communicated throughout the organization. In addition, staff with compliance respon-

Criteria	Auditor verification
A. *Pollution prevention programs*	
1. Programs should be in place to identify and reduce sources of pollution.	Determine what pollution prevention programs are in place.
2. Pollution prevention programs should be formalized in written procedures or plans.	Review pollution prevention documentation to ensure it contains the following elements: ■ Clearly defined responsibilities ■ Identification of sources ■ Applicable regulations ■ Training ■ Recordkeeping ■ Program evaluation
3. Pollution prevention programs should result in a reduction of pollutants.	Verify the reduction of pollutants for several programs.
B. *Preventive maintenance programs*	
1. Preventive maintenance programs should be in place to ensure proper operation of pollution control equipment.	Identify facility's pollution control equipment. Determine if preventive maintenance programs exist for identified pollution control equipment. Examine preventive maintenance documentation to ensure it contains the following elements: ■ Clearly defined responsibilities ■ Maintenance procedures ■ Applicable regulations ■ Training ■ Schedule ■ Record keeping ■ Program evaluation Verify preventive maintenance procedures are followed.
2. Effective preventive maintenance programs should result in increased availability of pollution control equipment.	Verify the availability of pollution control equipment.

Figure 17-8. Pollution prevention.

Criteria	Auditor verification
1. ECPs should be written and implemented for all regulatory areas.	Identify regulatory areas applicable to the facility. Verify that ECPs are in place for all areas. Ensure that each procedure: ■ Is issued by senior management ■ Is reviewed for technical adequacy ■ Contains clearly defined responsibilities ■ Is complete ■ Is accurate ■ Defines training requirements ■ Defines a recordkeeping system.
2. Employees should have access to and clear understanding of applicable ECPs.	Interview employees to verify they: ■ Have copies of ECPs ■ Understand their responsibilities and ECP contents.
3. A mechanism should be in place that ensures procedures are periodically reviewed and updated.	Interview environmental personnel on the process used to review procedures. In conducting audit, look for examples of where procedures are out of date.

Figure 17-9. Environmental compliance procedures (ECPs).

Criteria	Auditor instructions
1. The facility should have a system to track and interpret new/changed regulations.	Select a new or a recent change to a regulation, and determine how it gets incorporated into the ECPs and brought to the attention of appropriate facility staff.
2. Facilities should have a mechanism that factors new/changed regulations into the budget planning process.	Interview environmental personnel to determine how the additional resources for new regulations will be factored into the budget planning process. Verify how a new regulation was incorporated into this process.

Figure 17-10. Regulatory review.

Criteria	Auditor instructions
1. A record-keeping system should be developed that will ensure that all environmental requirements are verifiable.	Have the environmental manager describe the record-keeping process. Look for evidence that record-keeping system is: ▪ Complete ▪ Easily retrievable Verify that the record-keeping system is documented and includes document retention times.
2. There should be a mechanism that ensures all required reports and commitments are reported on time.	Ask the environmental manager how he/she knows reports or commitments will be reported on time. Look for evidence that all reports or commitments have been submitted on time.
3. Each facility should have a system to respond to environmental emergencies.	Examine the facility's emergency response procedure to ensure it includes the following actions: ▪ Investigate the cause ▪ Report to appropriate authorities. ▪ Correct the root causes ▪ Track incidents ▪ Monitor trends

Figure 17-11. Record keeping and reporting.

sibilities should be current on all new regulatory and policy and procedural requirements. It is also important that they be frequently reminded, orally and in writing, of management's commitment to compliance.

An auditor interviews management to determine how environmental matters are communicated. Employees are also interviewed to determine how communications occur and how effective they are. For example, the facility manager may inform the auditor of a performance goal for the year to eliminate 20 percent of the plant's chemical inventory. The auditor would inquire as to how this was communicated and then verify with the environmental manager and purchasing staff whether this goal was communicated and implemented. Effective ways to communicate management commitment include:

▪ Forming a management environmental review committee

▪ Including the environmental manager in management staff meetings

▪ Writing environmental information in plant newsletters

▪ Conducting management team environmental inspections

Figure 17-13 on page 473 provides an approach for the auditor.

Criteria	Auditor instructions
A. *Training program*	Determine if the facility has developed a formal training program.
1. There should be a process that determines personnel training needs.	Verify that the program satisfies all training needs.
2. The training program should ensure all employees are provided quality training in all identified areas.	Examine the training program to verify it provides the following: ■ Ensures all personnel receive mandated training ■ Ensures only qualified trainers are used ■ Training content meets the specific needs of the facility ■ Documents training attendees and course contents ■ Ensures new employees are promptly trained
B. *Program effectiveness*	
Training must be conducted so that personnel having environmental responsibilities are fully knowledgeable to carry out their assigned duties.	Interview selected personnel to ensure they have a clear understanding of: ■ Their environmental responsibility ■ Applicable regulations ■ How to complete required record keeping. Identify evidence where nonconformances are caused by lack of training.

Figure 17-12. Training.

Performance Evaluation. In addition to performing risk assessments and audits, another excellent control mechanism that ties back into the training program is the self-assessment or self-monitoring function. Facilities need more than audits to help them continuously improve and maintain compliance. Periodic compliance evaluations (self assessments) by central staff personnel or the facility environmental manager provide the facility with worthwhile information regarding the status of compliance and the effectiveness of controls. Because these evaluations are not audits and are informal, they are less disruptive to operations and problems can be addressed at lower levels of management. An auditor can verify the status of self assessments by using the approach shown in Figure 17-14.

Determining the Results of the Management Audit

When the audit team has gathered the required information needed to meet the audit objective and scope, the results must be compared to established criteria to

Criteria	Auditor instructions
A. *Management commitment to environmental excellence*	
Management should continually strive to create a culture conducive to environmental compliance.	Look for some of the following activities as evidence that management is communicating their commitment to the employees: ■ Frequent communication of goals ■ Environmental newsletters ■ Employee management environmental committees
B. *Solicitation of employee concerns*	
1. There should be a mechanism whereby employee environmental concerns are solicited and acted upon.	Interview the facility manager to determine: ■ The importance he/she places on getting employee input ■ How he/she actively solicits this input ■ What process is used to act upon input received
2. Such a channel of communication should create an awareness of senior management support for environmental matters.	Interview employees about their understanding of senior management's: ■ Acceptance of employees' input ■ Commitment to protecting the environment
C. *Environmental awareness*	
Environmental awareness should be continually reinforced throughout the facility.	Interview the environmental manager to determine how employees are made aware of the importance of environmental protection. Compare the environmental awareness program to the health and safety program.

Figure 17-13. Communication.

determine what and how information will be reported to management. The established criteria are:

■ Federal, state, and local regulatory requirements

■ Company policies and procedures

■ Effective environmental management system elements

The auditors must consider the probable causes of deficiencies, determine the significance of risks presented by deficiencies, and determine whether additional controls are needed. Is the deficiency simply due to a minor error and insignificant? Or is it due to lack of or a breakdown in management controls?

For example, if four of 20 PCB transformers observed by the auditor did not have labels, was the lack of labels due to an employee not understanding the PCB

Criteria	Auditor instructions
A. *Performance measurements*	
The facility should develop a set of environmental performance measurements to assess the effectiveness of the environmental management system.	Examine the facility's environmental performance measurements. Consider some of the following criteria: ■ Compliance measurements ■ Pollution prevention ■ Waste generated and released ■ Risk measurements ■ Cost measurements Determine how performance measurements help meet environmental goals.
B *Self-assessment programs*	
1. A formal self-assessment program should be implemented to assess compliance with environmental regulations, policies, and procedures.	Review the self-assessment program to ensure it contains the following elements: ■ Written and issued by the facility manager ■ Assessment schedules ■ Responsibilities for assessments ■ Assessment procedures ■ Root cause analysis ■ Formal reporting system ■ All regulatory areas assessed Ensure the program has been fully implemented.
2. A formal corrective action program must be developed and implemented to ensure that management is not at risk for deficiencies identified in self-assessments that have not been corrected.	Examine corrective action program to ensure it contains the following elements: ■ Written, formal policy issued by facility manager on the importance of taking prompt actions to correct deficiencies. ■ Deficiencies tracked to completion. ■ Corrective action assigned to specific individuals. ■ Reporting mechanism to keep management informed of status. ■ Documentation of corrective actions. Review records of deficiencies and ensure corrective actions are taken promptly and documented.
C. *Continuous improvement*	
A process should be in place where the environmental management system is continually evaluated for improvements.	Interview facility management and environmental staff to determine: ■ Process for reviewing program. ■ How self assessment and audit results are used to improve the program.

Figure 17-14. Performance evaluation.

labeling requirements (i.e., lack of training), or were the transformers painted five days before the audit, and staff simply had not gotten around to relabeling the units? A management auditor collects the evidence through interviews, observation, and sampling. The auditor then determines the probable cause of the problem, the level of risk, and then evaluates how best to report the information to management. In this case, if the lack of labels was a result of a recent painting, an experienced management auditor would probably choose to inform the facility manager and staff verbally of the nonconformance and document this in his/her audit workpapers. This information would not be included in the report as a nonconformance because of its relative insignificance. If formally presented in the report, this issue would probably be viewed by management as nit-picking, and the effectiveness of the audit to correct significant problems would be eroded. Also, the credibility of the audit team could be adversely affected.

A peer review process, in which audit results are evaluated by the entire audit team before reporting the information to plant staff and management, can be of significant help to the team in evaluating results. In this process, each auditor analyzes the results for his/her assigned audit scope areas and decides how items should best be reported. The auditor then presents the information to the team and defends it, explaining why it is worthwhile and ought to be reported. Basic questions to be answered to test whether information is worthwhile and should be included in the report are:

- Of what value is this information to the client?
- How will it help reduce risks and improve operations?
- If the auditor got this report, would he/she find the information useful?
- Does it fall within the objective and scope of the audit?

In management audits, only nonconformances that reflect significant compliance problems and/or identify weaknesses in the management system should be included in the report. Of particular concern are those items that expose the corporation, facility, and/or management and employees to risk.

As a final step, the audit team should rate the overall effectiveness of the facility's environmental management system and determine whether necessary management controls are in place to manage risks. A compliance status ratings scoresheet can be used by each auditor to record the rating of each regulatory area.[11] After each regulatory area is rated, the audit team can then collectively determine the client's overall compliance status using categories as shown in Fig. 17-15. The chosen category is then used as the basis for the executive summary, or briefing document, which is presented to facility management at the exit meeting (see Fig. 17-16) and in the final audit report (Fig. 17-17).

It is vital for facility management to know where the strengths and weaknesses are in its management system so that resources (i.e., staffing and budget) can be tailored to best control the facility's risks, as determined from the evidence found during the audit. Use of a clearly defined rating system is an effective method to meet this need.

| | Major compliance area | | | | |
Rating	Air pollution control[1]	Water pollution control[2]	Solid and hazardous wastes[3]	PCBs	Other[4]
1. Meets regulatory and internal requirements					
2. Substantially meets regulatory and internal requirements					
3. Generally meets regulatory and internal requirements except as noted					
4. Requires improvement to meet regulatory and internal requirements					
5. Requires substantial improvements to meet regulatory and internal requirements					

1. Includes asbestos, used-oil burning, minor sources.
2. Includes National Pollution Discharge Elimination System (NPDES), labs, SPCC, and sewage.
3. Includes USTs and special wastes.
4. Includes drinking water, community right-to-know, pesticides, and National Environmental Policy Act (NEPA).

Overall Compliance Status Categories:

- *Meets regulatory and internal requirements*—The facility/activity has a highly effective environmental management system. Controls necessary to manage risks are in place and functioning. The facility/activity is judged to be in compliance with applicable requirements included in the audit scope.

- *Substantially meets regulatory and internal requirements*—The facility/activity has an effective environmental management system. Controls necessary to manage risks are substantially in place. The facility/activity is in compliance with most requirements.

- *Generally meets regulatory and internal requirements except as noted*—There are weaknesses in the facility's/activity's environmental management system. Additional controls are needed to manage risks. Several concerns, some of which were judged significant, were observed.

- *Requires improvement to meet regulatory and internal requirements*—The facility/activity has an ineffective environmental management system. There is a comprehensive lack of controls necessary to manage risks. Many concerns, some which were significant, were observed.

- *Requires substantial improvement to meet regulatory and internal requirements*—The facility/activity has a highly ineffective or complete lack of an environmental management system. There is a significant lack of controls to manage risks. Numerous significant concerns were identified.

Figure 17-15. Compliance status ratings.

CPC

Smith Fossil Plant (SFP)
Environmental Management Audit
January 14, 1994
Exit Meeting Briefing Document

Audit purpose:
The purpose of this audit is to work with facility management and employees to identify environmental risks and appropriate corrective actions. The audit has been designed and conducted to provide a service to the management and staff of SFP.

Audit report schedule and distribution:
The audit report will be issued no later than March 8, 1994, 30 working days from today. Your response is requested no later than 30 working days from the date the report is issued. Your response, along with Environmental Audit's reply, will be incorporated into the report to form the final audit report. The final report will be sent to you with copies to senior management and the general counsel. It will be transmitted to you no later than 30 working days from receipt of your response.

Overall compliance summary:
SFP substantially meets regulatory and company requirements. The facility has an effective environmental management system (EMS). Controls necessary to manage risks are substantially in place. The facility is in compliance with most requirements.

Since the last audit, SFP has taken the following steps to implement an effective EMS:

- Developed and communicated an SFP mission statement to all employees that commits the facility to operate with a concern for the environment.
- Assigned an environmental technician to assist the environmental engineer two days a week.
- Established a risk assessment program.
- Developed and implemented pollution prevention programs for hazardous waste, chemical control, and oil releases.
- Established a management environmental committee which meets monthly to discuss environmental concerns.
- Conducts self-assessments of all regulatory areas and reports results to the environmental committee.

Figure 17-16. Example of a briefing document.

The audit revealed several opportunities for strengthening SFP's EMS which will assist the plant in controlling risks:

- Develop additional environmental goals and objectives to help meet the company's goal of being an environmental leader.
- Develop environmental performance measurements to track progress toward meeting established goals and objectives.
- Define environmental responsibilities for all employees.
- Complete risk assessments in all regulatory areas.
- Incorporate results from risk assessments along with established environmental goals and objectives into the budget planning process.
- Document and formalize the EMS.
- Develop a training plan that will ensure all employees having environmental responsibilities have been properly trained.
- Develop an annual self-assessment schedule and a corrective action program that will ensure and document that corrective actions have been taken on all deficiencies identified in self-assessments.

Audit findings:

Findings identify nonconformances with federal, state, or local environmental law or regulations; or with company environmental policy or procedure. You are requested to respond to all recommendations. Findings are tracked and closed by the auditors when confirmation is received that corrective actions have been completed.

Regulatory

1. Cenospheres are being discharged from the ash pond to the Clinch River.
   - Report the release of cenospheres to the state.
   - Develop a plan to prevent cenosphere releases.

2. Several requirements of the Georgia Waste Reduction Act of 1990 are not being met.
   - Develop an accounting system for waste management costs.
   - Set specific quantitative goals (in weight or volume) for hazardous waste minimization of each waste stream.
   - Incorporate the hazardous waste minimization plan into site-specific procedures and training plans.

3. PCB record-keeping and marking requirements are not being met.
   - Complete documentation of the August 1993 PCB spill event in accordance with spill cleanup requirements.
   - Perform PCB self-assessments to ensure that spill cleanup documentation is completed.
   - Mark doors or entryways into PCB transformer areas and the hazardous waste storage facility with PCB labels.

Figure 17-16. (*Continued*)

■ Survey other site areas to ensure marking requirements are met.

Policy/procedural

None.

Observations:

Observations may (1) identify a management control deficiency such as lack of procedures or training, (2) warn of situations that could develop into nonconformances if not addressed, (3) highlight exceptional environmental programs, (4) point out noteworthy accomplishments, or (5) document current conditions or project status. You are requested to respond to recommendations contained in observations. Observations requiring a response are tracked and closed by the auditors upon receipt of the audited organization's response to each recommendation.

Observations—response requested:

1. Environmental management system (EMS)
   ■ Document SFP's EMS and incorporate the elements in the EMS outline provided by the auditors.
2. Status of items identified by the 1990 NPDES Water Pollution Control Improvements Task Force
   ■ Document the status of items identified by the task force.
3. Superfund Amendments and Reauthorization Act (SARA) reporting
   ■ In consultation with emergency response authorities, consider using the regulatory provisions for alternative reporting of chemicals.
   ■ If alternative reporting is not used, establish a system to track issuance of revised MSDSs for chemicals reported under SARA requirements.
4. Opacity excuse codes
   ■ Establish controls that will ensure recording explanations of opacity exceedances.

Observations—response not requested:

1. Sewage treatment system alternatives
2. Chemistry laboratory quality program
3. New oil spill response shed
4. New weir cleaning method
5. Asbestos management
6. PCB risk assessment
7. Waste oil injection system
8. Status of findings from the previous audit

Figure 17-16. (*Continued*)

Overall compliance status categories:

- *Meets regulatory and CPC requirements.* The facility/activity has a highly effective environmental management system. Controls necessary to manage risks are in place and functioning. The facility/activity is judged to be in compliance with applicable requirements included in the audit scope.

- *Substantially meets regulatory and CPC requirements.* The facility/activity has an effective environmental management system. Controls necessary to manage risks are substantially in place. The facility/activity is in compliance with most requirements.

- *Generally meets regulatory and CPC requirements except as noted.* There are weaknesses in the facility's/activity's environmental management system. Additional controls are needed to manage risks. Several concerns, some of which were judged significant, were observed.

- *Requires improvement to meet regulatory and CPC requirements.* The facility/activity has an ineffective environmental management system. There is a comprehensive lack of controls necessary to manage risks. Many concerns, some which were significant, were observed.

- *Requires substantial improvement to meet regulatory and CPC requirements.* The facility/activity has a highly ineffective or complete lack of an environmental management system. There is a significant lack of controls to manage risks. Numerous significant concerns were identified.

Additional items:

The following are items which are judged as minor risks and will not be included in the final report. This judgment is based on the insignificance of the deficiency and the fact that it does not reflect a weakness in the environmental management system. Though not addressed in the audit report, these items warrant management attention and correction.

1. Signatory and certification requirements for reports required by the NPDES permit

2. Hazardous waste labels on oily rag containers

3. Errors on the land disposal restriction notification forms

4. Combustible materials near PCB transformers

5. Errors in 1992 PCB annual report

6. Signing and dating of SARA reporting forms

7. Certification of technicians for repair of equipment containment refrigerants

8. State approval of sewage sludge disposal

Figure 17-16. (*Continued*)

CLINCH POWER COMPANY
SMITH FOSSIL PLANT (SFP)
Environmental Management Audit Report
Audit Dates: January 11–14, 1994
Audit No. SF-94-01-11

Date of Issue_____

TABLE OF CONTENTS

Figure 17-17.

EXECUTIVE SUMMARY

The Smith plant substantially meets regulatory and company requirements (see Sec. IV for overall compliance status categories). The facility has an effective environmental management system (EMS). Controls necessary to manage risks are substantially in place, and the plant is in compliance with most requirements.

Since the last audit, the plant has taken the following steps to implement an effective EMS and reduce risks:

- Developed and communicated a mission statement to all employees that commits the facility to operate with a concern for the environment
- Initiated a risk assessment program
- Developed and implemented a pollution prevention program which addresses chemical control, hazardous waste, and oil releases
- Established an environmental management committee which meets monthly to discuss environmental concerns
- Conducted self-assessments of all regulatory areas and reported results to the environmental management committee
- Assigned an environmental technician to assist the environmental engineer two days a week

The audit revealed several opportunities for strengthening the EMS which will assist the plant in controlling risks:

- Develop additional environmental goals and objectives to help meet the company goal of being an environmental leader
- Develop environmental performance measurements to track progress toward meeting established goals and objectives
- Define and communicate environmental responsibilities for all employees
- Complete risk assessments in all regulatory areas
- Incorporate into the budget planning process the results from risk assessments, established environmental goals and objectives, and legislative and regulatory changes
- Develop a training plan that will ensure all employees having environmental responsibilities have been properly trained and proficiency is maintained
- Develop an annual self-assessment schedule
- Develop a plan that will ensure corrective actions are taken and documented for deficiencies identified in self-assessments
- Document the EMS

Figure 17-17. (*Continued*)

1. OBJECTIVE AND SCOPE

The objective of the audit was to work with plant management and staff to determine if the plant's system of controls is adequate to manage environmental risks. The auditors interviewed plant, partner, and regulatory personnel; reviewed documents and records; and inspected the facility. The scope of the audit included federal and state environmental regulations and company environmental policies and procedures. The audit team consisted of Jim Hall (lead auditor), Paul Hill, Jane Doe, David Ochs, and Susan Day.

II. FINDINGS—RESPONSE REQUESTED

A finding is a nonconformance with a federal, state, or local environmental law or regulation (Sec. A—Regulatory Findings) or with company environmental policy or procedure (Sec. B—Procedural Findings).

A. REGULATORY

1. *Cenospheres are being discharged from discharge serial number (DSN) 001.*

   The plant's National Pollution Discharge Elimination System (NPDES) permit prohibits the discharge of "floating solids or visible foam in other than trace amounts." On January 12, 1994, the auditors observed a large (200 square foot) mat of cenospheres in the intake channel about 50 yards from the intake. Further inspections along the channel (in the direction of DSN 001) revealed numerous 4- to 6-square inch patches of cenospheres along the shoreline. Also, cenospheres were observed flowing over the weir.

   Plant staff believe that a dredging operation, now in progress, is the cause for cenosphere accumulation around the weirs. On January 13, plant staff deployed booms to isolate the weirs from cenospheres. Release of cenospheres to waters of the state was also an issue in the March 1987 plant audit report (SFP-AWSO-87-03-24, Finding B-11). (Auditor: Jim Hall)

   Recommendations:

   The plant should:

   - Report the release of cenospheres to the state.
   - Implement additional controls as necessary to isolate the weirs from cenospheres.

   Client's response:

   The National Pollutant Discharge Elimination System (NPDES) noncompliance occurred on January 12, 1994, and was reported to the state in the plant discharge monitoring report for that month. The rate of

Figure 17-17. *(Continued)*

accumulation of cenospheres has increased significantly since dredging of the active ash disposal pond was begun. High surface winds during the period caused cenospheres to be suspended in the turbulent upper layer of water in the stilling pond. A 1.54-inch rainfall event on the previous day increased the flow through the skimming weirs sufficiently to carry some of these cenospheres through the weirs, after which they refloated in the calmer water of the plant intake channel. The risk of the discharge of cenospheres has been identified and assessed. Booms were deployed on January 13 to isolate the weirs from cenosphere accumulation. Floating cenospheres in the active and stilling ponds have been contained by booms and are being removed. A redundant skimmer has been installed at the active pond weir to reduce the potential of the material from entering the stilling pond. A skimmer has been installed at the dredge cell weir to prevent dredged cenospheres from reentering the active pond. Weekly inspections of cenosphere accumulation in the active and stilling ponds will be conducted by Yard Operations during the dredging operations. The inspection record will be tracked as an indicator in the NPDES process. Additional containment and removal activity will be conducted as more cenospheres accumulate. This finding should be closed.

Audits reply:

This finding is closed.

2. *Several requirements of the Georgia Hazardous Waste Reduction Act of 1990 are not being met.*

The Georgia Hazardous Waste Reduction Act of 1990 requires that all large quantity generators in calendar year 1989 develop a hazardous waste reduction plan by January 1, 1992. The plan must be designed to provide a 25-percent reduction in each hazardous waste stream at the facility by 1995. The Act outlines what the plan, at a minimum, should include. Some of the requirements are:

- A description of employee hazardous waste reduction awareness training programs.
- A description of methods to incorporate the plan into facility management practices and procedures.
- A description of the hazardous waste accounting system that identifies waste management costs.
- Specific quantitative performance goals, expressed in weight or volume per unit of production, for each hazardous waste stream. These goals are to be based on the 1989 hazardous waste stream generation rate or the first year the waste stream was generated after 1989. Goals should reflect the generator's planned progress each year to achieve the 1995 goal.

Figure 17-17. *(Continued)*

The Act also requires an annual review by analyzing and quantifying progress made toward meeting each performance goal established in the plan. Results of this review should be documented in an annual hazardous waste reduction progress report.

The plant is presently a small-quantity generator and has made commendable progress in reducing hazardous waste; however, it has not met all the requirements of the Act. Because the plant was a large-quantity generator in 1989, the facility prepared a hazardous waste reduction plan before January 1, 1992. The first annual hazardous waste reduction progress report was also completed on time. However, the plan and annual progress report do not contain all the information required by the Act. For example:

- Implementation of the plan has not been incorporated into existing hazardous waste training or plant's procedures.
- Although the plan describes a hazardous waste accounting system that would identify waste management costs, the system has not been implemented.
- Performance goals in the plan do not adhere to requirements outlined in the Act. Some are not based on 1989 generation rates, and none show planned reduction progress year by year to achieve the 25 percent or more reduction by 1995.

(Auditor: Susan Day)

Recommendations:

The plant should:

- Incorporate the hazardous waste reduction plan into site-specific procedures and training plans.
- Develop and implement an accounting system for waste management costs.
- Set specific quantitative goals for hazardous waste reduction for each hazardous waste stream.

Client's response:

The new hazardous materials and waste management process does incorporate the implementation of the hazardous waste minimization plan. The procedure requires input from each department in the development of waste reduction goals. The procedure also requires monthly monitoring of the hazardous waste generated to determine if minimization goals are met.

The hazardous waste reduction facility plan for the plant was revised as of March 1, 1994, to identify waste management cost and performance goals.

Waste transportation and disposal costs have been addressed for each waste which is transported off-site for treatment and/or disposal. Other

Figure 17-17. *(Continued)*

costs are assessed for the hazardous waste management program in general, which include personnel costs for inspections, hazardous waste, Best Management Practices (BMP), spill response, hazardous communication and waste minimization training, and accumulation and preparation for shipment.

Specific production ratio reduction goals have been identified in the plan for each active waste stream with the exception of stream nos. 2, 3, 13, and 30. These goals are based on 1989 generation levels or on the first year of generation where applicable.

Waste streams nos. 2 and 3 are acid and caustic wastes from the demineralizer which are neutralized and discharged through an NPDES-permitted outfall. Costs of managing the wastes are minimal because the waste acid and caustic are combined at the end of the process, and only minor adjustments are necessary to meet NPDES discharge limits. Because 1989 was such an atypical year, the drought improved intake water quality dramatically, and a portable demineralizer (regenerated offsite) was used for part of the year, the generation of waste was unusually low. A production ratio reduction goal based on 1989 would not be achievable under normal conditions and would be meaningless. The reduction goals for these streams are set at a challenging level of 25 percent of the 1991 generation level. Reduction efforts to date have resulted in maximum efficiency, and therefore minimal waste generation of the system. Waste production will vary with intake water quality, but further active reduction in these waste streams must be realized by reducing losses in boiler water by making major changes or equipment upgrades in the system.

Waste stream no. 13, mercury waste, has been recently revised to include the accumulation of fluorescent lighting tubes destined for recycling. The uncertain future of the classification of this waste and the lack of data on the quantity of generation make the setting of a reduction goal meaningless. The waste stream will be evaluated at the close of 1994 for the quantity of generation. Additional guidance from the Environmental Protection Agency could also be available at that time, potentially eliminating fluorescent tubes destined for recycling from the hazardous waste classification.

Waste stream no. 30 was opened in December 1993 for the accumulation of chargeable batteries containing cadmium, silver, and mercury. Insufficient time has elapsed to develop a rate of generation for this waste; therefore, no production ratio can be set. This stream will also be reevaluated at the close of calendar year 1994.

This finding should now be closed.

Audit's reply:

This finding is closed.

Figure 17-17. (*Continued*)

3. *PCB record-keeping and marking requirements are not being met.*
Environmental Protection Agency (EPA) regulations contain specific cleanup requirements for spilled PCB dielectric fluid. A PCB spill has occurred when PCBs have run off or are about to run off the equipment. Documentation requirements for each spill event include:

- The area of visible contamination, noting the extent of trace areas and the center of the area of contamination
- The cleanup method used
- Dates and times of discovery and cleanup

The above information was not available for an August 1993 spill from a PCB transformer (serial number G.E. L495006-74P). Since this spill event occurred, a spill cleanup documentation form has been developed by the environmental staff and made available to the site.

EPA regulations require doors and entryways to PCB transformers and areas used to store PCBs and PCB items be marked with PCB labels. The marking must be in a manner that can be easily read by emergency response personnel fighting a fire involving PCB equipment or PCB items.

The exterior doors on the west side of the powerhouse and stairwells leading to the powerhouse basement which contains PCB transformers were not marked. Also, the hazardous waste storage building, which is periodically used to temporarily store PCB wastes, was not marked with a PCB label. At the time of the audit, a container of PCB wastes was being stored in the building. (Auditor: Jane Doe)

Recommendations:

The plant should:

- Complete documentation of the August 1993 PCB spill event in accordance with the spill cleanup policy.
- Initiate self-assessments to ensure PCB spill cleanup requirements are met.
- Mark doors and entryways to PCB transformer areas and the hazardous waste storage facility.

Client's response:

Documentation of the August 1993 spill has been completed according to the plant's PCB spill cleanup certification, and cleanup has been verified by wipe test. Electrical maintenance activity item nos. 16 and 19 of the PCB process will significantly reduce the risk of recurrence of the documentation error, and technical services activity no. 8 provides adequate assurance that spills will be properly managed. All entrances to areas containing PCB equipment have been marked with PCB labels.

This finding should be closed.

Figure 17-17. (*Continued*)

Audit's reply:

This finding is closed.

B. PROCEDURAL

None.

III. OBSERVATIONS

An observation identifies a management control deficiency such as lack of procedures or training; warns of situations that could develop into nonconformances if not addressed; highlights exceptional environmental programs; points out noteworthy accomplishments; or documents current conditions or project status.

A. RESPONSE REQUESTED

1. *Environmental management system (EMS) improvement.*   The audit revealed that the plant has an effective EMS. There are, however, several opportunities for improvements that would strengthen the EMS and assist the plant in controlling environmental risks. The following are suggestions for improving the EMS:

- *Management commitment.*   Each organization should demonstrate its commitment to environmental management by developing and communicating a formal environmental policy. This policy should outline upper management's support and line management's accountability for environmental performance and include a framework for the organization's environmental goals, objectives, targets, and programs.
    The plant has communicated a vision statement to all of its employees which states that "SFP will be operated with a concern for the safety of the employees, the public, and the environment." The statement also commits the plant to compliance with all federal, state, and local environmental regulations and company policies and procedures. The plant has one environmental goal for 1994—to operate with no notices of violation (NOVs). SFP should consider developing additional "beyond compliance" goals.
- *Organization.*   Clear lines of authority and responsibility should be defined, documented, and integrated into appropriate operating staff functions.
    This has been done for the environmental engineer. In addition, the facility has recently redefined responsibilities (including environmental) for shift engineers and assistant unit operators. The audit revealed, however, that some environmental responsibilities of a few engineers have not been defined in writing. This should be accomplished when the new procedures are in place.
- *Environmental risk assessments.*   Environmental risk assessments should be conducted periodically to identify and prioritize conditions which present the corporation with the greatest risk and highest probability of occurrence.

Figure 17-17. *(Continued)*

The plant has performed formal risk assessments for PCBs and asbestos. The environmental engineer has performed informal risk assessments for environmental training, oil management, and hazardous waste minimization. Additional risk assessments should be conducted for compliance with the new Clean Air Act requirement and implementation of Fossil and Hydro's new environmental processes. SFP has committed to conduct periodic risk assessments for each regulated area.

- *Operational planning and budgeting.* The business plan should take into account the impact of current and pending environmental regulations and results of risk assessments. In addition, environmental goals and objectives should be integrated into the business planning process.

  As stated above, SPF has incorporated an environmental goal of no NOVs into its business plan. It should consider establishing "beyond compliance" environmental performance goals which will help the company meet its environmental leadership goal. These goals should be factored into the business planning process to ensure adequate resources are available to meet the goals.

- *Development and implementation of environmental programs.* A site pollution prevention program (source and waste reduction) should be developed and implemented addressing each regulatory area (i.e., air, water, and solid and hazardous waste).

  The plant has initiated some elements of a pollution prevention program such as reduction of hazardous waste inspections to prevent releases and a chemical control program. Additional pollution prevention measures should be considered in other regulatory areas.

- *Employee awareness and training.* The plant should have an environmental training program which identifies and provides necessary environmental training needs for all personnel. The environmental engineer has provided environmental training for PCBs, hazardous waste, hazardous materials transportation, and air monitoring. A training matrix should be developed that will help ensure all employees having environmental responsibilities receive required training and that proficiency is maintained.

- *Internal and external communication.* The organization should have a process that provides for periodic communication, both internally and externally, on matters related to environmental commitments, goals, performance, and concerns. The plant has met this element of an effective EMS by establishing an environmental committee. The committee consists of the management team plus the plant environmental engineer. Monthly meetings are held to discuss environmental concerns, track corrective action progress, and analyze results of risk assessments. In addition, the environmental engineer discusses environmental concerns with all employees in quarterly informational meetings.

Figure 17-17. (*Continued*)

- Environmental performance evaluation. The organization should have a self-assessment process which evaluates environmental performance measurements, compliance with regulations, and adherence to company policy and procedures. Corrective actions to address deficiencies should be developed and implemented in a timely manner and tracked and reviewed by management to ensure completion.

  The environmental engineer performs monthly self-assessments for each regulatory area, and results of these assessments are presented to the environmental committee each month. This activity has not been formalized, and there is no documented corrective action program to ensure deficiencies identified in assessments are corrected.

(Auditor: David Ochs)

Recommendations:

To improve its EMS, SFP should document its EMS and evaluate practices in each of the above areas and make necessary improvements.

Client's response:

The goal of SFP is to be a world leader in producing efficient and reliable electric power at a competitive cost while protecting the safety of its employees, the public, and the environment. The continued commitment of the SFP management team and plant employees to its EMS system directed toward compliance with federal, state, and local environmental regulations and company environmental policy is a necessary element in achieving this goal.

The immediate goals of the EMS are (1) to deploy and implement each of the revised environmental procedures by May 31, 1994, (2) to communicate to all SFP managers and employees their environmental responsibilities and ensure that they are adequately trained to perform those duties by October 1, 1994, and (3) to ensure that adequate resources are made available so that the environmental impact of SFP is minimized and regulatory risks are assessed. The long-range goal of the EMS is to further develop and refine the pollution prevention program in the context of the revised environmental procedures and consistent with the environmental leadership goals of the plant.

Process deployment: The deployment of each process is a risk assessment in the concerned media, and compliance with the requirements of each procedure will provide indicators for the evaluation of SFP's performance in that media. The procedures contain the means to track and evaluate plant performance in each regulatory, procedure, and media area. The evaluation of the performance indicators will identify areas of existing potential regulatory and/or environmental risk for SFP, provide a framework for improvement activities, and identify the priorities for further pollution prevention programs.

Figure 17-17. *(Continued)*

Communication: The members of the EMS Environmental Management Committee (SFP-EMC) meet monthly. The meeting agenda allows for identifying environmental concerns, tracking existing or resolved concerns, analyzing the results of risk assessments, and discussing the status of environmental process indicators and regulatory or policy changes. Employees with specific environmental responsibilities are provided with their job descriptions which have those responsibilities identified. Training for those employees with specific responsibilities is current. Required training and awareness training for all employees will be provided, updated, and tracked through TS/ENVR/GEN/FOS/1.2, Environmental Training, as it is implemented in April 1994. Employees are made aware of current environmental information meetings attended by all employees.

Resource allocation: The plant manager is ultimately responsible for resource allocation at SFP. The plant manager is made aware of environmental risk assessments, performance indicators, and the status of corrective and preventative actions through his participation in the Environmental Management Committee. The FY 1995 business plan is being developed and will include at least one additional performance indicator associated with TS/ENVR/GEN/ALL/1.1, Environmental Process Management System. Capital, job order, and operation and maintenance (O&M) projects with environmental justification are prioritized and considered with other resource demands and are included and funded appropriately in the business plan. The plant manager has accepted the mission statement of SFP, which includes environmental commitments and acts in the best interest of plant and the environment.

Audits reply:

This observation is closed.

2. *Status of NPDES water pollution control improvements.* The plant has not documented the status of items identified by the NPDES Water Pollution Control Improvements Task Force. The plant environmental engineer stated that all items have been or are being addressed. (Auditor: David Ochs)

Recommendation:

SFP should update the status of these items in writing to document plant actions to address and/or correct each item.

Auditee's response:

All items of concern raised by the task force have been addressed and documentation is on file at the plant. Action on four issues has been delayed until the plant is issued a renewal of its NPDES permit. Two of these concern redwater seeps and two concern nonchemical metal

Figure 17-17. (*Continued*)

cleaning wastes. It is expected that these issues will be addressed in the new permit, and the permit will provide regulatory guidance for managing them.

Audit's reply:

This observation is closed.

3 *Superfund Amendments and Reauthorization Act (SARA).* SFP currently has no system in place to ensure that significantly revised material safety data sheets (MSDSs) received for chemicals subject to SARA reporting requirements are sent to the state, fire department, and local emergency planning committee within three months of receipt. This supplemental reporting is not necessary if SFP follows steps for alternative reporting as set forth in 40 CFR Part 370.21. Alternative reporting does not require submittal of MSDSs and may be preferred by emergency response authorities. (Auditor: Paul Hill)

Recommendations:

The plant should establish controls to ensure SARA reporting requirements are met.

Client's response:

SFP used the alternative method, as described in 40 CFR 370.21, for the 1993 calendar year report. The report requirements are scheduled in the plant MPAC system.

Audit's reply:

This observation is closed.

4. *Opacity excuse codes.* Explanations for opacity exceedances are not always recorded on opacity charts by the shift engineer. The shift engineer is responsible for ensuring that reasons for opacity exceedances are recorded on the opacity recorder chart; occasionally, this is not done. If the reason is not recorded on the chart, Technical Services has to send the chart back to Operations for an explanation for the exceedance. Then excuse codes can be assigned and results reported to the state. It is often difficult for operators to reconstruct circumstances concerning the exceedance, and therefore, incorrect excuse codes may be assigned. (Auditor: Jim Hall)

Recommendation:

SFP should establish controls that will ensure that shift engineers record explanations for exceedances.

Client's response:

Figure 17-17. (*Continued*)

Shift supervisors have been informed of their responsibilities concerning the recording of the reasons for exceedances of the 20-percent opacity limit. The revised media procedure TS/ENVR/GEN/FOS/1.0, Air, will be amended to reflect this responsibility, and the codings will be tracked as a performance indicator in this process until the new continuous emissions monitoring system data tracking system is operating.

Audit's reply:

This observation is closed.

B. *RESPONSE NOT REQUESTED*

1. *Sewage treatment system alternatives.* The sewage treatment system presents the plant with risk because of the potential for bypasses and other nonconformances. The wetland system is underdesigned, and the drainfield becomes saturated due to hydrologic overloading. Water is not percolating, vegetation has died, and replanting efforts have failed. There have been two notices of noncompliance in the past year, and bypasses have nearly occurred on several occasions.

   The plant is reviewing the situation and considering alternatives under a capital project. This project will include a feasibility study for tying the plant into the Uptown Utility sewage system. (Auditor: Jim Hall)

2. *Chemistry laboratory quality program.* The plant and Fossil & Hydro Water Chemistry are commended for quality assurance and quality control initiatives relative to the plant NPDES chemistry program. Examples of controls that have been initiated to help ensure quality laboratory results include:

   - A cross-training program including rotating staff into the plant chemistry laboratory analyst position for one year. This will help ensure backup for analysts that are on leave or leave the organization.
   - Submittal of unknown samples to periodically assess the accuracy of plant laboratory results.
   - The new chemical laboratory training plan developed by Water Chemistry.
   - The new performance audits developed by Water Chemistry and the fossil plants.

   The new training and audit programs are noteworthy controls and should help ensure the plant maintains an effective NPDES laboratory compliance program. (Auditor: Jim Hall)

3. *New oil spill response shed.* The plant is commended for placing a new oil spill response equipment shed at the discharge. This should allow

Figure 17-17. (*Continued*)

for quick response to spills that may occur in the discharge area. (Auditor: Paul Hill)

4. *New weir cleaning system.*   The plant is commended for developing a new way to clean weirs using an air compressor. The weirs are cleaned twice per year, and compressed air has proven to be an effective way to clean weirs and improve their operation. (Auditor: Susan Day)

5. *Asbestos management.*   SFP has an effective asbestos management program. Regulatory reporting requirements have been met, and there is a positive working relationship between company and subcontractor staff. Extensive asbestos marking was observed in the powerhouse. A comprehensive asbestos survey was performed in late 1992 by XYZ, Inc. On the basis of data collected, ABC, Inc. prepared a risk assessment report in March 1993. This report outlines a formal plan for repair and replacement of damaged asbestos-containing materials at SFP. Company staff indicated that SFP intends to follow recommendations in the ABC report, and the facility is currently doing *Priority 1* renovation work. Additionally, the fiscal year 1994 budget will allow SFP to complete this phase and significantly address those areas identified as *Priority 2*. (Auditor: Susan Day)

6. *PCB risk assessment.*   Fossil Operations has initiated a PCB risk assessment at its facilities. This assessment is a commendable project that should assist plant management in determining priorities for future equipment retrofills and replacements. A survey of SFP's PCB equipment was done in December 1993. The survey focused on past leak and spill history, the presence of secondary containments, and equipment location. This information will be included in a systemwide data base that is scheduled to be presented to Fossil Operations management by February 1, 1994. (Auditor: Jane Doe)

7. *Waste oil injection system.*   STP currently burns its waste oil by applying it to the coal pile. This practice is questionable from an environmental standpoint, and it is labor intensive. In addition, oil is being spilled onto the pavement at the loading dock as it is being added to the waste oil accumulation trailer. The trailer has a sandbag containment, but spilled oil is leaking from the containment and moving to a storm drain about 30–40 feet away. This drain leads to the plant ash pond. SFP has budgeted for an oil injector system and is currently obtaining price quotes for its purchase. An injection system would reduce risks and improve housekeeping. (Auditor: David Ochs)

8. *Status of findings from the previous audit.*   Finding B-2 from the previous audit report (audit no. SFP-91-03-12) was open at the time of the

Figure 17-17. (*Continued*)

audit. The audit verified that corrective action has been taken, and this finding is now closed. (Auditor: Jim Hall)

_____

Lead Auditor

_____

Manager
Environmental Audits

_____

Vice President
Environmental Management

IV. SUPPLEMENTAL INFORMATION

1. *Response.* A written response from the client addressing recommendations for each finding and observation in Secs. II and IIIA is due within 30 working days after the report is issued. This response should be sent to the vice president, environmental management, indicating corrective actions taken or scheduled. The client's response will be included in the final environmental audit report. Environmental Management's reply to the client's response will be included, and the final report will be issued within 30 working days from receipt of the auditee's response. Observations will be closed upon receipt of the auditee's response to each recommendation.

2. *Overall compliance status categories*

   - *Meets regulatory and company requirements.* The facility has a highly effective environmental management system. Controls necessary to manage risks are in place and functioning. The facility is judged to be in compliance with applicable requirements included in the audit scope.
   - *Substantially meets regulatory and company requirements.* The facility has an effective environmental management system. Controls necessary to manage risks are substantially in place. The facility is in compliance with most requirements, only a few deficiencies were observed. These deficiencies represent isolated exceptions.
   - *Generally meets regulatory and company requirements except as noted.* There are weaknesses in the facility's environmental management system. Additional controls are needed to manage risks. Several nonconformances, some of which were judged significant, were observed.
   - *Requires improvement to meet regulatory and company requirements.* The facility has an ineffective environmental management system. There is a comprehensive lack of controls necessary to manage risks. Many nonconformances, some of which were significant, were observed.
   - *Requires substantial improvement to meet regulatory and company requirements.* The facility has a highly ineffective or complete lack of an environmental management system. There is a significant lack of controls to manage risks. Numerous significant nonconformances were identified.

Figure 17-17. (*Continued*)

Conclusion

Recent changes in environmental regulatory requirements, along with more aggressive enforcement policies and stricter federal regulatory guidelines, indicate a need for facilities to have effective environmental management systems. Environmental management audits are an excellent tool for assessing managers at both the corporate and facility level and for determining whether adequate management controls are in place to protect the company, its managers, and its employees from risks.

The management audit approach can be used to conduct audits of facilities, programs, and contractors that pose potential environmental risks. The environmental management audit provides management with the answers to five basic questions that, from a business perspective, are critical for effective long-term compliance:

1. What is the compliance status?
2. Is the environmental management system functioning as intended by management?
3. Are adequate controls in place where risks and potential liability exist?
4. What are the implications or possible outcomes if controls are not added or improved?
5. What can be done to improve the management system and compliance programs?

The environmental management audit approach provides the corporation with a greater level of risk protection. It can help significantly in pointing the company toward a position of environmental leadership, and it provides potential advantages for competing in the world economy.

References

1. U.S. Environmental Protection Agency, *Enforcement Accomplishments Report FY 1992,* Office of Enforcement, Washington, D.C., April 1993.
2. U.S. Environmental Protection Agency, press release, Dec. 9, 1993.
3. International Organization for Standardization, "Environmental Management," ISO/TC207, U.S. TAG Meeting, Philadelphia, August 1993.
4. British Standards Institution, *Specification for Environmental Systems,* British Standard 7750, London, 1992.
5. U.S. Department of Justice, *Factors in Decisions on Criminal Prosecution of Environmental Violations in the Context of Significant Voluntary Compliance or Disclosure Efforts By the Violator,* Washington, D.C., U.S. DOJ, 1991.
6. U.S. Sentencing Commission, *Organizational Sentencing Guidelines,* May 16, 1991 (56 *Federal Register* 22762).

7. U.S. Environmental Protection Agency, *Environmental Auditing Policy Statement,* Washington, D.C., 1986.

8. Institute of Internal Auditors, *Standards for the Professional Practice of Internal Auditing,* Altamonte Springs, Fla., May 1985.

9. J. Ladd Greeno et al., *Environmental Auditing, Fundamentals and Techniques,* Center for Environmental Assurance, Arthur D. Little, Inc., Cambridge, Mass., 1987.

10. Environmental Auditing Roundtable, *Standards for Performance of Environmental, Health, and Safety Audits,* February 1993.

11. Tennessee Valley Authority, *Environmental Auditing Procedure,* 8th ed., vol. I, *The Audit Process,* December 1993.

18

Environmental Health and Hygiene Services at Noranda

E. O. Villeneuve

*Director, Environmental Audits
and Special Projects, Noranda*

Over the last few years, public opinion polls have shown that the environment is the public's number one or two concern. Green products have become a marketing coup, packaging a priority. Politicians have included environment on all their agendas, and labor is becoming actively involved in promoting environmental causes. The environment has emerged as a key strategic issue of the 90s, not just the flavor of the month.

When it comes to the environment, natural resource companies such as Noranda have a high profile. Noranda, a highly diversified natural resource company, is active in mining, forestry, oil and gas, and manufacturing. These activities, by their nature, are intrusive on the environment, and this intrusiveness carries responsibilities. As a result, five responsibilities were identified as the basis for the environmental management system that Noranda has implemented. These are:

- Treat the environment as a finite and precious resource
- Recognize that the environmental challenge is global in scope
- Understand the environmental impacts of each of the activities carried out
- Provide full and honest disclosure on environmental matters
- Fix the problems we helped to create

Company Background

Noranda operates plants in four major divisions: Forest Products, Minerals, Energy, and Manufacturing. The company operates mostly in North America but has interests in Europe, South America, and Australia. Annual revenue totals about $10 billion, and the company has about 50,000 employees.

Accountability and responsibility for management is very much decentralized. Individual plant managers are fully accountable for:

- Maintaining and improving productivity, efficiency, and competitiveness
- Reducing costs
- Protecting the environment, the health, and safety of their employees and the public
- Maintaining credibility and good relations with government agencies, the public, employees, and their customers

Noranda's Environmental Management System

Noranda's first environmental policy dates back to 1965. Twenty years later, in 1985, the company adopted a new, progressive, and proactive environmental policy (see Fig. 18-1). This version is ready for an update and will include two essential components that are now missing: communication and research.

At Noranda, it is a fundamental belief that both proper organization and an environmental management system (EMS) are necessary to implement the company's environmental policy—and to reach a high level of environmental excellence. The elements of the Noranda EMS are:

- Direct assistance to operations/companies/divisions
- Environmental auditing
- Communications
- Legal and regulatory support
- Capital expenditure review
- Education and training
- Environmental research

In addition, the EMS must include a component dealing with outside company relations involving government, business, environmental groups, and universities, to name a few. A more detailed description of each of these elements is provided in Fig. 18-2.

Noranda Inc.

Environmental Policy

noranda operations will strive to be exemplary leaders in environmental management by minimizing the environmental impact on the public, employees, customers and property, limited only by the technological and economic viability. The following principles are basic to achieving this environmental objective.

1. The potential risks of new projects to employees, the public and the environment must be assessed so that effective control measures can be foreseen and taken and all parties made aware of these facts.

2. Noranda Group operations will implement site specific environmental, health, hygiene, safety and emergency response policies in the spirit of guidelines issued by Noranda Inc. as well as in conformation with applicable laws and regulations.

3. Noranda Group operations will constantly evaluate and manage risks to human health, the environment and physical property.

4. Noranda Group operations will be subject to periodic environmental, health, hygiene, safety and emergency preparedness audits.

5. A report on environment, health and hygiene, safety and emergency preparedness will be presented annually to the Board.

ALFRED POWIS
CHAIRMAN & CHIEF EXECUTIVE OFFICER

Figure 18-1. Noranda's environmental policy statement.

ENVIRONMENTAL MANAGEMENT SYSTEM

Both proper organization and an environmental management system are necessary in the implementation of the Environmental Policy and in reaching a high level of environmental excellence. The key elements of Noranda's system are:

1.   DIRECT ASSISTANCE TO OPERATIONS/COMPANIES/DIVISIONS

Plant managers and environmental coordinators at each facility must have access to specialized technical, scientific and political expertise. They must have a back-up team to whom they can turn in order to do their jobs effectively. Support may be required for environmental impact assessment, risk evaluation, the development of compliance and commissioning plans, responses to new and emerging issues, negotiations with governmental agencies, and so forth. Within Noranda, this expertise is provided by the corporate office and by external consulting services.

2.   ENVIRONMENTAL AUDITING

Although the auditing program developed and introduced at Noranda in 1985 is still an evolving and dynamic system, some general objectives have been developed, including compliance with regulatory requirements and Noranda guidelines, the application of best management practices, the minimization of potential risks and liabilities, and the application of good management systems to the environmental program. This is an essential element of our Environmental Management System. It is later described in more detail.

3.   COMMUNICATIONS

Employees of the corporation at all levels -- from the Chief Executive Officer to the plant maintenance person -- must carry with them the Board of Directors'

Figure 18-2. Noranda's environmental management system.

responsible and environmentally-sensitive vision. This cultural framework becomes firmly rooted when the commitment of senior officers is communicated effectively throughout the organization and when the commitment is supported by credible and progressive decisions on the plant floor. Communication, of course, is a two-way street. The environmental concerns of individual employees must flow freely to the decision-makers if a progressive program is to be maintained. There can be no environmental excellence without environmental communication. This vital element is discussed in more detail later in this paper.

4.   LEGAL AND REGULATORY SUPPORT

It is difficult to keep abreast of the detailed and complex regulations continuously emanating from federal, provincial, state and municipal agencies. The environmental coordinator at each operating plant, the managers and, in fact, each employee, must understand the law and understand it in its basic, every-work-day terms. Interpretative back-up must be available. AT Noranda, this is provided by the corporate legal advisor and by staff of Environmental Services. External legal expertise is retained in the event of enforcement and judicial actions.

5.   CAPITAL EXPENDITURE REVIEW

The capital expenditure control system for the Noranda Group requires a sign-off signature by the proponent of the Appropriation Request and by the business unit/division or corporate office for environmental considerations. A checklist provides a series of questions for the project managers. These include such items as environmental assessment, compliance and approvals, public information, reclamation/decommission planning, potential liabilities, hazard/operability studies and environmental protection considerations (control systems, risks, hazardous wastes, chemical storage, emergency response). All expenditures requiring approval by various levels within the corporation, as specified by internal directives,

Figure 18-2. (*Continued*)

are subject to an examination of environmental relationships. Each project manager must address these items before an Appropriation Request is approved.

6.   **EDUCATION AND TRAINING**

Employees at all levels must have the proper tools to advance the environmental program. Awareness of environmental issues on the global and local scale is essential for employees if they are to comprehend the relationships between their plant and their individual activities and the various components of the external environment.

A broad spectrum of training needs must be fulfilled to ensure that employees are equipped to effectively carry out their daily responsibilities. These include training for spill control, operation of treatment plant systems, monitoring and sampling, emergency response and so forth. Noranda has organized speciality workshops dealing with hazard operability studies, environmental auditing, legal issues and emergency response. These are offered on a continuing basis, depending on the need.

7.   **ENVIRONMENTAL RESEARCH**

Many of our complex environmental problems will be solved through research and technology. A major technology research program is examining the current processes being used in the search for cleaner and more efficient processes. The program's objectives are to:

•   improve productivity while creating less waste by using raw materials more efficiently;

•   eliminate the use of hazardous chemicals wherever possible;

Figure 18-2. (*Continued*)

- eliminate or minimize emissions and discharges of contaminants;

- improve effluents and emissions control processes to meet our long-term goal of virtually eliminating discharges;

- promote recycling; and

- reduce energy consumption.

In addition, environmental research includes a significant ecological component.

Figure 18-2. (*Continued*)

Noranda's Environmental Audit Program: Principal Elements

Environmental auditing was formally introduced in Noranda in 1985, the result of an extensive consultation process—with several major chemical producers—to investigate the safe storage, handling, processing, and transportation of hazardous chemicals. One of the results of this consultation was the strong conviction that an environmental audit program would help manage the issues more effectively.

The program elements were developed in 1985 and 1986, and the first audits took place during the second half of 1986. Figure 18-3 shows the number of audits conducted since 1986.

Definition of Environmental Audits

An environmental auditing program is the key element of any environmental management system, and several definitions of environmental auditing have been suggested. But in setting up the Noranda program, a special effort was made to agree on a definition that would account for the format, scope, and objectives of the audit program. At Noranda, we have defined environmental audits as a systematic and objective method of verifying that standards, regulations, procedures, and corporate guidelines in the following areas are being followed:

- Environmental
- Health

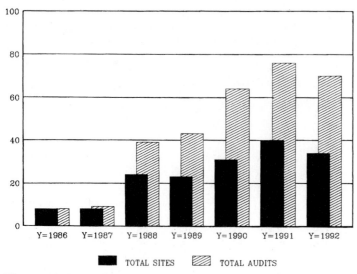

Figure 18-3. Total sales audited.

- Industrial hygiene
- Safety
- Emergency preparedness

The examination involves:

- Analysis
- Testing
- Confirmation of procedures/practices at an operating plant/facility

The environmental process also evaluates the adequacy of the environmental management system, including:

- Communications
- Training
- Risk assessment and management
- Application of best management practices at plants

Environmental Audit Objectives

From the outset, the intent of the audit program was to provide operating managers with an accurate and concise assessment of their compliance status vis-à-vis the applicable standards. It is a strong belief that by clearly identifying areas of noncompliance, the audit program provides some of the essential elements required for the development and implementation of corrective action plans, the purpose of which is to eliminate the deficiencies.

This ongoing activity can provide management with the assurance that environmental, health, and safety risks have been identified and that systems are in place for their control.

Also, Noranda personnel participating in the audit program are exposed to technologies and environmental management systems at other facilities. This learning function facilitates technical networking and information transfer.

Through the exercise of identifying and correcting environmental deficiencies, the company benefits by the continuous improvements to its environmental management system.

The key objectives of a properly designed environmental auditing program could be summarized as follows:

- To ensure compliance with regulatory requirements and corporate guidelines
- To ensure the application of best management practices
- To minimize environmental liabilities and risks

On meeting these objectives, potential risks and both corporate and personal liabilities are minimized.

Environmental Audit Goals

Specific goals of Noranda's environmental program are to:

- Correct all deficiencies and findings in a timely and cost-effective manner
- Reduce liabilities and risks to a minimum by improving previous practices and engineering designs, by implementing process modifications, and by substituting chemicals
- Improve awareness and understanding of environmental regulations, standards, guidelines, and codes of practice among Noranda operational staff at all facilities
- Transfer technology and improve awareness of good environmental management systems among Noranda plants
- Improve the efficiency and the cost-effectiveness of the environmental audit program
- Audit all operations at least once every four years

In order to meet these goals, it has been necessary to implement effective reporting procedures, which will be discussed later.

The Environmental Audit Process

The three basic phases of the environmental audit process are:

- Preaudit activities
- Activities at site
- Postaudit activities

Each phase encompasses several essential activities:

- Preaudit activities
  Selection and scheduling
  Contacting facility
  Planning
  Selection of audit team
- Activities at site
  Plant familiarization/internal controls
  Assessment of internal controls
  Evidence gathering
  Evaluation of findings
  Presentation of findings
- Postaudit activities
  Draft report
  Final report
  Action plan
  Follow-up on implementation of action plan

Critical to the process are the preparation of both an audit report and the action plans and the follow-up activities. The action plan closes the loop and ensures that all deficiencies and findings are corrected in a timely and cost-effective manner. It represents the due-diligence element of the program. Figure 18-4 provides a graphical illustration of the three phases and related audit activities.

Audit Teams, Auditor Training, and Continuous Improvement

In developing its program, Noranda determined that the use of internal auditors, primarily trained specialists selected from operating facilities in each business unit, would best serve its objectives and goals. The primary reason is that the extensive knowledge, experience, and environmental awareness developed during the auditing exercise remains within the corporation. In fact, the in-depth understanding of regulations and the exposure to outstanding management sys-

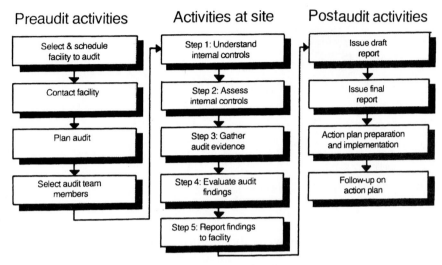

Figure 18-4. Basic steps of an environmental audit.

tems and environmental controls at other plants are valuable assets that the auditors transfer to their own and other facilities.

Auditor candidates have been selected from plants within each business unit, supplemented by staff from the corporate environmental services. The professions include:

- Chemists
- Engineers
- Environmental scientists
- Industrial hygienists
- Medical doctors and nurses
- Process, transportation, and emergency response experts

Extensive training for auditors has been provided by a specialized consulting firm, which, in addition to teaching the techniques and process, leads the participants through a trial review. Figure 18-5 illustrates how the audit resources within Noranda are distributed by area of specialty.

All environmental, occupational health, industrial hygiene, emergency response, and closure audits are conducted with internal personnel. Teams of three to six auditors are assigned to a specific task by the director—environmental audits, based on the scope of the review and the expertise required.

For small facilities, environment, occupational health, and industrial hygiene audits may be conducted at one time, in which case the team will have expertise from each area. However, in the case of large plants, a review will generally be limited to one protocol requiring three or four experts in that particular field.

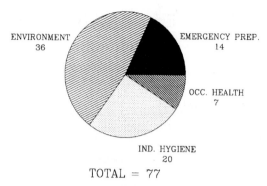

ENVIRONMENT
36

EMERGENCY PREP.
14

OCC. HEALTH
7

IND. HYGIENE
20

TOTAL = 77

Figure 18-5. Noranda audit resources by specialty (as of December 1992).

Each auditor may be assigned to two or three audits per year. Currently, Noranda has 77 trained auditors who completed 308 audits through the end of 1992.

Third-party, external auditors are used for acquisition, divestiture, and other special reviews, where an unbiased assessment of potential liabilities and the costs of upgrading environmental systems is required.

Noranda personnel involved in the program are invited to participate in regular workshops to discuss the audit program, review protocols, and provide input to help with the continuous improvement of the program. Such workshops also provide an opportunity for additional training aimed at upgrading auditor skills and proficiency.

Types of Audits

From a single type of audit—emergency response—in 1986, the program has evolved to eight distinct types of audits:

- External environment
- Occupational health
- Industrial hygiene
- Emergency response
- Environment and land management
- Acquisition
- Divestiture
- Plant closure (intended primarily for mine closures; the focus is on reviewing closure plan implementation)

Figure 18-6 provides the table of contents for six protocols used internally. As was noted, for acquisitions, divestitures, and closures, there is greater emphasis on the identification of liabilities, the costs associated with upgrading the facility

NORANDA INC.

Environmental Review Program

ENVIRONMENT

REVIEW ELEMENTS

o **Policy & Responsibilities**

o **Communications**

o **Training**

o **Risk Assessment, Environmental Impact Studies & Research**

o **Monitoring Program**

o **Equipment Operating & Maintenance**

o **Air Emission Control**

o **Water Effluent Control**

o **Pipelines**

o **Land Impact**

o **Solid/Hazard Waste Management**

o **PCB Program**

o **Transportation of Dangerous Goods**

o **Fuel and Chemical Storage**

o **Water Supply**

o **Security**

o **Groundwater Protection**

Figure 18-6.

NORANDA INC.

Environmental Review Program

INDUSTRIAL HYGIENE

REVIEW ELEMENTS

o **Policy**	o **Equipment**
o **Regulation Compliance**	o **Health Hazard Evaluation**
o **Staffing**	o **Record Keeping**
o **Facilities**	o **Health Hazard Control**
o **IH/OH Interface**	o **Training**
o **Health Hazard Recognition**	o **Right to Know**

Figure 18-6. (Continued)

to today's standards, and the issues that must be addressed in a closure or recla-mation plan. Acquisition and divestiture audits usually involve the participation of third-party auditors to lead and execute the required activities.

Audit protocols are divided into two distinct areas:

- Compliance with federal, provincial, state, and municipal regulations, stan-dards, and bylaws, as well as corporate standards, policies, and guidelines
- Best management practices

Prevention of potential environmental impacts through the application of pre-cautionary procedures is an intentional, more cost-effective strategy.

Audit protocol development is a strong feature of the audit program, as it requires the input of all those participating in the program as auditors. Figure 18-7 illustrates the steps used in protocol development.

NORANDA INC.

Environmental Review Program

OCCUPATIONAL HEALTH

REVIEW ELEMENTS

o   **General**

o   **Company Health Department**

o   **Medical Surveillance**

o   **Co. Health Relation with Other OH Specialists**

o   **Workplace Contamination**

o   **Meetings**

o   **Medical Surveillance Jobs**

o   **Equipment**

o   **Program**

Figure 18-6. (*Continued*)

Audit Report Security and Postaudit Activities

Final audit reports and action plans are distributed to the specific plant manager, his or her superior in that particular business unit, the senior vice president of environment, and the director of environmental audits. Postaudit activities are outlined in Fig. 18-8.

The question of report accessibility by governmental agencies and by court-directed disclosure is one that industry and legislators struggle with. Noranda, however, has concluded that the benefits flowing from the auditing program far outweigh any negative concerns associated with disclosure.

NORANDA INC.

Environmental Review Program

EMERGENCY PREPAREDNESS

REVIEW ELEMENTS

o   **Policy & Responsibilities**   o   **Training**

o   **Risk Evaluation**   o   **Community Response &
                                                   Public Relations**

o   **Emergency Plan Framework**   o   **Legal**

o   **Emergency Equipment &   o   **Plant Inspections**
      Physical Facilities**

o   **Security**   o   **Internal Communications**

o   **Plant Maintenance**   o   **Miscellaneous**

Figure 18-6. (*Continued*)

A due-diligence court defense, consisting of an active audit program that critically and independently (independent to the particular plant in question) identifies and corrects environmental deficiencies and findings represents the best defense.

Priority Setting and Audit Scheduling

Acquisition, divestiture, and closure audits are obviously scheduled on an as-needed basis. The director of environmental audit plans a four-year schedule. Priorities are based on the input of Corporate Environmental Services and the factors listed on pages 516 and 518.

NORANDA INC.

Environmental Review Program

FORESTRY OPERATIONS

REVIEW ELEMENTS

PART I : DOCUMENTATION REVIEW

o   **Policy & Responsibility**

o   **Documentation and Permits**

o   **Conservation Policies & Practices**

PART II : FIELD REVIEW

o   **Road Building**

o   **Water Quality**

o   **Harvesting**

o   **Aesthetics & Recreation**

o   **Site Management**

o   **Environmental Communications**

o   **Site Protection**

o   **Forest Planning**

Figure 18-6. (*Continued*)

NORANDA INC.

Environmental Review Program

PLANT CLOSURE (MINES)

REVIEW ELEMENTS

o   **Environmental Policy, Closure Plan & Responsibilities**

o   **Communications**

o   **Training**

o   **Environmental Assessment**

o   **Monitoring Program**

o   **Post Closure Operation & Maintenance**

o   **Legal Compliance; Water Effluents**

o   **Land Issues**

o   **Solid Waste Management**

o   **PCB Program**

o   **Transportation of Dangerous Goods**

o   **Fuel and Chemicals**

o   **Water Supply**

o   **Safety and Security**

o   **Special Wastes**

Figure 18-6. (*Continued*)

- New or modified legislation
- The size of the facility
- The processes and the characteristics of the chemicals and raw materials used and the volumes stored
- Employee exposure to in-plant chemicals and process by-products
- The emission effluent and waste volumes and characteristics

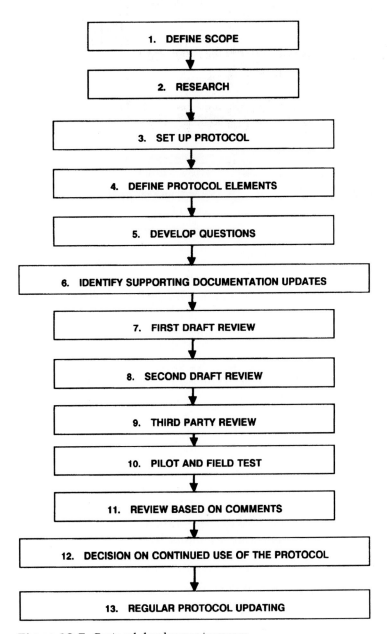

Figure 18-7. Protocol development process.

Reporting and Follow-up

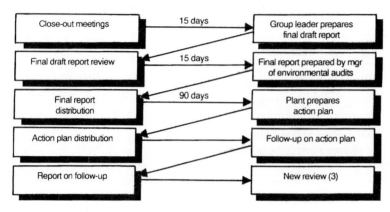

Figure 18-8. Environmental auditing (Noranda).

- The sensitivity of the environment surrounding the facility and the nature of the receiving environment
- The proximity of public residences to the plant

Although the intent is to audit each facility once every four years, operations that have a higher rating in terms of the above-noted criteria are scheduled more frequently.

Program Results

The results and benefits from implementing a formal environmental audit program are numerous. These benefits apply both to the individual operations as well as to the corporation as a whole, and they impact on the individual employee, plant managers, and officers of the corporation.

In general, these results and benefits can be summarized as follows:

- Increased focus on environmental issues and higher priority level assigned to activities having an environmental component
- More focused approach to environmental control strategies
- Development of a tool for assigning levels of priority on the many activities involved in managing an operation in the 1990s
- Clearly defined responsibilities for the different activities under the environmental program allowing commitments to be made with more assurance
- Higher level of awareness at all levels for environmental matters because in most cases the solutions involve the input and efforts of more than one person or department

- More efficient management of the related documentation such as permits, records, manifests, etc.

Conclusion

Environmental protection is an integral part of our business. Today it is unthinkable to build a plant:

- Without assessing the environmental impact
- While knowingly or willingly violating environmental regulations
- Without communicating environmental issues with employees and the public
- Without cooperating with governments
- Without planning for better processing and environmental technologies

This concern to do better is tangible evidence that corporate culture is changing.

Within this context, the environmental audit program plays a major role. At Noranda, active participation in the program from each operation, either as a supplier of auditing resources or as an audited facility, emphasizes the basic commitment to the audit process and a better environment.

19

The WMX Environmental Audit Program

John Nagy

Director, Environmental Audit Department,
WMX Technologies

The WMX Technologies and Services, Inc. (WMX), environmental audit program was established in 1983 to verify the environmental compliance status of active company facilities and operations and generate information on a formalized, continuing basis. The information developed through audits is intended to assist division, region, group, and WMX management in determining the adequacy of the systems in place for assuring environmental compliance. The audit process is also designed to ensure that specific issues that are identified during audits are fully addressed and that management systems are modified as necessary to improve environmental compliance performance.

Objectives of the Environmental Audit Program

The objectives of the WMX environmental audit program are to:

- Provide assurance to WMX management that systems and controls are in place and being implemented at company facilities and operations to ensure continuing environmental compliance

- Assist facility and region managers in the identification and resolution of specific environmental compliance issues

- Evaluate environmental compliance trends across regions and business groups and work with the business groups to assess the need to strengthen company environmental policies and management systems.

Audit Scope and Planning

Requirements reviewed during environmental audits include federal and state environmental statutes and regulations that address solid and hazardous waste management, air and water pollution control, and other environmental-related areas. In addition, requirements established in environmental permits, administrative rulings, contractual requirements, local ordinances, and company environmental policies are also reviewed. The subject areas covered in environmental audits are shown in Fig. 19-1.

---

1. Permits, authorizations, consent agreements, and contracts
2. Construction
3. Operations (landfill, transfer, hauling, incineration)
4. Leachate management
5. Water management and sewer use
6. Landfill gas management
7. Groundwater monitoring
8. General facility standards, including security and access controls and emergency management procedures
9. Tank management
10. Container management
11. Treatment and stabilization
12. Customer and site-generated special waste
13. Land disposal restrictions
14. Generator standards
15. Air emissions and monitoring
16. Employee and community right-to-know
17. Closure, postclosure, and financial assurance
18. Compliance management programs

Figure 19-1. Audit areas reviewed during WMX environmental audits.

The environmental audit department maintains a "master audit schedule" of all facilities that are currently subject to a WMX environmental audit. The schedule includes basic information regarding the subject facilities, including dates of previously conducted audits and the next projected audit. Waste Management Inc. (WMI) sanitary landfills are audited once every 3 years. Audits of WMX medical services facilities are conducted once every 2 years. Initial audits of WMI hauling divisions and transfer stations are currently being conducted. Individual facilities are audited on a more frequent basis as circumstances warrant.

Notification that an environmental audit is being planned is provided to corporate, region, and facility managers at two distinct points in time. Initial notification letters are sent to the division president, with a copy to the group vice president of environmental management, approximately 2 months prior to the audit. Shortly after this notification, a member of the department initiates contact with facility and region management to collect information used for quarterly audit scheduling and audit preparation purposes.

After specific dates for an audit are established, more detailed notifications are provided to corporate, region, and facility management. This notification occurs approximately 4 to 6 weeks prior to the audit team's arrival on site. Advance notification is intended to minimize the disruption of facility operations and ensure that all information needed for the audit either is sent to the environmental audit department or is available on site.

Audit Preparation

Environmental audits are usually conducted in teams or two to four auditors, with all auditors sharing responsibility for completion of the project. The audit team is responsible for preparation activities, on-site auditing, and reporting and follow-up activities. Individual auditors are typically allocated 1 to 2 weeks for audit preparation. Each auditor on an audit team is assigned a specific set of audit areas for which he or she is responsible.

Audits are conducted in accordance with established audit protocol (see sample protocol in Fig. 19-2). Standard audit area assessment guides are maintained by the department; these assessment guides include references to all federal and company environmental compliance requirements (see sample assessment guide in Fig. 19-3). During the audit preparation stage, the assessment guides are annotated and expanded to reflect all state, local, permit, legal, and contractual environmental requirements applicable to the subject facility. Audit preparation also includes obtaining and reviewing copies of site-specific plans and documentation and reports from the previous environmental audit (if applicable). During the audit preparation stage, the audit team's objectives are to acquire a working knowledge of the subject facility's operations, permits, and environmental regulatory framework, and to prepare the assessment guides that are used during the on-site audit for evaluation of the subject facility's management systems and compliance status.

The following landfill construction protocols were designed to be used by fully trained auditors with auditing experience. Such protocols are of limited use without training and direction. Landfill construction is just one of 15 to 25 "management areas" that are reviewed during a WMX audit.

LANDFILL CONSTRUCTION TECHNOLOGY REQUIREMENTS
SUBSURFACE (Construction—Subsurface)

NOTE: The letters A, B, and C in the left column refer to the relative level of significance in terms of managing compliance.

NA Phasing (sequence of construction in accordance with design plans)

B   Unstable area demonstration

B   Controlling leachate during construction
  - Toe containment berms constructed
  - Leachate lines between active cells and cell excavations plugged

B   Physical separation from groundwater
  - Below or above the liner
  - Vertical distance from top of liner to groundwater table

B   Dewatering (control of surface water during construction)

B   Liner type (options)

B   Liner subbase system
  - Material
  - Thickness
  - k
  - Evaluation

A   Clay bottom-liner system
  - Spreading and scarifying
  - Material (depth, quality)
  - Compaction
  - Lift thickness
  - Permeability
  - Base grades

A   Synthetic liner system
  - Materials
  - Type of seams

Figure 19-2.

- Integrity of seams
- Bond to clay (composite liners)
- Chemical properties and resistance to waste
- Base grades

A   Leachate collection and detection system(s)

- Installed
- Compatibility
- Headers and laterals
- Leachate access risers
- Sumps
- Pumps
- Drainage blanket system

    Installed
    Material
    Thickness
    k
    Gradation
    Compatibility and composition

B   Protective cover system and liner protection

- Material
- Thickness
- Permeability

A   Leachate conveyance and storage system

- Pipes outside limits of liner
- Collection tanks
- Force main

A   Gas management system

- Wells
- Headers
- Condensate knockouts and tanks
- Flares
- Vents

NA Surface impoundments—CWM only

Figure 19-2. (*Continued*)

LANDFILL CONSTRUCTION TECHNOLOGY REQUIREMENTS
ABOVE GRADE (Construction—Above grade)

B   Phasing (filling and capping areas as required)

NA Hazardous waste cell grid markers (covered under Hazardous Waste Operations)

A   Intermediate cover
- Material
- Thickness
- Timely placement and leachate minimization

B   Type of final cover system (options)

B   Soil final cover system
- Materials (depth, quality)
- Compaction
- k
- Timely placement and leachate minimization

B   Synthetic final cover systems
- Materials
- Stability calculations
- Type of seams
- Integrity of seams

B   Vegetation
- Scarifying
- Timely placement

B   Top slopes and side slopes
- Maximum elevation
- Minimum top slope
- Maximum side slope
- Horizontal limits of waste<horizontal limits of liner

B   Surface water control system (construction—surface water)
- Control structures (berms, benches, ditches, downslope pipes, culverts)
- Retention structures (ponds and basins)

Figure 19-2. (*Continued*)

      • Discharge structures (outfalls)

      • Runoff and run-on control system

      • Flood control and levels

NA Erosion controls (covered under Surface Water Discharges)

B   Maintenance activities

      • Vegetative cover maintained

      • Settlement repaired

      • Ponding of water prevented

      • Erosion minimized

C   Surface mining, coal mining, and mining reclamation activities

C   Stockpiles and borrow areas (location and activity authorized)

C   Roads (authorized)

Figure 19-2. (*Continued*)

On-Site Auditing Activities

At the subject facility, on-site auditing consists of an evaluation of management system effectiveness through environmental compliance verification. The on-site audit is typically 3 to 5 days in duration. The audit team evaluates the facility's compliance with environmental requirements, using a combination of facility inspections, document reviews, and interviews of facility personnel to compile the information needed to assess compliance.

The on-site audit is conducted with a closing conference, during which the audit team presents all audit findings to facility and region management. A preliminary draft of the audit report is distributed at the closing conference.

Audit Follow-up Activities

Audit Reports

The findings of an environmental audit are documented in a formal audit report, which is prepared by the audit team immediately upon return to the office. The audit report lists specific compliance deficiencies identified during the on-site audit; the audit team also prepares a narrative audit summary to be sent out with the report. At the time of report issuance, all compliance issues are entered into the company's issue-tracking system, the compliance action reporting system (CARS). The audit report is distributed to key facility, region, group, and corporate managers.

NOTE: Definition of symbols for each of the following assessment guides:

PA—preliminary assessment
FA—final assessment
I—interview, i.e., name of person interviewed, title, where, when
D—document, i.e., plans, monitoring data
O—observation, i.e., what was observed

AUDIT AREA: CONSTRUCTION

Elements and facility information	Citation	PA	Testing and conclusions	FA
II. TECHNOLOGY REQUIREMENTS SUBSURFACE (Construction—Subsurface) I: D: O:				
B Phasing (sequence of cell construction)				
B Unstable area—demonstration of engineering measures to ensure structural integrity (258.15[a])				
■ Placed in O. R.				
■ Notify state director of placement in O. R.				
B Controlling leachate during construction				
■ Toe containment berms constructed				
■ Leachate lines between active cells and cell excavations plugged				

Figure 19-3. Landfill construction assessment guides

AUDIT AREA: CONSTRUCTION

Elements and facility information	Citation	PA	Testing and conclusions	FA
B Physical separation from groundwater				
■ Below or above the liner				
■ Vertical distance from top of liner to groundwater table				
B Dewatering				
■ Control of surface water during construction				
■ Control of groundwater during construction (OSHA)				
B Liner subbase system				
■ Material (soil type, quality)				
■ Thickness				
■ Permeability (k)				
■ Evaluation for stability, settlement, consolidation, heave, and constructability (LDD) (LMP)				
A Synthetic liner system				
■ Materials, (HDPE, geotextiles; quality)				
■ Type of seams				
■ Integrity of seams				
■ Bond to clay (composite liners)				

Figure 19-3. (*Continued*)

AUDIT AREA: CONSTRUCTION

Elements and facility information	Citation	PA	Testing and conclusions	FA
■ Chemical properties and resistance to waste				
■ Base grades				
Minimum 1% slope toward LCS (LDD)(LMP)				
A Leachate conveyance and storage system [D AS-BUILT OF SYSTEM SERVICING ONE CELL PER AREA]				
■ Pipes outside limits of liner (NOTE: If underground pipes must be double-lined or inspected to ensure no leak-age) (LDD)				
■ Collection tanks (location and size) (PLANS)				
Inside landfill liner				
Outside landfill liner				
■ Force main (location and size) (PLANS)				
III. TECHNOLOGY REQUIREMENTS—ABOVE GRADE (construction—above grade)				
I:				
D:				
O:				

Figure 19-3. (*Continued*)

AUDIT AREA: CONSTRUCTION

Elements and facility information	Citation	PA	Testing and conclusions	FA
B Phasing (sequence of filling and closing cells)				
A Intermediate cover				
■ Material				
■ Thickness				
■ Timely placement				
To minimize leachate (LMD)				
w/in 180 days (LMD)				
B Synthetic final cover systems				
■ Materials (HDPE, geotextile, geonet; quality)				
■ Stability calculations completed and documented (LDD)				
■ Type of seams				
■ Integrity of seams				
B Surface water control system **[D MOST RECENT TOPO OR AS-BUILT]** (construction—surface water)				
■ Control structures				
Berms				
Benches				

Figure 19-3. (*Continued*)

AUDIT AREA: CONSTRUCTION

Elements and facility information	Citation	PA	Testing and conclusions	FA
Ditches				
Downslope pipes				
Culverts				
■ Retention structures				
■ Ponds and basins				
■ Discharge structures				
Outfalls				
■ Run-on and runoff control system (258.26[a])				
Design for 24-hour, 25-year storm				
Construct				
Maintain				
■ Flood controls and levels (258.11[a]; 257.3-1[b])				
Demonstration of unrestricted flow of 100-year flood				
Placed in O.R.				
Notify state director of placement in O.R.				
Unreduced temporary water storage capacity of floodplain				
Washout of solid waste prevented				

Figure 19-3. (*Continued*)

Facility Responses

Within 30 days of the issuance of the final report, facility management (with group management approval) must formally respond to each compliance issue presented in the audit report. The written response is a plan describing the required corrective actions and a schedule for implementation of these measures; responses must also address preventive actions that will be taken to ensure that deficiencies will not recur.

Close-out of the Audit Process and Audit-Issue Tracking

Approved responses that have been entered into CARS are reviewed by staff in the environmental audit department as a "quality check" on the corrective and preventive actions that have been planned. Feedback is provided to the business groups, as appropriate, as part of the QER process. Upon confirmation that appropriate and complete corrective and preventive actions have been entered into CARS, the environmental audit process is closed out. Compliance issues continue to be tracked through CARS until they are fully resolved.

Department Organization and Staffing

The environmental audit department director reports to the WMX director of environmental compliance, who in turn reports to the senior vice president, law and compliance. This reporting relationship assures that the audit program is organizationally independent from the business groups being audited. The director of environmental compliance routinely submits reports to the WMX chairman of the board and chief executive officer and the WMX president and chief operating officer regarding auditing activities and audit findings.

The planning and completion of audits and the preparation of reports and work papers are supervised by environmental audit department managers. The managers report to the director and manage the activities of the environmental audit staff. There are three job classifications for members of the audit staff:

- Environmental auditor
- Lead environmental auditor
- Senior environmental auditor

Approximately 50 percent of the auditors are engineers (chemical, civil, or environmental); the remainder have backgrounds in geology, chemistry, or other environmentally related fields. Training and professional development are emphasized for the audit staff to assure that audits are completed efficiently and effectively and that findings thoroughly address regulatory and technical issues.

Details on training activities are provided in the Environmental Audit Department Training and Employee Development Plan.

Information Management

A substantial amount of information on technical issues and company and regulatory requirements is needed for auditors to properly conduct environmental audits. Reference information is maintained in the environmental audit department reference center and state regulatory files. Regulatory and company policy information that is routinely used during audits is compiled in policy and regulation reference books. Maintaining complete, up-to-date information related to environmental requirements is an essential element of the audit program.

All information collected during an audit, including completed assessment guides, is compiled in audit work papers. Work papers are assembled and approved by the audit team following the completion of on-site audit work. The environmental audit department maintains audit work papers from a particular audit until the next audit at the same facility. At that time, any relevant documents are incorporated into the new audit work papers, and the remaining records are destroyed.

All information related to findings from environmental audits, including work papers, is managed in a confidential manner. Work papers and audit reports are maintained in locked files in the environmental audit department. After response approvals, all copies of audit reports are required to be returned to the environmental audit department; related correspondence is required to be removed from region, facility, and other corporate files. Thus only one copy of audit documentation (i.e., correspondence, the audit report) is maintained (by the environmental audit department) for confidentiality. Documentation related to individual audits is retained in accordance with the department's records retention schedule.

20

AT&T's Environmental and Safety Audit Program

Ronald DiCola

Environmental and Safety Compliance
Assessment Technical Manager, AT&T

AT&T's worldwide environmental and safety audit program consists of due diligence; disposal and recycling facility; management systems; pollution prevention; energy; OEM; and compliance assessment-assurance audits. This chapter, however, deals only with company audits that measure compliance against laws and regulations at AT&T's U.S. and non-U.S. factories. These compliance audits are full scope—environment and safety. Before describing the company's compliance audits, however, some brief information about the nature of the company will prove useful.

Organization of the Company

Before divestiture in 1984, AT&T was similar to a holding company. It coordinated the efforts and financial management of Bell Telephone Laboratories, Inc. (research and development); Western Electric Company (the manufacturing and supply unit); The Telephone Operating Companies; The Long Lines Division (later called AT&T Communications, Inc.); and second-tier subsidiaries such as Nassau Recycle Corporation and Teletype Corporation.

Because of Western Electric's role as manufacturing and supply unit to the company, it was a leader in the development of safety programs and environmental engineering controls. Bell Telephone Laboratories also developed strong safety and environmental programs for its own unique needs. AT&T itself, as a holding company, had little reason to develop safety or environmental programs.

With the January 1, 1984, divestiture, everything changed, including structure of the AT&T companies. The Telephone Operating Companies became independent, competing Regional Bell Operating Companies. AT&T reorganized into current major units: Communication Products, Communication Services, Financial Operations, and Network Systems. In 1992, the company acquired NCR. Manufacturing and research and development (Bell Laboratories) functions were assigned to the appropriate business units.

Today, AT&T comprises some 20 business units and divisions based on individual product or service offerings. These range from computer chips and components to voice and data transmissions worldwide through the AT&T Network. Each business unit is responsible for its own financial performance. The company operates more than 40 U.S. and 20 non-U.S. manufacturing locations worldwide, owns over 5000 buildings, service centers, and laboratories, and employs over 315,000 employees in all 50 states and 130 different countries.

Corporate Environmental and Safety Engineering

In 1986, the Superfund Amendments and Reauthorization Act (SARA) made postdivestiture AT&T more aware of its environmental liabilities. As AT&T gathered data on emissions in compliance with SARA-313, the company was able to see how many pounds of materials were going out its smokestacks. The company wanted those numbers reduced dramatically.

To reach this goal and to provide an integrated, comprehensive global approach to environmental and safety management, the company reorganized its corporate staff to create an Environmental and Safety Engineering Center (E&SEC) under vice president David R. Chittick. The group, headquartered in New Jersey, has overall corporate responsibility worldwide. Field Environmental and Safety Engineering Center professionals are deployed in four regional areas around the United States to support AT&T's domestic service and salespeople.

The group's mission is to establish the direction, the standards, and the engineering support for all of AT&T's operations and OEM suppliers necessary to protect AT&T employees, customers, shareowners, the public, and the natural environment worldwide. This organization is responsible for auditing the quality of AT&T's environmental and safety performance.

Environmental and Safety Engineering Center Responsibilities

- Compliance assessment
- Employee transportation
- Energy
- Environmental and safety engineering
- Industrial hygiene and ergonomics
- Product liability
- Real estate assessment and remediation
- Record keeping and reporting
- Superfund

Business Unit Environmental and Safety Officers

In daily operations, environmental and safety officers for each business unit and division are key to maintaining sound performance worldwide. These officers are responsible for coordinating the performance of their individual units. The Environmental and Safety Engineering Center supports and supplements their efforts. For example, it provides information and advice to business units about government regulations and other issues of the global marketplace. In addition, because business units are encountering a proliferation of regulations from all levels of government all around the world, E&SEC monitors and helps interpret these regulations to ensure that business initiatives will not be frozen out of a given market or found in noncompliance.

AT&T's Compliance Audit Program

The Role of the Law Division

Some companies think of compliance as a risk management issue. AT&T does not. In fact, the group executive of AT&T Network Systems, who has top-level responsibility for environmental and safety concerns at AT&T, has stated, "Compliance with law is not a risk issue. It is absolute, and it is mandatory."

AT&T compliance audits are done by the Environmental and Safety Engineering Center at the direction of the law department to help it counsel management on compliance in the plants and other AT&T facilities. The law department also communicates audit results to business unit lawyers so that they, in turn, can advise business unit management concerning the unit's compliance status, legal exposure, needs for compliance planning, and any necessary steps to correct compliance weakness. Since the audit findings are used as the basis for providing this legal advice, results of the audit are covered by the attorney-client privilege. Distribution is limited accordingly. However, to provide necessary information to people in the plant who need to act on them, E&SEC prepares an action plan that is separate from the audit report. The action plan doesn't draw conclusions but rather provides a list of recommended actions to assure continuing compliance with both AT&T and legal requirements. The law department reviews the action plan but doesn't need it as a basis for legal advice because that basis is in the audit report analysis. Consequently, the action plan is treated as proprietary but is not considered to be legally privileged.

Once a company establishes an audit program, it is of paramount importance to follow up on deficiencies discovered by the audit. If not, the audit and legal time spent are both wasted. Moreover, if a federal or state agency investigating a violation finds that a company discovered a problem long ago and did not fix it or at least make good faith attempts to fix it, the agency is likely to penalize that company more severely, arguing that the violation is a "willful" violation.

AT&T audits against legal standards and its own written practices. It also audits against good management practices, practices not necessarily required by law and perhaps not explicitly required by company policy, but practices that enable it to more efficiently meet both legal and company requirements. For example, labeling drums is a legal requirement, but developing a system for labeling drums is a "good management practice." Audits also encompass pollution prevention efforts that have many benefits, including reducing compliance costs, staying ahead of federal and state laws, decreasing documentation work, etc.

The scope of AT&T's compliance audits includes:

- Clean Air Act
- Comprehensive Environmental Response Compensation and Liability Act
- Federal Water Pollution Control Act
- SARA Emergency Planning and Community Right-to-Know Act
- Toxic Substance Control Act
- Safe Drinking Water Act
- Resource Conservation and Recovery Act
- Hazardous Materials Transportation Act
- Occupational Safety and Health Act
- Nuclear Regulatory Commission
- Federal Insecticide, Fungicide and Rodenticide Act
- Delivery, storage, and handling of regulated materials (chemicals)

Globalizing AT&T's Audit Program

AT&T's globalization efforts are relatively new. Historically, U.S.-based, the company has expanded into Europe, Asia, and Latin America. The goal has been to start up these operations correctly at the beginning. The company wants the people operating local plants to know that it is concerned with compliance and that it will be auditing their operations.

When an audit program is globalized to cover factories in Mexico, in the European Community, or in Asia, the biggest difficulty is identifying the applicable laws. It is sometimes difficult to get specifics to audit against. Auditors often have to tap every resource, every component of their network to determine relevant laws and regulations and to get them accurately translated.

In many non-U.S. situations, AT&T's own standards are stricter than the relevant local laws. In all cases, company audits are based on both AT&T's standards and the laws of the land in which the factory is located.

The Philosophy behind AT&T's Audits

The audit program exists to enable the Environmental and Safety Engineering Center to be of service to its customers. Audits help to identify needs and help facilities to be better prepared to comply with applicable laws, regulations, and policies.

The Evolution of AT&T's Audit Program

The compliance auditing program at AT&T has evolved over the last several years and continues to evolve. Perhaps the biggest evolution has been the change from being "report cards" on compliance to being management tools.

Incorporating the Principles of TQM

An integral part of that change has been the incorporation of the principles of total quality management (TQM) into the auditing program. With the incorporation of these principles, audits evolved from being technical checklists—laundry lists of noncompliance—with little long-term "value added." Environmental and Safety Engineering Center auditors were unable to answer questions such as:

- "What are the causes of our problems?"
- "Why are we out of compliance?"
- "Why is the fire extinguisher mislabeled?"

- "Why isn't that drum labeled?"
- "What's wrong with our system?"

Using TQM methodology, auditors are now able to answer such questions. In addition, they are able to identify both root causes of compliance problems and the processes that need to be improved. In order for the auditors to find root causes, however, they must have an understanding of human factors such as organizational behavior.

By structuring the audits to fit within the TQM pattern, results become much more useful to the heads of the business units, who now use quality processes to increase customer satisfaction, to increase productivity, to improve product quality, and to achieve the company's environmental goals.

Quality management means management by fact—-measuring actuality against goals and showing the gaps that exist between the two. Gaps in compliance cannot be improved unless they can be measured. By doing a management-system and root-cause analysis, E&SEC audits provide a means for their customers to identify gaps, develop plans, and support closure of the gaps. Once this is accomplished, E&SEC is then able to help facilities develop their strategic plans. This is part of the "value-added" the organization provides its customers, and as such, the audit is perceived as a tool for the facilities themselves.

Shortening the Audit Cycle

A second evolution occurred when E&SEC realized that the audit cycle was taking longer and longer to complete. One reason for the increasing length of the process: Facility personnel viewed the auditors as consultants and would look to them for answers to detailed operational questions.

Even after the audit, plant personnel continued to call the auditors for consultation. This slowed the audit process significantly, and because of it the process was changed. Now, when field people have questions on the technical requirements resulting from the audits, they are referred to environmental engineering staff for follow-up. This has greatly improved the efficiency of the audit program. Today, the entire audit cycle takes approximately 60 days.

Frequency of Audits and Evolution of the Audit Team

A third and interrelated evolution in the audit program concerned the frequency of audits and the makeup of the audit team. Originally, each plant was to be audited once a year. Depending on how one defined a plant, that might mean from 40 to 90 audits a year—a physical impossibility given available resources.

Today, AT&T audits are based on need. Facilities with the highest risk or the largest compliance gap are audited more frequently. The company established 2 years as the average frequency for the audit program; however, an audit may be conducted every year if needed, or every 3 or 4 years for low-risk and high-performance facilities. To meet the increased audit workload, both in-house auditors and consultants are used in the program.

The On-Site Audit Team

The audit team includes one E&SEC engineer who directs the activities of environmental and safety experts provided by the outside consulting firm. The consulting firm then completes much of the report preparation that consumed so much staff time in the past. However, when the audit concerns a "sensitive" facility such as AT&T Bell Laboratories, the audit team consists entirely of company professionals: five engineers from E&SEC and a "borrowed" AT&T industrial hygienist and safety engineer.

This new approach has been found to work well, has enabled E&SEC to do more audits more efficiently, and has freed up the time of company engineers. Overall, the number of audits has increased from 5 to 20 audits per year over the last several years, and the cost per audit and the time per audit have been reduced drastically.

Evolution of the Audit Report

The format of the audit report also evolved to become more customer-focused. Instead of reporting a technical checklist of violations, the revised report provides an executive summary and a program-by-program analysis of a facility's status. A separate action plan provides the plant with individual action items. The plan assigns responsibilities, projects timetables, and establishes due dates for completion of the action items.

Evolution of the Measurement Scale

The measurement scale of compliance audits has also evolved. Originally, the number of action items was the measurement. Now compliance is measured on a scale of 1 to 5: A plant receiving a grade of 1 requires major improvement in the way it handles environmental and safety issues; while a grade of 5 means the plant is managing its environmental and safety programs very well.

This scale is quality-driven, for in quality terms everything is a process. If drums are unlabeled, the labeling process must be analyzed. And once the process is fixed, every drum is expected to be properly labeled every time that facility is audited.

Rating	AT&T Audit Rating System Definition
5.	MEETS governmental and AT&T internal compliance requirements, where all (or virtually all) requirements are satisfied.
4.	SUBSTANTIALLY MEETS governmental and AT&T internal compliance requirements, where most of the requirements are satisfied and a high degree of compliance is substantiated.
3.	GENERALLY MEETS governmental and AT&T internal compliance requirements, where many of the requirements are satisfied, especially the more significant requirements, but weaknesses in design and/or implementation of compliance programs were noted.
2.	REQUIRES IMPROVEMENT to meet governmental and AT&T internal compliance requirements, where a number of exceptions to applicable requirements were noted, reflecting the absence of some required programs and/or some significant departures from established criteria.
1.	REQUIRES SUBSTANTIAL IMPROVEMENT to meet governmental and AT&T internal compliance requirements, where many of the requirements are not satisfied, including the absence of a number of required programs and/or significant departures from established criteria.
0	NO DATA AVAILABLE

The Audit Process

Preaudit Preparation

It is not the intent of the audit team to spring an inspection on a day's notice to "catch" facilities (E&SEC customers) in the act of doing wrong so as to penalize them. E&SEC believes such an approach offers no value to customers. Instead, a few months before an audit, E&SEC alerts the facility that it is going to be audited. There is no surprise when the audit team arrives.

Audit Protocols

Originally, each area of specialty (air, water, handling of chemicals, etc.) had its own protocol. Consultants helped E&SEC combine the individual checklists into a standard set of protocols in the form of questions and answers. There is one protocol for environment and another for safety. The protocols are modified to keep current with changing laws and regulations. Before an audit team visits a facility, it adds applicable federal, state, and local regulations to the protocols, indicating which requirements are governmental and which are AT&T practices, or good management practice. Six weeks before the audit, the protocols are sent to the facility that is to be audited.

The On-Site Audit

On-Site Opening Meeting.   When the audit team arrives at a facility, usually on a Monday, the first step in the audit is an on-site opening meeting. Because every plant is organized and managed differently, opening meetings range from 5 to 30 people. At the meeting, the audit team describes the audit process and enlists the help and support of the facility.

An opening meeting might include the following personnel:

- General manager of the facility
- Environmental and safety manager
- Chief engineer
- Medical doctor
- People from training, and others

Documentation Review

Next, the team begins a documentation review. The team analyzes the facility's permits, chemical control documents, inventory documents, training records, notifications, etc., to see if the facility is meeting AT&T's written practices and federal, state, and local requirements. It also looks for good management practices, and the absence of them. The team checks that all documentation is complete and correct, that it is signed by the appropriate people, and that all entries are supported with sufficient backup information.

Technical Audit

In addition to documentation review, the team does a complete technical audit of the facility. The technical review covers areas such as:

- PCPs
- Underground tanks
- Use of chemicals
- Chemical handling

- Chemical substance control
- Substance waste and solid waste management
- Water quality
- Transportation
- Air pollution control
- Emergency planning

When noncompliance is found, the audit team looks for its root causes.

Management Systems Audit

The technical audit is supplemented by a management systems audit that examines how well the facility's management systems run and how well they work together. The auditors look for accountability, especially regarding safety, and they ask questions such as:

- "Does management recognize its responsibility and accountability?"
- "Does management have a meaningful way of approaching environmental and safety as part of doing business?"
- "How is accountability disseminated around the floor?"
- "Are safety meetings held?"
- "If safety meetings are held, are the proceedings documented?"

Catastrophic Event Analysis

The audit also involves catastrophic event analysis. Here auditors review how a facility uses compressed gases, flammable liquids, and other chemicals such as corrosives and oxidizers. They examine how these chemicals are regulated by the local jurisdiction as well as by federal and/or state regulations. The auditors are looking at the potential for catastrophe such as a large release of toxic gas or a spill of a flammable liquid that could cause a conflagration. The point is to anticipate what the root causes of such potential catastrophes might be. For instance:

> During the inspection of a pipeline system that pressurizes methanol, a flammable liquid, a senior E&SEC engineer determined that hundreds of gallons of methanol would spill on the floor if one of the pipes ruptured, creating a potentially catastrophic event. The engineer noted this to management as an "action item," and the situation was corrected.
>
> During another audit, auditors located incompatible chemicals stored next to each other. The audit action plan showed how a common spill might prove disastrous and required that the situation be corrected. The audit report examined the root causes of the potentially dangerous situation. In the analysis presented to management, the auditors asked:

- "Do you need to use this material?"
- "Do you need to have this gas under pressure?"
- "Do you need to pipe it?"
- "Is there a better way of moving it?"

"Boarding"

By Thursday, the walk-through, inspections, interviewing, and documentation review are essentially complete. The audit team meets in private to "board" the major issues affecting the facility. They rank these issues on the basis of potential risk—in terms of both liability and personal risk. Team members discuss priorities for the facility and determine which action items are most important. Root causes and trends are analyzed. For instance, instead of listing what was done incorrectly on 45 out of 50 air permits, "themes" are discussed.

The Action Plan

As part of this effort, the team prepares a written action plan—a "deliverable"—for the plant management so that it can begin making improvements and correcting problems uncovered by the audit. The action plan is presented at a final meeting with the plant personnel who were at the opening meeting. While this final meeting is often painful, it is nevertheless the meeting top unit managers find most valuable. This is the meeting that reveals the gaps in compliance that need to be addressed.

Quality Improvement Teams

To close the gaps, personnel at all levels of the plant must be involved with the improvement efforts. Consequently, as part of the audit procedure, the audit team helps the facilities to establish quality improvement teams. Auditors tell plant personnel that: "In order for you to close this gap, you need to have a quality improvement team. That team needs to have a broad spectrum of participation, and it needs to have management support."

Action Plan Updates

Each facility has the ultimate responsibility for being in compliance. The environmental and safety officer, however, must stay in contact with the facility, making sure that the improvements called for in the action plan are done. To help E&SEC auditors follow up on a plant's action plan, each audited facility does a quarterly update of its action plan. The data submitted are entered into a software application custom-developed by E&SEC engineers. The software allows staff auditors to accurately track progress and to analyze results by subject area.

The Audit Report

After completion of the on-site audit, auditors prepare the audit report for the law department. The report, approximately 10 pages in length with a one-page executive summary, discusses findings, management systems, and root causes. It also discusses "why" and includes exemplary findings.

Report Distribution

The audit report is a privileged and confidential report—as such it is distributed to select upper management by the law department. After the audit report is complete, the audit team meets with appropriate personnel from environmental engineering, safety engineering, fire protection, chemical substance control, safety and industrial hygiene engineering, and the law department. The audit team shares information about the audit and about items requiring the help of these professionals.

During the 6-week audit cycle, auditors, business unit environmental and safety officers, and facility managers form a three-way team. Once the audit report and action plan are released, environmental engineering experts replace the auditors as the third leg of the team.

Conclusion

AT&T environmental and safety audits are directed by the law department and carried out by the company's Environmental and Safety Engineering Center (E&SEC). The audits are full-scope, covering applicable environmental, health, and safety laws, regulations, and policies at U.S. and non-U.S. facilities.

The compliance audit program has evolved and continues to evolve. The biggest change has been the shift in the way the compliance audits are perceived—from report cards used to judge a facility to a cooperative effort by auditors, business unit management, and facility managers. Another important change has been the incorporation of total quality management principles in order to help management by measuring gaps in performance and by providing an independent assessment of the root causes of the gaps.

On average, a facility is audited once every 2 years. By using consulting experts under direction of E&SEC engineers, the number of audits performed in a year has increased significantly.

The on-site audit includes a documentation review, a technical audit of the facility, an audit of management systems, and a catastrophic event analysis. The auditors develop an action plan which helps the facility correct noncompliance issues. The audit team provides additional value by facilitating the formation of quality improvement teams to focus on closing the gaps. A more detailed audit report is also prepared and is used by the law department to counsel management on compliance issues.

The evolutionary changes in the company's audit program reflect the needs of E&SEC customers. As these needs change, the program will continue to evolve so that it remains a useful tool for business unit management and helps them ensure that their facilities are in compliance.

21

Environmental Assessments at a Major Money Center Bank

Judd Bernstein

Vice President, Chemical Bank

The dramatic increase in environmental laws and regulations over the past 15 years has created additional risk for real estate lenders. They are threatened with the following:

- That environmental liabilities will compromise the borrower's creditworthiness
- That contamination will negate the value of the borrower's collateral
- That the lender itself will become liable for cleanup costs

While the financing of real estate is unique and each loan represents its own set of issues, lenders have responded to this increased exposure by developing various environmental due diligence techniques. The environmental site assessment (ESA), a full-blown Phase I assessment, or some sort of screening is now customary commercial practice prior to most decisions to finance commercial real estate. But there is no master formula for how to conduct an environmental site assessment.

Numerous organizations have attempted to define and standardize the environmental site assessment. The American Society for Testing and Materials (ASTM) has undertaken one of the most ambitious efforts that involves an attempt to standardize both the on-site investigations and the reporting of the results. In 1993, it published the following standards for environmental site assessments:

- Phase I Environmental Site Assessment Process (ASTM designation E1527-93)
- Transaction Screen Process (ASTM designation E1528-93)

The purpose of these standards is to define good commercial and customary practice for conducting environmental site assessments. The standards are intended to permit a user to satisfy one of the requirements of the "innocent landowner defense" and "secured creditor exemption" under the Comprehensive Environmental Response, Compensation and Liability Act (CERCLA) and Superfund Amendments and Reauthorization Act (SARA).

With this in mind, what follows is an outline of what a major bank would consider a minimum guide or scope of what prudent lenders should do to protect their institution from environmental liabilities, such as placing a mortgage on an environmentally impaired property or just plain lending to a company that is in an environmentally sensitive business without knowing what environmental liabilities that company has and what they have done, if anything, to mitigate them (see Fig. 21-1).

As a general rule, environmental site assessments are conducted in phases. The initial phase is called the Phase I assessment. It is intended to identify potential sources of environmental problems or liability. The objective of a Phase II environmental assessment is to confirm the presence or absence of a potential environmentally hazardous material or liability through on-site sampling. This sampling might be extensive—with borings and soil and groundwater samples—or just preliminary soil gas investigations. There are two stages of Phase II assessments—the initial stage and a more advanced stage that includes extensive soil and groundwater sampling. In general, lenders should not be involved in Phase II investigations unless they are the owners of the property. The Phase II work should be done by the borrower or owner of the property with the lender only reviewing the scope of work that is to be performed to verify that it will satisfy the lender's requirements.

Once the existence of a problem has been confirmed, a Phase III investigation may be needed to quantify the magnitude of the problem and develop a plan for establishing a cleanup cost estimate.

The actual cleanup work can be called the Phase IV assessment. In most instances lenders should not get involved in detailed Phase III or Phase IV of the environmental site assessment process unless through unfortunate circumstances—their foreclosure or taking the deed in lieu of foreclosure—they have become the owners of the property.

Once the transaction closes, the purchaser or borrower assumes operational control of the property that was the subject of the site assessment investigation and can take action to ensure that on-site activities do not contaminate the property. Most lenders, however, will have little or no operational control of environmental issues and will only have the rights and remedies (to monitor or

1. Does the borrower keep environmental records for the facility?

2. Has there been an environmental audit of the borrower's facility?

3. Are there any federal, provincial, or municipal environmental agreements with respect to the borrower's facility?

4. Are there underground storage tanks, either in use or no longer in use?

5. Are there chemicals on the site?

6. Is there asbestos on the site?

7. Has there ever been a spill at the facility, such as an oil spill, for example?

8. Is there a spill control and emergency response plan?

9. Is there a system for reporting environmental accidents such as spills or other releases of contaminants?

10. Are there PCBs or PCB-contaminated equipment on the site?

11. Are there any pesticides on the site?

12. Have there been any warnings or notices of violation or orders from regulatory inspectors or agencies?

13. Have there been any complaints from residents or workers?

14. Are they any discharges of waste water from the borrower's facility?

15. If so, is there any monitoring of the quality of the waste water discharged into a body of water or the sewer or sanitary sewer system?

16. Are there any municipal agreements with respect to discharges into sewers?

17. Have there been problems with the municipality and discharge exceedences?

18. Are there any air or noise emissions?

19. If so, is there any monitoring of the air emissions and the effectiveness of the ventilation?

20. Are there any certificates of approval or permits for emission sources or applications for such permits?

21. Is or was there any hazardous and/or solid waste storage and/or disposal at the borrower's facility?

22. Is there a solid-waste management plan for nonhazardous waste?

23. Is there off-site waste disposal or hazardous waste treatment storage?

24. Are there any waste manifests on the site?

25. Is there waste processing equipment such as incinerators, lagoons, or solvent stills on the site?

26. Does the borrower transport dangerous goods?

27. Are there activities generating environmental contamination on adjacent properties?

Figure 21-1. A sample checklist for the lender.

investigate on-site activities) that are provided for in the loan documents. Further, lenders must consider whether:

- Lenders should not have any day-to-day operational control of a borrower's business or property if they wish to claim protection as a "secured creditor" under CERCLA and SARA and do not wish to be considered or be named as an "operator." This would mean they could be liable for cleanup costs.
- Furthermore, lenders must consider whether potential cleanup costs may adversely affect the borrower's ability to pay back the loan.
- The priority of the lender's lien position (mortgage) can be adversely impacted by state environmental superlien provisions.

Of course, the lender has the option of walking away from the property. All these factors and many others must be balanced during the underwriting and due diligence process.

Most lenders have developed site-assessment procedures or policies that are based on the various regulatory programs referred to in Table 21-1. These policies may be highly structured or informal. However, owing to the dynamic nature of environmental law and regulations, constant diligence is necessary to keep this type of material current (see Fig. 21-2). Also, an environmental attorney (in-house or outside) should be made a member of the overall environment risk management team. This team is made up of the internal staff, account officer, environmental consultant, and the environmental attorney.

Table 21-1 Major Federal Laws That Govern Lenders' Environmental Concerns and Phase I Assessments

Name	Citation	Comments
Comprehensive Environmental Response, Compensation, and Liability Act	42 USC S9601-9675	Established national scheme for liability; authorized EPA to adopt National Contingency Plan
Resources Conservation and Recovery Act (Solid Waste Disposal Act)	42 USC S6901-6992K	Established comprehensive scheme for management of hazardous wastes; established program for state management of underground storage tanks; authorizes EPA to require corrective action at hazardous waste sites
Emergency Planning and Right-to-Know Act	42 USC S1100-11050	Requires routine reporting to state and local committees
Water Pollution Control Act	33 USC S1251-1387	Established permit standards and requirements for discharges to surface waters
Clean Air Act	42 USC S7401-7671q	Requires EPA to set air quality standards to be enforced by state permit programs

I. Introduction

As it is increasingly important for real estate lenders to protect themselves against becoming involved with environmentally sensitive properties, an environmental due diligence policy will assist a lender in minimizing the environmental liability associated with existing and potential environmental hazards by delineating the due diligence procedures to be followed. In addition, the policy is to be in conformance with the Guidelines for an Environmental Risk Program issued by the Federal Deposit Insurance Corporation (FDIC) in 1993.

The lender should maintain a technical services staff to assess the existence and magnitude of environmental hazards and to manage the environmental due diligence process.

The policy of the lender should be that the technical consultant, as hereinafter defined, be engaged by the lender to perform the required environmental site assessment or any other environmental investigation as deemed necessary by the technical services staff for its own benefit.

The lender should also form an environmental policy committee, whose members should be drawn from the various divisions of the lender most directly affected by the environmental policy.

The policy applies to "covered transactions," which are defined as any of the following:

1. Any extension of credit (e.g., loans, lines of credit, and letters of credit herein referred to as "loans") secured by:
   a. Commercial real estate
   b. Interests in commercial real estate
   c. Interests in entities owning commercial real estate (e.g., closely held corporations and partnerships)
2. Any business loan to a customer operating an environmentally sensitive business, whether or not commercial real estate forms part of the collateral
3. Real estate to be acquired in satisfaction of debts
4. Real estate acquired by purchase or net lease for use as blank premises
5. Those instances when, in a fiduciary capacity (e.g., executor, testamentary trustee, or *inter vivos* trustee), a bank holds interests of the type described in (1) *supra.*

II. Definitions

The following definitions are used throughout this policy:

1. **Environmental site assessment Phase I ("Phase I Assessment")**—This is a nonintrusive survey of real property for purposes of determining envi-

Figure 21-2. Sample basic environmental due diligence policy.

ronmental concerns resulting from the generation, transportation, usage, storage, and/or disposal of hazardous materials. In addition, issues such as asbestos-containing materials, wetlands, lead paint, and radon shall be investigated as determined appropriate by either the technical services staff or the consultant.

Environmental site assessment Phase II ("Phase II Assessment")—This is the assessment that determines the actual presence or absence and the extent of hazardous substances and contamination.

2. **Technical consultant**—A qualified external consulting firm acceptable to and retained by the technical services department.

3. **Technical services manager**—The individual at each bank who is responsible for applying the policy and granting exceptions thereto.

4. **Environmentally sensitive business**—A business involved in the manufacture, handling, transportation, storage, or removal of hazardous materials.

III. Transactions Requiring an Environmental Site Assessment

A Phase I assessment is required prior to entering into a covered transaction. A Phase I assessment should be requested through the technical services staff, which will arrange for the engagement of the technical consultant.

Prior to entering into a covered transaction, a Phase II assessment or environmental site assessment will also be undertaken, as required, based on the recommendation of designated members of the technical services staff and legal counsel. For example, if at the conclusion of a Phase I assessment, it appears that more information is required to explain such things as soil discolorations, large earth disturbances, and nearby or surrounding property concerns (e.g., well-water odors or contamination, colored seepages, and abandoned dump sites), then a Phase II assessment should be conducted. Ordering and paying for a Phase II assessment will be the responsibility of the obliger.

1. **New transactions**—A Phase I assessment must be conducted for each covered transaction as defined herein.

2. **Renewals, extensions, modifications, and restructuring**—Updated (or new) Phase I assessments may be required where the business or property is vulnerable to environmental hazards or when the transaction in question was "booked" prior to the establishment of this policy. The technical services staff is authorized to determine the need for an updated or new environmental site assessment in such instances.

Figure 21-2. (*Continued*)

IV. Minimum Scope of the Phase I Assessment

A Phase I assessment is performed to determine whether a Phase II assessment is needed before the transaction can go forward. Types of environmental problems that would require further investigation include:

- Polychlorinated biphenyls (PCBs)—contaminated soil used as fill material

- Lead contamination of subject site (in soil, in paint, in water)

- Presence of underground storage tanks on-site or on adjacent sites

- Evidence that a solid or hazardous waste landfill has been located on or adjacent to the subject property or that a toxic chemical lagoon, pit, or site registered as a hazardous waste site by any governmental agency is on or located near the property.

- Manufacturing processes that might have caused a hazardous material discharge to the air, soil, or water.

At the sole discretion of the technical services manager, a modified environmental site assessment may be performed. This may include, but not be limited to, an environmental transaction screen modeled on the ASTM standard E152B-93, environmental data bank reviews, update environmental site assessments (no historical data), checklist site assessments, manufacturing loan environmental survey, or any combination of the above.

For nonindustrial type loans under $500,000 a waiver of the environmental site assessment (Phase I environmental assessment) can be granted by the technical services manager and the appropriate M.D., S.V.P., or E.L.L. (both must approve) upon the submission of a fully complete site assessment waiver form.

The technical consultant selected to complete the Phase I assessment shall perform such investigations as are established by the technical services unit. These shall include but not be limited to the following:

1. Site visit to identify all obvious visual signs of contamination and/or potential sources of environmental or regulatory concern

2. Identify uses of the property, currently and for the last 50 years.

3. Review of various documents and records of the U.S. Environmental Protection Agency and the appropriate state and local agencies for the purpose of identifying any history of hazardous waste activity that is known to those agencies (enforcement actions, consent orders, and reported spills) to have occurred at the subject property or at any site within a prudent ($1/2$ to 1 mile depending on the site) radius of the subject property.

Figure 21-2. (*Continued*)

The current compliance status of such activity with respect to applicable laws and regulations will be reviewed. A list of all identified activity along with its compliance status will be included in the final report.

The documents will also be reviewed to determine if the facility has the required air emission and wastewater discharge permits and a Spill Prevention Control and Countermeasures (SPCC) plan, if appropriate.

4. Specific report concerning the present or past existence of any storage tanks (above and below ground), wells, and waste disposal facilities to determine the potential for leakage by review of available information on age, composition, contents, water table, and monitoring or inspection records.

5. Specific report concerning electrical transformers and fluorescent light ballasts regarding the use or containing of PCBs.

6. Activity review of all adjacent businesses, land, and cross-property easements to include both current and prior use.

7. Proximity to any sensitive ecological area, i.e., wetland or floodplain

8. In existing buildings:

   a. Identification of the presence of any probable asbestos-containing material or other toxic material such as radon gas; for residential buildings, lead paint, if required.

   b. Followed up by a sample testing of the material to measure its asbestos content for comparison with allowable and regulatory levels.

   c. If asbestos is identified, an estimate of the extent of asbestos-containing material should be made and the cost of its removal, plus estimate of operations and maintenance plan, if required.

9. The findings should be summarized in a written report (four copies) that will contain color photographs and may contain an estimate of the cost to remove hazardous material, if any, found at the project site and to restore any damage caused by such removal, inclusive of all professional and/or supervisory (construction management) fees and air monitoring and testing if necessary or appropriate. The body of the report will be signed and dated by a officer of the technical consultant.

V. Documentation

1. **Report preparation and review**—A Phase I assessment report is to be prepared by a technical consultant. The technical services department staff will prepare a written review detailing the findings of the assessment and a summary of pre- and posttransaction closing items requiring attention, if any. To minimize the lender's exposure to environmental liability, Phase I assessment recommendations should be incorporated into the process for approving the covered transaction.

Figure 21-2. (*Continued*)

2. **Environmental certification**—As a condition to close a covered transaction, a certificate will be prepared by the appropriate technical services unit staff member for each property upon its determination that the assessment is acceptable and that the minimum requirements of this policy have been met. If a review indicates that items need to be addressed prior to closing, then the certification will not be issued until the technical services staff is satisfied that all open issues have been satisfactorily addressed.

3. **Legal documents**—All legal documents including any contracts of sale, trust agreements, or loans documents must include appropriate protective representations and warranties regarding environmental responsibility. All loan documents must contain appropriate protective representations and warranties requiring that, if hazardous substances are used in the ordinary course of an obliger's business, such substances are handled and disposed of according to applicable laws and regulations. The handling of such materials outside of the ordinary course of a customer's business should be prohibited. Loan documents should also include the following:

   a. Indemnities from the obliger or guarantor to the lender in the event that public or private third parties assert environmental or hazardous waste claims against the lender.

   b. A provision giving the lender the right to undertake subsequent environmental assessments including tank testing once the loan has been made.

   c. Remedial or other actions required to be taken prior to or following closing of the loan.

Under no circumstances may these covenants or representations be modified or waived without discussion with in-house counsel and a managing director or equivalent-level officer.

All agreements should allow the bank to withdraw from any transaction if the results of an environmental assessment or other environmental site assessment disclose environmental conditions that, in the sole discretion of the lender, make the transaction inadvisable.

VI. Customer Submitted Phase I Assessment

When a customer submits a Phase I assessment that is not more than 6 months old and was commissioned and performed prior to the covered transaction, the technical service unit or equivalent will then review it to see if it meets the lender's requirements. At the same time as the submission of the environmental site assessment, the customer must also supply the following:

1. A signed, valid environmental certification letter from the technical consultant

Figure 21-2. (*Continued*)

2. A qualification and background package of the technical consultant (if it has not already been determined to be acceptable to the lender), which shall also include three references and an insurance certificate indicating what professional liability insurance is carried

3. A letter from the borrower giving the lender permission to speak to the technical consultant and directing and instructing the consultant to fully cooperate and answer all of the questions

The acceptance of the borrower-submitted Phase I assessment is at the sole discretion of the technical services manager.

VII. Release of Environmental Assessments

No environmental site assessment may be released to any party without the consent of the head or the deputy of the appropriate group credit officer, or the equivalent, and the technical consultant. The recipient will be required to provide a signed release letter before a copy of it is furnished if consent is obtained.

VIII. Duty to Disclose

In the course of conducting an environmental site assessment, the lender may gain information it may be required to disclose to governmental or other appropriate authorities. The determination to make such disclosure will be made by the environmental policy committee after consultation with counsel and technical services staff.

IX. Billing and Reimbursement

The cost of environmental site assessments and its review undertaken prior to or subsequent to the closing of a covered transaction is the sole responsibility of the customer. It is the obligation of the lender's officer to ensure reimbursement and to collect all consulting fees associated with due diligence prior to or at closing of the covered transaction regardless of whether the transaction is completed.

Figure 21-2. (*Continued*)

Environmental Policy

To protect themselves prior to providing funds where properties have potential environmental problems or where properties may be contaminated with materials defined as "hazardous substances," lenders should have in place a policy that will help to minimize the environmental liability associated with existing and potential environmental hazards. To assist lenders in establishing such a policy, the FDIC published "Guidelines for an Environmental Risk Program," dated Feb. 25, 1993, and sent the document to all chief executives of FDIC-insured institutions. The document noted that, "Examiners will review the institution's compliance with its own environmental risk program as part of the examination of lending and investment activities." Banks without environmental due diligence programs must establish them, and banks with such a program must achieve compliance. (See Fig. 21-2.)

Here are repeated from above what the FDIC guidelines identify as the three major environmental risks facing lenders:

- The borrower's environmental liabilities could compromise its creditworthiness.
- Contamination could negate the value of the borrower's collateral.
- The lender itself could become liable for cleanup costs.

To limit exposure to these risks, the FDIC prescribes a program with the following elements:

1. A formal procedure to identify and evaluate the environmental risks associated with a transaction. The procedure should be approved by the board of directors and implemented "by a senior officer knowledgeable in environmental matters."

2. Training to assure that employees have sufficient knowledge to assess environmental risks themselves or to retain outside counsel or consultants where appropriate.

3. Written policies for evaluating environmental risks in various situations, such as loan origination, credit monitoring, workout, and foreclosure.

4. Environmental risk analysis prior to lending.

5. A more structured risk assessment (such as a Phase I site assessment) where appropriate.

6. Loan documentation provisions designed to safeguard the lender against environmental liability.

7. Monitoring the borrower's activities during the loan terms to identify any environmental problems that arise.

8. Avoiding such participation in management of the borrower's affairs as would render the bank liable.

9. Carefully evaluating environmental risks prior to foreclosure.

Site Environmental Risk

The prudent lender must not assume that any site, however pristine, is free of hazardous materials unless it has been sufficiently investigated by qualified professionals. The list of potential site contaminants is long. The most common sources of contamination are petroleum-product storage tanks, dry-cleaning plants, gasoline stations, electroplating businesses, machine shops, electrical substations, and accidental or illegal discharge of waste oils.

Because a great number of hazardous materials are potentially present on any site, a comprehensive preliminary site assessment report (see Fig. 21-3) should

The written report should address the scope of work requirements and include the following items:

A. Title page
  - Project identification and date of report.
B. Introduction
  - Description of scope of work.
C. Executive summary
  - Brief description of findings and recommended actions, if any.
D. Site description
  - Describe site or building and/or facility use and surrounding properties.
E. Record review
  - Summarize federal, state, and/or local database search.
  - Analyze historical sources.
F. Site reconnaissance and interviews
  - Describe existence and location of any hazardous materials or visible indications of other environmental concerns. Also confirm the absence of such materials and/or concerns.
  - Contact with employees, neighbors, local government officials, past owners, etc.
  - Analyze other records as appropriate.
G. Conclusions and recommendations
  - Analyze the information and provide a definitive recommendation as noted in Conclusions and Recommendations sections.
  - Signature of report auditor or writer.
  - Signature of report reviewer.
  - Signature of an officer of the firm.
H. Appendixes
  - Location map and site diagram or facility sketch
  - Photos and color copies
  - Sampling location map, if available
  - Sampling results, if available
  - Database reports
  - Additional correspondence
  - Résumés of personnel who performed the assessment
  - Chain-of-title listing

Figure 21-3. Sample Phase I environmental assessment report format.

be a basic requirement in every commitment letter as a condition of making the real estate loan. The site investigation and assessment should be conducted by a professional firm that specializes in this type of work and possesses the requisite geotechnical, hydrologic, and chemical analytic skills and resources (see Fig. 21-4).

The comprehensive preliminary site investigation and assessment should include an on-site inspection of the buildings and grounds of the subject site as well as inspection of abutting properties. Historical records sufficient to identify all previous owners and uses of the site or adjacent sites must be obtained and reviewed. The location of all proximate sites where toxic discharges have occurred must be examined in relation to groundwater flow patterns. Interviews with on-site, and possibly off-site, management and maintenance personnel should be conducted. Plans of existing or proposed improvements should be reviewed. The records of regulatory agencies should be searched for evidence of any hazardous discharge. Any evidence that suggests the potential for contamination by hazardous materials or for the release of previously contained hazardous materials must be followed up by subsurface investigation and analysis. Any underground tank is suspect, regardless of contents, age, or material of construction. It is virtually impossible to say that an underground tank and its associated piping is not leaking or has not been overfilled without subsurface investigation.

Remedial Action

Once the nature and extent of present or potential site contamination from hazardous materials has been determined, the procedures for its removal or man-

---

The environmental consultant should furnish the following before being considered to become an approved environmental consultant:

1. A sample report of an environmental assessment conducted by the consultant
2. All qualifications (i.e., licenses, certificates, and education) and experience
3. Proof of liability insurance and amount of coverage (this is important not only because of the amount of liability or errors and omissions insurance but also because it confirms that the environmental consultant is deemed to be an insurable risk by the insurance company)
4. Name or names of testing laboratory affiliations
5. Name or names of references for which work has been done
6. Contact name, address, and phone number

Figure 21-4. Sample requisites for approval of an environmental consultant.

agement can be prescribed and the associated costs estimated. If the lender decides to continue its involvement, indemnifications must be obtained from the borrower to the limit of the lender's exposure and liability. In all cases, the lender should seek the advice of counsel as to this specific concern.

Environmental Risk Procedure

To facilitate the engagement of the environmental consultant, the technical services department has set up a basic outline memo of information that is to be completed by the originating officer (see Fig. 21-5). After receiving the memo, the technical services department contacts a preapproved consultant to obtain a cost quote to perform the required Phase I assessment. The originating officer is informed of the quote, and he or she in turn informs the borrower and receives acceptance from them, preferably in writing. At the time of acceptance the consultant is contacted to start the assessment and to submit an agreement letter. (A sample agreement letter is given in Fig. 21-6, which outlines the services that the environmental consultant is to provide the lender. This letter is modified as needed to meet specific project conditions.)

An average time frame to complete a Phase I assessment would range from 4 to 6 weeks and occasionally longer, depending on the size and complexity of the property in question.

After receiving the assessment, the technical services department reviews it and makes comments. In the event the assessment finds an environmental risk, remedial recommendations are given to the account officer. At this time, the account officer makes the necessary underwriting decisions. Payment for the consultant's services is required either prior to or at the closing of the loan.

Structured Risk Assessment

Phase I Site Assessments

A Phase I environmental assessment is required prior to entering into a transaction involving real estate as security or with an environmentally sensitive business. For a detailed description of the Phase I assessment see Figs. 21-7, 21-8, and 21-9.

Phase II Site Assessments and Other Assessments

Phase II site assessments and other environmental assessments will also be undertaken as required, based on the recommendation of the internal technical staff, the environmental consultant, and the legal counsel. It should be noted that any Phase II site assessment will be the responsibility of the borrower, unless the lender is considering taking title to the property (see Fig. 21-10).

FROM: _____   TEL. _____

_____

TO: _____

Please order an environmental property assessment based on the information presented below:

1. Name of project: _____
    Property address: _____
    _____
    _____

2. Borrower: _____

3. Property type: _____

4. Building: _____
    Description: _____
    Year built/Age: _____

5. Bldg. Size (sq. ft.): _____

6. Land area: _____

7. Vacant land, if any: _____

8. Site contract: _____
    Name: _____
    Title: _____
    Tel.: _____

9. Chain of title attached: Yes:_____   No:_____

Approved by: _____

Figure 21-5. Sample environmental property assessment request form.

(Consultant's Letterhead)

Mr._____

(Lender)

RE: (Project name/Property location)

Dear Mr._____:

Submitted herewith is our proposal to do a Phase I environmental hazard assessment report applicable to the above-referenced project:

I. Based on information provided to us, we understand the project to be as follows:

   A. (Brief description of project)

II. (Name of consultant) shall perform the following Phase I environmental hazard assessment for (name of lender).

   A. Site visit to identify all obvious visual signs of contamination and/or potential sources of environmental or regulatory concern.

   B. Identify uses over the past 50 years and current uses of the property.

   C. Review of various documents and records of the U.S. Environmental Protection Agency and the appropriate state and local agencies for the purpose of identifying any history of hazardous waste activity that is known to those agencies (enforcement actions, consent orders, and reported spills) to have occurred at the subject property or at any site within a prudent ($^{1}/_{2}$ to 1 mile depending on the site) radius of the property. The current compliance status of such activity with respect to applicable laws and regulations will be reviewed. A list of all identified activity along with its compliance status will be included in the final report.

     The documents will also be reviewed to determine whether the facility has the required air-emission and wastewater-discharge permits and a Spill Prevention and Control and Countermeasures (SPCC) plan, if appropriate.

   D. Specific report concerning the present or past existence of any storage tanks (above or below ground), wells, and waste-disposal facilities to determine potential for leakage by review of available information on age, composition, contents, water table, and monitoring or inspection records.

   E. Specific report concerning electrical transformers and fluorescent light ballasts regarding the use or containing of PCBs.

Figure 21-6. Sample environmental hazard assessment agreement letter.

F. Activity review of all adjacent businesses, land, and cross-property easements to include both current and prior use.

G. Proximity to any sensitive ecological area, i.e., wetland, floodplain.

H. In existing buildings:

1. Identification of the presence of any probable asbestos-containing material or other toxic material, such as radon gas; for residential buildings, lead paint, if required.

2. Followed up by a sample testing of the material to measure its asbestos content for comparison with allowable or regulatory levels.

3. If asbestos is identified, an estimate of the extent of asbestos-containing material should be made and the cost of its removal, plus an estimate of the cost of an operations and maintenance plan, if required.

I. Our findings will be summarized in a written report (five copies each) that will contain color photographs and an estimate of the cost to remove hazardous material, if any, found at the project site and to restore any damage caused by such removal, inclusive of all professional and/or supervisory (construction management fees and air or monitoring testing if necessary or appropriate). The body of the report will be signed and dated by an officer of (name of consultant).

III. Compensation

(Name of consultant) fees for services relative to the subject project are as follows:

For the Phase I environmental hazard assessment under Section II, A through I, $_____, which includes all expenses.

In the event that additional work is required, we assume this request shall be accommodated at the reasonable and standard cost.

It is mutually agreed by all parties that (name of consultant) has been engaged by (name of lender) as its independent environmental consultant for this project. (Lender) shall have the right to provide copies of the materials you prepared to both participants and investors in this transaction and to potential investors or participants in a subsequent financing of these properties. All reports, both verbal and written, are for the benefit of (name of lender) and its agents, employees, participants, and assigns.

We will perform the work in a professional manner, consistent with customary standards and practices for work of this nature.

Figure 21-6. (*Continued*)

(Name of consultant) shall have no obligation to owner/seller, purchaser, or borrower, or any agents thereof. Services performed herewith are to be performed for (name of lender) as its independent consultant.

(Name of lender) acknowledges that (name of consultant) may look to it for payment of fees. Any requirements by (name of lender) for payment by owner/seller, purchaser, or borrower shall in no way alter the above relationship.

After reviewing this proposal, kindly signify your acceptance by signing at the designated place and returning the original to us. The additional copy is for your files.

Very truly yours,

AGREED AND ACCEPTED

By: _____

Date: _____

Figure 21-6. (*Continued*)

A. Site visit to identify all obvious visual signs of contamination and/or potential sources of environmental or regulatory concern.

B. Identify the past 50 years and current uses of the property.

C. Review of various documents and records of the U.S. Environmental Protection Agency (USEPA) and the appropriate state and local agencies for the purpose of identifying any history of hazardous waste activity which is known to those agencies (enforcement actions, consent orders, and reported spills) to have occurred at the subject property or at any site within a prudent ([$^1/_2$] to 1 mile depending on the site) radius of the subject property. The current compliance status of such activity with respect to applicable laws and regulations will be reviewed. A list of all identified activity along with its compliance status will be included in the final report.

   The documents should also be reviewed to determine if the facility has the required air emission and wastewater discharge permits, and a Spill Prevention Control and Countermeasures Plan (SPCC), if appropriate.

D. Specific report concerning the present or past existence of any storage tanks (above and below ground), wells, and waste disposal facilities to determine potential for leakage by review of available information on age, composition, contents, water table, and monitoring or inspection records.

E. Specific report concerning electrical transformers and fluorescent light ballasts regarding the use or containing of PCBs.

F. Activity review of all adjacent businesses, land, and cross-property easements to include both current and prior use.

G. Proximity to any sensitive ecological area, i.e., wetland, floodplain.

H. In existing buildings:
   1. Identification of the presence of any probable asbestos-containing material or other toxic material, such as radon gas; for residential buildings, lead paint, if required.
   2. Followed up by sample testing of the material to measure its content for comparison with allowable or regulatory levels.
   3. If asbestos is identified, an estimate of the extent of asbestos-containing material should be made and the cost of its removal, plus estimate of an operations and maintenance plan, if required.

I. The findings should be summarized in a written report which will contain photographs and an estimate of the cost to remove hazardous material, if any, found at the project site. The body of the report will be signed and dated by an officer of the consulting firm.

Note: Subject to the results of Phase I, a Phase II investigation, such as soil and groundwater sampling, air samples, tests, precision-type tank test in accordance with NFPA recommended practice 329, may be required to verify the scope and cleanup cost of any potential toxic conditions discovered in Phase I.

Figure 21-7. Sample outline of minimum acceptable scope of work for a Phase I environmental hazard assessment. Standard scope.

A. Site visit to identify all obvious visual signs of contamination and/or potential sources of environmental or regulatory concern.

B. Identify the past 50 years and current uses of the property.

C. Review of various documents and records of the U.S. Environmental Protection Agency (USEPA) and the appropriate state and local agencies for the purpose of identifying any history of hazardous waste activity which is known to those agencies (enforcement actions, consent orders, and reported spills) to have occurred at the subject property or at any site within a prudent ($1/2$ to 1 mile depending on the site) radius of the subject property. The current compliance status of such activity with respect to applicable laws and regulations will be reviewed. A list of all identified activity along with its compliance status will be included in the final report.

   The documents should also be reviewed to determine if the facility has the required air emission and wastewater discharge permits, and a Spill Prevention Control and Countermeasures Plan (SPCC), if appropriate.

D. Specific report concerning the present or past existence of any storage tanks (above and below ground), wells, and waste disposal facilities to determine potential for leakage by review of available information on age, composition, contents, water table, and monitoring or inspection records.

E. Specific report concerning electrical transformers and fluorescent light ballasts regarding the use or containing of PCBs.

F. Activity review of all adjacent businesses, land, and cross-property easements to include both current and prior use.

G. Proximity to any sensitive ecological area, i.e., wetland, floodplain.

H. In existing buildings:

   1. Identification of the presence of any probable asbestos-containing material or other toxic material, such as radon gas; for residential buildings, lead paint, if required.

   2. Followed up by sample testing of the material to measure its content for comparison with allowable or regulatory levels.

   3. If asbestos is identified, an estimate of the extent of asbestos-containing material should be made and the cost of its removal, plus estimate of an operations and maintenance plan, if required.

I. Environmental compliance status: Check compliance, including compliance status, general reputation, and history of violations of the facilities, by contacting responsible regulatory agencies.

J. Occupational safety compliance: Confirm status of compliance with OSHA requirements, "right to know" rules, emergency procedures. This

Figure 21-8. Sample outline of minimum acceptable scope of work for a Phase I environmental hazard assessment. Manufacturing scope.

should include review of company documentation and check with agencies. (Note: worker health and safety compliance problems also suggest a potential for worker health liability claims.)

K. If applicable, determine identity of hazardous and waste haulers and recyclers or disposal facilities, both present and past. Check reputation and potential Superfund involvement of haulers and recyclers or disposal facilities.

L. Verify RCRA hazardous waste generator status and compliance.

M. The findings should be summarized in a written report which will contain photographs and an estimate of the cost for remediation, if applicable. The body of the report will be signed and dated by an officer of the consulting firm.

Note: Subject to the results of Phase I, a Phase II investigation, such as soil and groundwater sampling, air samples, tests, precision-type tank test in accordance with NFPA recommended practice 329, may be required to verify the scope and cleanup cost of any potential toxic conditions discovered in Phase I.

Figure 21-8. (*Continued*)

Transactions Requiring Environmental Assessments

All transactions primarily secured by real estate deserve special attention since any contamination may render the property valueless. Additionally, businesses involved in the handling, transportation, storage, or removal of hazardous materials are clearly areas where environmental liability may arise.

New Transactions.   Phase I assessments must be conducted for:

- All real estate transactions that are defined as indebtedness "primarily secured by real estate" with the exception of residential mortgage loans (one-to four-family dwellings) and consumer loans

- Highly leveraged transactions in the manufacturing, industrial, and transportation areas, or in those instances involving entities with significant reliance on real property for the generation of cash flow.

- Those instances involving obligers engaged in activities concerning petroleum products, waste disposal, the handling of nuclear materials, or other hazardous materials

- Those situations in which the bank or lender is involved in the management of trust and estate accounts involving land, other real estate, and corporations in environmentally sensitive situations

- Other transactions to entities engaged in activities that are likely to raise environmental concerns should also be referred to the internal technical staff.

Sample outline of minimum acceptable scope of work for a Phase I environmental hazard assessment

A. Site visit to identify all obvious visual signs of contamination and/or potential sources of environmental or regulatory concern.

B. Identify the past 50 years and current uses of property.

C. Review various documents and records of the U.S. Environmental Protection Agency (USEPA) and the appropriate state and local agencies for the purpose of identifying any history of hazardous waste activity which is known to those agencies (enforcement actions, consent orders, and reported spills) to have occurred at the subject property or at any site within a prudent ($^{1}/_{2}$ to 1 mile depending on the site) radius of the property. The current compliance status of such activity with respect to applicable laws and regulations will be reviewed. A list of all identified activity along with its compliance status will be included in the final report.

   The documents should also be reviewed to determine if the facility has the required air emission and wastewater discharge permits, and a Spill Prevention and Control and Countermeasures Plan (SPCC), if appropriate.

D. Specific report concerning the present or past existence of any storage tanks (above and below ground), wells, and waste disposal facilities to determine potential for leakage by review of available information on age, composition, contents, water table, and monitoring or inspection records.

   1. Gasoline filling station sites (existing or past) protocol.

      a. Initial phase:

         ▪ If tanks are 1 year old or less—review plans and specifications.

         ▪ If tanks are older than 1 year—perform field photoionization or gas chromatograph tests of soil vapors (a minimum of 5 test holes). The test holes shall be a minimum of 3.5 ft from the surface or below the frost line.

      b. Advanced phase (if initial phase discloses problems—soil contamination):

         ▪ Perform borings (min. 30 ft deep or to groundwater or bedrock) and analysis to substantiate findings.

         ▪ Analyze soil and groundwater samples to define extent of contaminations.

         ▪ Submit report—synopsis of remedial action required with cost estimate.

Figure 21-9. Sample outline of minimum acceptable scope of work for a Phase I environmental hazard assessment. Known gas station scope.

E. Specific report concerning electrical transformers and fluorescent light ballasts regarding the use or containing of PCBs.

F. Activity review of all adjacent businesses, land, and cross-property easements to include both current and prior use.

G. Proximity to any sensitive ecological area, i.e., wetland, floodplain.

H. In existing buildings:

1. Identification of the presence of any probable asbestos-containing material or other toxic material, such as radon gas; for residential buildings, lead paint, if required.

2. Followed up by sample testing of the material to measure its content for comparison with allowable or regulatory levels.

3. If asbestos is identified, an estimate of the extent of asbestos-containing material should be made and the cost of its removal, plus estimate of an operations and maintenance plan, if required.

I. The findings should be summarized in a written report which will contain photographs and an estimate of the cost to remove hazardous material, if any, found at the project site. The body of the report will be signed and dated by an officer of the consulting firm.

Figure 21-9. (*Continued*)

Renewals, Extensions, Modifications, and Restructuring. Updated Phase I environmental audits may be required where the business or property is vulnerable to environmental hazards. The technical staff should be authorized to determine the need for an updated assessment in such instances.

Conveyance of Title. A Phase I environmental assessment must be undertaken and reviewed prior to the commencement of a foreclosure, taking of a deed in lieu, or similar action that may result in the title's being conveyed to the lender, its subsidiaries, or affiliates.

Documentation

Report Preparation and Review

Phase I environmental reports are to be prepared by a qualified external consulting firm acceptable to and retained by the lender. The technical staff will prepare a written review detailing the findings of the assessment and a summary of pre- and postcredit closing items requiring attention, if any. Under certain exceptions, a borrower-submitted Phase I report will be accepted upon review

1. If the results of the Phase I assessment are inconclusive or disclose the possible existence of contaminants, additional investigations and sampling will be implemented in a Phase II environmental assessment. Testing efforts that might be involved (given the nature and location of the property in question) include:

   - Collection and analysis of soil, surface water, storm sewer, sediment, or groundwater samples
   - Testing of underground storage tank(s)
   - Analysis of indoor air samples or building materials for chemical composition

2. For Phase II investigation work:

   - Installation of necessary wells must be in conformity with required state and local regulations and U.S. EPA protocols.
   - Analytical work must be done by testing laboratories certified by the EPA or the state where the work is being performed.
   - Underground tank tests must be conducted by a certified tank-testing firm.

3. Utilizing the information obtained from these sampling and testing programs, the environmental consultant will prepare a written report. The report will ascertain whether the site is free from contaminants or, if not free from contaminants, what remedial measures can economically be employed to clean up the site. These measures are to be incorporated into a remedial action plan that will be part of the report—the plan will detail the estimated costs of the necessary remedial measures.

4. Following implementation of the remedial measures, the environmental consultant will inspect the remediated site using a follow-up testing and analysis program that will verify successful implementation of the remedial plan.

5. The written report should expressly be for the benefit of the lender but should allow the lender to distribute copies of the report to whomever the lender deems appropriate (for example, to the borrower and to the principals of the borrower and to any loan participant or assignee).

Figure 21-10. Basic outline of Phase II environmental assessment.

and approval by a designated member of the technical staff. To minimize the bank's exposure to environmental liability, Phase I recommendations should be incorporated into the underwriting process.

Environmental Certification

As a condition to close a transaction, a signed environmental certificate will be required. The certificate will be prepared by the appropriate technical staff member for each property upon his or her determination that the assessment is acceptable and that the minimum environmental policy requirements have been met. If a review indicates that items need to be addressed prior to closing, the certification will not be issued until the technical staff is satisfied that all open issues have been satisfactorily addressed. Phase II analyses will be undertaken at the discretion of the technical staff and legal counsel.

Loan Documents

All loan documents must contain appropriate protective representations and warranties requiring that if hazardous substances are used in the ordinary course of a borrower business, such substances are handled and disposed of according to applicable law and regulations. Handling of such materials outside of the ordinary cause of a customer's business should be prohibited. Loan documents should also include the following:

- Appropriate protective representations and warranties requiring that, if hazardous substances are used in the ordinary course of the borrower's business, such substances can be handled and disposed of according to applicable laws and regulations. The handling of such materials outside of the borrower's ordinary course of business is prohibited.

- Indemnities from the borrower and/or guarantor to the lender in the event that the public or third parties assert environmental or hazardous waste claims against the lender.

- A provision giving the lender the right to undertake subsequent environmental assessments any time after the credit agreement has become effective.

- Providing for remedial or other actions required to be taken prior to or following closing of the loan.

- Allowing the lender to withdraw if the results of an environmental assessment or other information disclose environmental conditions that, in the sole discretion of the lender, make continuation of the relationship inadvisable.

Under no circumstances may these covenants or representations be modified or waived without the express written consent of counsel and a senior credit officer or the equivalent.

Release of Environmental Assessments

Phase I environmental assessment reports are not to be released to the borrower or other interested parties without the consent of the head or deputy of the technical staff and the appropriate group credit officer or the equivalent. The borrower will be required to provide a signed release letter before a copy of the assessment is furnished (see Fig. 21-11).

Phase I Environmental Assessment

A Phase I environmental assessment is a nonintrusive survey of prior and current uses of real property for purposes of determining environmental concerns resulting from the generation, transportation, usage, storage, and/or disposal of hazardous materials. In addition, non-CERCLA issues such as asbestos-containing materials, wetlands, lead paint, and radon are investigated as determined appropriate by either the bank or its consultant.

Phase II Environmental Assessment

A Phase II environmental assessment is the assessment that determines the actual presence or absence and the extent of hazardous substances and contamination. If, at the conclusion of Phase I, it appears that more information is required to explain such things as soil discolorations, large earth disturbances, and nearby or surrounding property concerns (e.g., well-water odors or contamination, colored seepages, abandoned dump sites), then Phase II assessment should be conducted. As a minimum, this assessment should include the following:

- Conducting an analysis on soil borings, building materials, and air and water samples, if appropriate. (The number and location of samples taken and types of analyses conducted are determined solely by the consultant in consultation with the borrower and acceptance of the scope of work by the lender, based on the Phase I assessment.)

- Conducting whatever other tests are necessary to establish either the absence of hazardous substances or the nature and extent of contamination.

- Preparing a written report to the lender certifying the site does not represent a significant environmental risk to the lender, or recommending remedial action if contamination is determined to be present and the estimated costs. This report should be a stand-alone document which includes a summary of all observations, laboratory results, data evaluations, and conclusions of the site characterizations.

(Lender's Letterhead)

Date:

_____

_____

_____

_____

Dear —————————:

As requested, I am enclosing a copy of a _____ report of the _____ dated _____, prepared by ___(consultant's name)___ . Please be advised that the report being forwarded has been prepared solely for the benefit of ___(lender's name)___ and is being furnished to you solely as a courtesy. The report cannot be furnished by you to any other person and should not be relied on, published, quoted, copied, or disseminated by you or any person other than ___(lender's name)___ , its successors, and assigns.

___(lender's name)___ and ___(consultant's name)___ make no warranties or representations that the report has identified all potential problems that exist at the property, or as to the accuracy, completeness, or thoroughness of the assessment and provide no other assurances or guarantees regarding the report. ___(Consultant's name)___ is intended to be a third party beneficiary of this letter.

The report is solely intended to meet the loan requirements of ___(name of lender)___ . The enclosed report is being forwarded at your specific request and ___(name of lender)___ and ___(name of consultant)___ assume no obligation to provide you with any further information.

AGREED AND ACCEPTED:

By: ———————————————   By:_____

Title: ——————————————   Title:_____

Date: ——————————————

Figure 21-11.   Sample assessment report release letter.

Phase III Work

Phase III environmental work entails the preparation of the design and the implementation of a remedial program and postremedial sampling to document cleanup, which includes obtaining required governmental approvals as required.

Internal Environmental Staff

It is my opinion that the lender should have an internal staff capable of engaging, monitoring, coordinating, and reviewing the work of the external environmental consultant that is engaged by and for the benefit of the lender. This internal staff would also be a source of information on environmental matters. It should also have the responsibility for providing the guidance and training for account officers and/or relationship managers.

The internal staff's function is to protect the lender from potential liability, protect the value of the collateral, and verify the borrower's compliance with environmental regulations. This staff can be called "technical service," "environmental risk management," "environmental compliance unit," or whatever is deemed appropriate. This unit can be part of the real estate lending, corporate credit, risk and insurance management departments, or it can stand alone, reporting to the lender's compliance officer or senior credit officer.

The following are sample job descriptions for internal environmental staff:

Senior Staff Environmental Specialist

Staff at this level work under the direction of the manager of technical services. Senior staff are responsible for reviewing the soundness of environmental issues of proposed real estate lending, investment, and development projects; reviewing outside environmental consultants' Phase I and Phase II reports and advising lending officers; providing bankwide, in-house consulting services pertaining to environmental issues; and assisting in the selection, monitoring, and coordination of outside environmental consultants and contractors.

The incumbent typically has 8 to 10 years of related work experience, a bachelor of science degree in one of the environmental sciences, and a master's degree in environmental engineering or in one of the related sciences. Background should include work experience in a city, state, or federal environmental protection agency.

Staff Environmental Specialist

Staff at this level works under the manager of technical services and senior staff specialist. Their responsibilities are the same.

The incumbent typically has a minimum of 4 years of related work experience, and a bachelor of science degree in one of the environmental sciences. Background should include work experience at a city, state, or federal environmental protection agency.

Conclusion

It is prudent for a lender to have an environmental policy and to have Phase I environmental site assessments performed for the following reasons:

Environmental conditions can impact a lender's position in several ways. Contamination of real estate may completely devalue a parcel and expose the borrower to cleanup expenses that far exceed the apparent appraised value of the parcel. A lender's intimate involvement in the management of a borrower's operations may expose the lender to direct liability. Contamination that remains upon default and foreclosure may expose the lender to direct liability for the borrower's cleanup obligations.

Contamination of a parcel or environmental noncompliance by a borrower may also severely impact creditworthiness. Environmental contamination at or near a parcel that serves as the lender's collateral may dramatically impair the borrower's ability to make payments as required, and it may also severely devalue a parcel upon which a lender most foreclose. Finally, environmental contamination may also lead to third-party lawsuits from residents or neighbors based upon the health impacts of environmental hazards.

By taking a proactive role in the management of environmental risk, the lender minimizes exposure to environmental liability and avoids losing the viability of the collateral.

22

Environmental Auditing at General Mills

Scott M. Brown
P.E., General Mills, Inc.

General Mills, Inc. (GMI) is a Fortune 70 food company that manufactures and markets many widely recognized consumer food products such as Cheerios and Wheaties cereals (among many other brands), Gold Medal flour, Yoplait yogurt, Gorton's seafood products, and Betty Crocker cake mixes and other packaged food products. In all, GMI has over 40 manufacturing and operational facilities that are included in the environmental auditing program.

GMI's Environmental Program

As a food company that sells products directly to consumers, it is important to GMI to avoid negative publicity concerning environmental issues. GMI's written corporate environmental policy is to comply with all applicable laws and regulations. The corporate Health and Human Services Department is charged with developing specific programs and guidance to implement this policy. As part of the implementation of the corporate policy, an environmental auditing program was initiated over two years ago, and each manufacturing facility appointed an environmental coordinator for its facility.

The environmental coordinators are responsible for overseeing the implementation of the environmental program at their facility. They ensure that their facility is complying on a day-to-day basis and that new environmental regulations are communicated to the plant manager, plant engineering staff, department managers, and wage employees. At most plants, the environmental coordinator position is not full time but is done in addition to the person's other normally assigned duties. Some facilities, however, are environmentally complex enough that they need and have a full-time environmental coordinator.

GMI's philosophy is to keep the corporate environmental staff as small as possible without sacrificing integrity or effectiveness. This is accomplished by making each plant responsible for its own environmental compliance, and by using computer coordination of environmental issues as much as possible.

GMI's Environmental Auditing Program

Objective of GMI's Auditing Program

The objective of GMI's auditing program is to achieve and maintain compliance at all GMI facilities by identifying environmental compliance issues and resolving them expeditiously. A side benefit of environmental auditing is that the environmental coordinators become trained on environmental regulations and good management practices for environmental management.

Personnel

Environmental audits at GMI are conducted completely internally. Outside consultants are never used for two reasons: (1) to maintain better control on the information that comes out of the audits and (2) to be cost-effective. A typical quote from a reputable consulting firm for conducting a rigorous environmental compliance audit, such as those being conducted internally at GMI, was $30,000, not including consulting services for resolving audit items.

GMI's corporate auditing staff consists of one person, the environmental assessor. The environmental assessor reports to the director of safety and environmental management, who in turn reports to the vice president of health and human services.

The auditing program is managed by the environmental assessor, who is responsible for:

1. Selecting and developing audit protocols appropriate for the facilities

2. Scheduling audits

3. Conducting the audits as the audit team leader

4. Writing up the audit exception reports, which include recommendations for achieving compliance and a timetable for item resolution

5. Following up on audit item resolution with the facilities

6. Providing conceptual engineering support in resolving the audit issues

7. Training facility environmental coordinators in environmental regulatory issues

8. Periodically reporting to top management on the compliance status of GMI facilities

A typical GMI audit team consists of the environmental assessor and the facility environmental coordinator. Occasionally, the environmental coordinator from another facility in the same division will participate in the audit. Other facility personnel may also participate in portions of the audit.

Approach

GMI believes that the best way to solve problems is to encourage its employees to work together, regardless of business division, to come up with solutions. The approach taken in the auditing program reflects this philosophy.

The GMI approach to auditing is analogous to that of an internal consultant. The environmental assessor works with the plant in identifying noncompliance areas and in developing and implementing solutions. This fosters a relationship of trust between the plant and the environmental assessor that is crucial for a successful and thorough audit. The plants see the audit as a function to help them rather than judge them. Because of GMI's decentralized approach where each plant is responsible for compliance, the facilities generally do not view the audit as an intrusion because they realize that it helps them accomplish their environmental compliance mission.

In some companies, the auditor reports only exceptions and does not make recommendations or assist in implementation of solutions. The reported concern is that the auditor may lose his or her objectivity. This is the bureaucratic approach to auditing.

In an audit, most things are black and white—something is either in compliance or it isn't: the plant either has a permit or they don't; they either have cathodic protection on new USTs or they don't; they either have all signed hazardous waste manifests on file or they don't; they either have an SPCC on a 150,000-gallon fuel oil tank or they don't. All of these are objective assessments of the plant's compliance with unambiguous regulations and require that the auditor have a detailed working knowledge of all environmental regulations.

The "gray" areas in an audit are in areas of equipment performance assessment and scenario assessment. For example, the augers beneath a baghouse leak filter cake fines that pile up beneath the augers. Is this a problem? It depends on whether the baghouse is outside or inside and how the fines are being managed. Is the baghouse outside situated next to a navigable water body and are the fines allowed to accumulate? If so, this is a violation of stormwater regulations because the fines, even though they are just cereal or grain dust fines, are considered "materials associated with industrial activity" and not allowed to leave the facility with stormwater runoff. However, making recommendations and even assisting in implementing the solution (e.g., helping to justify capital to replace the baghouse or assisting in procuring environmental permits for the construction and operation) does not interfere in any way with an objective assessment of the original problem during the audit or in subsequent audits.

Environmental auditing is an eminently practical activity that requires a solid engineering background, a detailed working knowledge of the regulations, the ability to communicate succinctly with plant personnel and corporate and divisional management, both in writing and verbally, and a willingness to literally get your hands dirty. An effective and worthwhile audit requires that the auditors walk and crawl around areas that may be dusty because this is the nature of areas where pollution control equipment is operated.

Tools

GMI uses two important computer tools to prepare for and conduct environmental audits: Enflex Info and Audit Master. These are not tools that are simply nice to have—without these two tools, a larger staff would be required to effectively implement the audit program. By effectively using these two tools, GMI is saving a lot of money that would otherwise be spent on extra staffing without sacrificing the quality or integrity of the auditing program.

Enflex Info is a CD ROM-based system that carries almost all environmental, safety, and transportation regulations at the federal level and environmental and safety regulations for many states. It is used extensively in both preparing for the audit by reviewing relevant state regulations and writing up the final audit report. For example, in explaining the regulatory basis of an audit item in the audit report, the salient regulations can be quickly located in Enflex Info and excerpts can be electronically "pasted" into the report being prepared in Audit Master.

Audit Master is a PC-based auditing system from which the audit is conducted. Audit Master is installed on a notebook computer and taken to the plant. The environmental assessor and the plant's environmental coordinator sit down together and work through hundreds of questions on various environmental program areas. Comments are entered as appropriate so that the audit report is partially written when the plant visit is completed.

While computerized auditing programs like Audit Master are a valuable tool for conducting a thorough audit, they are not a substitute for an experienced auditor with an engineering background and a thorough working knowledge of environmental regulations. However, using a computerized auditing tool like Audit Master greatly improves the thoroughness and efficiency of the audit. For a paper-based audit to be as thorough as a computer-based Audit Master audit, the auditor would have to carry bulky notebooks filled with audit questions to the facility. Then, either the auditor or a secretary would need to go through all the auditor's handwritten notes and transcribe them into a draft report that the auditor would go through, adding recommendations and additional descriptions. And if regulatory excerpts were to be added, they would have to be typed in by hand.

Quality Control

Using a computerized auditing program minimizes subjectivity in evaluating compliance items at a facility. Furthermore, references and regulations are cited for each audit question in the program and can be instantly reviewed. This often provides additional valuable information needed to evaluate the compliance of a situation with applicable regulations.

Auditor certification and registration are also important in ensuring the integrity of the auditing program. Professional engineering registration (P.E.) and the Certified Hazardous Materials Manager (CHMM) are two examples of professional designations that each have a code of ethics. Most states have specific regulations that forbid a P.E. from engaging in unethical or unscrupulous professional conduct, such as falsifying an audit report. If it can be proved that a P.E. willfully falsified an audit report, his or her engineering license can be revoked.

The Auditing Process

Preparation

The environmental assessor and the facility environmental coordinator must work together in preparing for an audit. Each has specific tasks that must be accomplished. Preparation consists of the following activities:

1. The environmental assessor mails audit screening questions and materials-to-be-gathered lists to the environmental coordinator of the facility to be audited. These items are typically sent 3 to 4 weeks prior to the plant visit.

2. The environmental coordinator works through the screening questions and returns them to the environmental assessor at least 1 week prior to the plant visit.

3. The environmental assessor reviews the screening questions, enters the coordinator's answers, and creates Audit Master files for the facility based on the answers. The answers to the screening questions help select audit questions that are appropriate to the facility. For example, if the facility does not have underground storage tanks (USTs), then questions on USTs will not be selected for use during the audit.

4. The environmental coordinator uses the materials-to-be-gathered lists to pull together all appropriate documentation, plans, drawings, permits, etc., for review during the audit.

5. The environmental assessor reviews relevant and appropriate regulations promulgated by the state in which the facility is located.

Conducting the Audit

The scope of a typical audit includes questions in the following program areas, referred to in Audit Master as modules:

- Air quality
- Water quality

- Emergency Planning and Community Right-to-Know (EPCRA)
- Solid and hazardous waste
- Storage tank management
- Spill prevention
- Hazardous materials transportation

The audit team usually consists of the environmental assessor and the facility environmental coordinator, but other people at the facility may participate in portions of the audit. For example, if the safety coordinator is in charge of the MSDS sheets and is involved in the Tier II and Form R reporting, he or she would participate in the EPCRA portion of the audit.

All audits at the plants are conducted using a combination of the three standard auditing techniques: document review, physical inspection, and personnel interviews. The audit operations at the plant are usually based out of a conference room that the facility environmental coordinator has reserved for the audit. All the documentation to be reviewed during the audit is spread out in piles on the conference room table.

The duration of the plant visit portion of the audit varies with the size of the facility and the type of operations. Flour mills can be audited in a day and a half, cereal plants take from a minimum of two days up to a week, and the facilities in the other divisions each take about two days. A typical audit day at a plant is about 12 hours long.

After all appropriate audit modules have been completed, a red flag report is printed out at the facility and reviewed with the environmental coordinator. This report is basically a rough draft of the final audit report and lists all exceptions noted during the audit. The environmental coordinator and the environmental assessor review the red flag report and discuss the oral summary of the audit that will be presented to the plant manager at the wrap-up meeting.

The wrap-up meeting is where the environmental coordinator and the environmental assessor present the major findings to the plant manager. Other personnel present at the wrap-up might include the plant engineering manager, the safety coordinator, and the plant sanitation manager, depending on their involvement with the plant's environmental program.

Final Audit Report

A GMI audit report is divided into environmental program area sections (e.g., air quality, water quality, etc.). Each section is comprised of audit items, which are exceptions noted during the audit. These exceptions can be either regulatory compliance issues or good management practice (GMP) issues. The audit items reported in the final audit report are of two types: trackable and FYI.

Trackable items are entered into the Corrective Action Tracking Database (CATD) as part of the final report preparation and are assigned a target implementation date for tracking purposes. The target implementation date is determined by the environmental assessor and is based on item priority and resource requirements. Audit items that are regulatory compliance issues are higher priority than GMP items. Resource requirements are items such as labor hours, capital, contractors, and design work. Resource-intensive audit items, such as those requiring high capital or labor hours, are generally assigned a later target implementation date.

FYI items are included as either purely informational items or for elaboration on other audit items which may or may not be trackable items.

The audit report format is very much like a detailed punch list. A narrative report format simply is not useful for the environmental coordinator for implementation or to the environmental assessor for tracking. Each item reported consists of the following information:

- The exact audit question
- The response to the audit question
- Comments entered during the plant visit
- Additional descriptions of the situation, usually consisting of excerpts of salient regulations on which the audit item is based
- Recommendations for corrective actions
- The corrective action database entry listing information such as a brief description of the item, the item location at the plant, the responsible person, and the target implementation date

Typically, it takes about 10 working days (allowing for other work to be accomplished during that time) for the final audit report to be written up and issued.

Only two copies of the final audit report are printed. One is sent to the facility environmental coordinator and the other stays in the environmental assessor's files. The summary transmittal memo, which consists of a narrative summary of the audit and an abbreviated punch list of all trackable audit items, accompanies the final report sent to the plant and gets wider distribution, including the plant manager, plant technical manager, divisional environmental contact, and GMI corporate environmental legal counsel.

Corrective Action Tracking

Corrective action tracking is crucial for protecting GMI from liability associated with conducting environmental audits. Since the audit report documents items of noncompliance, it is important that these items are tracked to ensure that they are satisfactorily corrected in a timely manner.

It is also a good idea to establish an item resolution paper trail. At GMI, this is accomplished by sending a quarterly audit follow-up report to each facility that still has unresolved audit items. Plant management as well as divisional management is copied on this report. The environmental coordinator responds in writing to the quarterly tracking report, documenting audit actions taken on audit items that were resolved during the quarter and indicating the status on the unresolved audit items. Throughout the whole corrective action tracking process, there is active telephone communication between the facility environmental coordinator and the environmental assessor about various engineering solutions for audit items at the plant.

Summary

GMI has implemented an effective environmental auditing program that has low overhead compared with programs at other companies and that efficiently identifies and tracks audit items to resolutions. The entire program is operated and managed by the environmental assessor, who relies heavily on a CD ROM-based regulations database and a computer-based auditing system that allows for computer coordination and tracking of audit items.

GMI uses the internal consultant approach in conducting its auditing program to encourage all employees to work together as a team in maintaining environmental compliance.

PART 7

The Global Interest in EHS Auditing

23

Environmental, Health, and Safety Auditing in Germany

Christopher J. Keyworth

*Senior Project Manager, ENSR,
Consulting and Engineering*

The development of environmental, health, and safety auditing in the Federal Republic of Germany has followed a remarkably different pattern from that experienced in the United States. Variations between the two nations in terms of their legislative framework, regulatory agencies, corporate cultures, and social attitudes, among others, have contributed to the divergence in the paths followed. The resulting differences may be illuminating to those familiar with U.S. attitudes and practices, and particularly those involved in multinational corporate environmental, health, and safety management. As regards developments prior to reunification in 1990, the discussion below relates to the then existing states ("Bundesländer") of the Federal Republic of Germany, or former West Germany.

The German population has one of the highest levels of environmental consciousness in the world. Opinion polls in Germany have consistently shown the environment to be placed high on the list of public concerns. The high level of public awareness dates back at least to the 1970s, when "Waldsterben," or the dying forests, became a national concern. The Green environmental political party, Die Grünen, has been represented in federal and state parliaments and has participated in state government. In the international arena, Germany has frequently been at the forefront of moves to address global environmental issues. Germany is second to none in the percentage of gross national product (GNP) dedicated to investment in environmental protection.[1]

The strong feelings prevalent in Germany about the protection of the environment have been reflected in the development of environmental law. German industry has also long been accustomed to strict health and safety requirements.

The Water Management Act of 1976, for example, set down stringent requirements for the protection of surface water and groundwater. The 1980s saw the enactment of a large body of environmental regulation, which has continued to be strengthened and augmented in the early 1990s. The passage in 1991 of the pioneering packaging ordinance, designed to reduce the generation of packaging waste, was only possible given the concern for the environment shared by lawmakers, industry, and the public.

One visible effect of, and also evidence of, the high public level of awareness of environmental issues is the strong emphasis seen in advertising on the "green" aspects of products, packaging, and also production processes. German industry, conscious of this awareness, has established a reputation for taking environmental, health, and safety issues seriously, reflected in the design and engineering of products and production facilities.

In the mid and late 1980s, however, when the practice of environmental auditing was emerging rapidly in the United States, few corresponding developments were visible in Germany. Initial activity could be identified mainly within German companies that were part of U.S. multinational corporations. Only in the early 1990s, partly on account of the promotion of auditing by the International Chamber of Commerce, did German companies in significant numbers begin to implement environmental or environmental, health, and safety auditing programs.

More in evidence in Germany have been technical studies of specific areas or units of a facility in order to assess their current status relative to regulatory requirements. Such studies are usually performed or commissioned because an awareness exists that some type of upgrading is required, and a baseline analysis is considered a desirable starting point for planning purposes. Although some characteristics of audits can be identified in these analyses, they are usually focused engineering studies bearing little overall resemblance to full, routine audit programs.

If Germany is indeed such an environmentally conscious nation, why have German companies been slow to pursue the idea of auditing? A simple answer is that most of the specific factors[2] that served to initiate and promote auditing in the United States did not have parallels in Germany at that time. These factors were largely absent in Europe as a whole, and Germany has not lagged noticeably behind other European countries, with the exception of the United Kingdom, in the introduction of auditing.

As an example of the lack of motivating influences, German companies experienced no imposition of audit requirements such as those mandated by the U.S. Securities and Exchange Commission (SEC) on a company failing to identify environmental liabilities. In fact, no detailed reporting of environmental liabilities, as required by the SEC for the "10-K" annual reports, is necessary in Germany.

Also absent in the early 1980s in Germany were (1) impositions by governmental agencies of million-dollar fines on environmental offenders and (2) the issuance of a statement by the U.S. Environmental Protection Agency promoting (voluntary) auditing.[3] The potential for environmental liability resulting in significant adverse financial impact to a corporation was at that time of considerably lower concern to a German company compared with its U.S. counterpart. These were all important reasons for the delayed development of auditing in Germany. The complete answer is, however, more complex.

The German Attitude toward Auditing

An examination of why auditing did not develop in Germany at the rate of the United States involves the consideration of a number of interwoven factors. Two key contributors are the differences between Germany and the United States in the development of corporate environmental, health, and safety management and the differing perceptions of the nature and the role of auditing. Influences present in Germany to a lesser degree than in the United States, such as liability potential, have already been mentioned. Some of the most significant additional considerations are discussed below.

The factors involved can be divided into two distinct categories. The first category includes those factors that are of general validity and the second category contains those that have specifically arisen since 1990 when the European Commission[4] first disclosed draft proposals for an "eco-auditing" requirement. The debate on auditing in Germany has, since 1990, largely been framed by the long and eventful saga of the development of the European Union (E.U.) Eco-Management and Audit regulation.

The Traditional Approach to Auditing in Germany

The general category of factors affecting attitudes relating to auditing is discussed below.

External Oversight.   German industrial facilities have generally been subject to close scrutiny by supervisory agencies. Inspections are made by a number of parties, including:

- The commercial supervisory authorities (Gewerbeaufsicht)
- The local water authorities (Untere Wasserbehörde)
- The worker safety association (Berufsgenossenschaft)

A notable feature of regulatory supervision in Germany is the empowerment of nongovernmental bodies to perform some supervisory functions. For example, regular inspection by one of the Technical Supervisory Associations (Technische Überwachungsvereine, or TÜV) or equivalent experts is required for certain components of a facility, such as storage facilities for "substances hazardous to water." Regulations also provide for the installation and maintenance of these components by officially approved organizations.

In addition to site inspections, supervisory bodies are active in review of documentation submitted in accordance with reporting requirements. Air emissions reporting requirements exist and each type of hazardous, or "special," waste shipment for treatment or disposal must be specifically approved. Additional reporting requirements might be stipulated in the permit conditions of larger facilities.

Much of the responsibility for enforcement of regulations lies with the individual Bundesländer, and the degree of supervision can therefore vary somewhat between the states. The amount of supervision also is dependent on factors such as the size of a facility, the activities conducted, and the level of risk that these activities present. The largest facilities have become accustomed to inspections from supervisory bodies on an almost monthly basis, with the frequency of inspections being more infrequent and more irregular for smaller firms. As a result, the larger, publicly visible corporations have generally experienced a high level of external supervision.

Insurance coverage has been widely available to German companies against liability for adverse impacts on surface or groundwater (Gewässerhaftpflichtversicherung). Application for such coverage normally triggers an inspection of the facility to determine the degree of risk posed by facility operations, which includes the general level to which the facility is in compliance with relevant regulatory requirements.

Internal Supervision.   A particular feature of German law is the requirement for environmental officers at many industrial facilities (Umweltschutzbeauftragte). With regard to their legal status, rights, and responsibilities, these positions have no equivalent in U.S. legislation. The stipulated duties of the appointees go well beyond ensuring that the facility is in compliance with the relevant regulatory requirements. Such officers are mandated for facilities conducting certain types of activities, such as refineries and cement plants, or where stipulated thresholds, such as volume of wastewater discharge, are exceeded. Requirements exist regarding officers for waste management, water protection, process safety,[5] and air quality protection.

These employees are afforded special statutory protection relating, for example, to access to the necessary resources to perform their duties and limitations on termination of employment. Annual reports to the company management are prepared by these officers detailing the implementation of their duties over the year. Management is also required to solicit the input of the appropriate officer on investment decisions that could impact his or her area of responsibility.

Federal Environmental Management Requirement.   With the revision in 1990 of the Federal Air Quality Control Act[6] (Bundes-Immissionsschutzgesetz), which subjects a wide range of industrial facilities to permitting requirements, a far-reaching new provision was introduced. Those facilities subject to permitting were required to report to the authorities on the manner in which the observance of the environmental protection requirements arising from the statue was ensured.[7] Significantly, in view of the following discussion on the E.U. audit regulation, no attempt was made in the statute to define how the assurances were to be achieved. However, the legislative intent of this provision is clearly for a notification to be made to the authorities describing the company's environmental management system related to the permitted installations.[8] The statute also calls for the name of the member of the company management responsible for environmental protection to be notified to the authorities. This highly significant requirement has been recognized as representing a powerful catalyst for the stimulation of formalized corporate environmental management systems. Some larger companies were able to provide documentation of their existing systems for the notification, while others needed to augment or refine their organization, responsibilities, procedures, and other elements.

Owing to the three groups of factors described above, the view became not uncommon among large companies, such as large chemical producers and power plant operators, that the reporting procedures and the agency inspections, combined with the existing technical and administrative measures, systems, and controls in place at their facilities, provided an appropriate level of supervision of environmental, health, and safety issues.[9] Auditing was considered to represent a superfluous additional level of control. However, these large German companies represent just those industry leaders that would have needed to take the initiative in the 1980s for any substantial development of audit programs to take place at that time. Similarly, the concept of auditing became successful in the United States only once the pioneering efforts were made by larger companies.

Other, more subtle factors that have been given less attention involve the German corporate culture. As environmental audit programs began to make headway in the United States, an association was created between an audit program and the use of external consultants as auditors. Little or no auditing expertise was available in most companies, and other arguments, such as independence of judgment, spoke for the use of external specialists. In the meantime, many large U.S. corporations have established audit programs primarily utilizing in-house resources.

In Germany, the use of consultants has traditionally been more sparing than in the United States. Consultants may be called in to solve a specific technical problem where they can provide expertise lacking within the company. The idea, therefore, of bringing in a consultant to investigate the intimate workings of the company have hitherto been somewhat of an anathema for many corporations. The more common attitude is to ensure that internal staff are performing the necessary monitoring and supervisory functions.

Certainly some attitudes and concerns varied from the position discussed above in the late 1980s and particularly the beginning of the 1990s. German firms observed the positive auditing experiences reported by some U.S.-owned companies operating in Germany.[10] Awareness of the benefits of auditing was increased by the publication in 1989 of a *Position Paper on Environmental Auditing* by the International Chamber of Commerce. On the whole, however, initial reception in Germany of the concept of auditing could at best be described as mixed.

Germany and the E.U. Eco-Management and Audit Regulation

The discussion below considers the factors affecting German attitudes toward auditing that relate directly to the E.U. Eco-Management and Audit scheme (EMAS).

In 1990, the European Commission disclosed initial proposals for an "eco-auditing" directive. After many further revisions, an Eco-Management and Audit regulation[11] was finally adopted in 1993. Germany is one of the (currently) 12 member states of the E.U., and regulations issued by the E.U. become legally binding in member states.[12] The objective of EMAS is to promote continuous improvement in the environmental performance of industrial activities by:

- The establishment and implementation of environmental policies, programs, and management systems
- Regular environmental auditing
- The provision of information on environmental performance to the public

From the first announcement of the commission's intentions, auditing, not hitherto at the center of corporate attention, became an intense subject of debate in Germany. Many companies, particularly those not previously familiar with the concept of auditing, did not differentiate between the establishment of an audit program and participation in EMAS. Therefore, the arguments for and against the concept of auditing per se began, in Germany, to be lost in the debate on the merits of the commission's proposals for EMAS. The reservations held by many representatives of German industry about the moves in Brussels had a profound impact on German perceptions of auditing. Misgivings on the part of industry about the commission's proposals were, it should be noted, by no means restricted to Germany.[13] The wisdom was questioned of introducing a concept at the E.U. level that had not been broached at a national legislative level in any of the member states.[14]

Between 1991 and 1993, many modifications were made to the commission's proposals, some of which respond to an extent to the German concerns, but perceptions had by then to a large degree already taken hold. Some of the many concerns voiced, publicly or privately, are briefly mentioned below.

In the United States, the subject of public disclosure of corporate environmental information took a separate path from that of auditing. Auditing has generally been accepted to be an internal mechanism for corporations to use for their own purposes without any compulsion existing for companies to publicize the results of audits. On the other hand, the concept of "right-to-know" has become well entrenched in the United States. Extensive reporting of facility releases to the environment and waste disposal is required in the United States under federal right-to-know legislation (referred to as SARA Title III), and some state laws require further disclosures.

By contrast, the process of developing the E.U. regulation had the effect of throwing the two issues of auditing and public disclosure into one pot. The public disclosure or "environmental statement," as initially envisioned by the commission, was to have largely reflected the findings of an audit. Since then, the environmental statement has been "decoupled" from the specific audit findings, but it still contains considerable detail on environmental performance, including "an assessment of all the significant environmental issues of relevance to the activities concerned." The environmental statement is prepared following the completion of each audit and thereby retains a strong association with the audit. Because of the nature of the environmental statement, especially as it was originally intended, the perception of auditing as a public process took hold in Europe. Therefore, the willingness of corporations to commit to establishing an audit program became linked with the willingness to publicize detailed information on corporate environmental performance.[15]

The perceived public nature of the audit process clashes strongly with the fact that German industry has traditionally been shrouded in a greater cloak of mystery than its U.S. counterpart. In this respect Germany does not differ greatly from most European countries. German facilities are, depending on type and size, required to submit various information to the supervisory authorities, such as biannual air emission statements and hazardous-waste generation data. These reports, however, generally went no further than the authorities, who themselves were previously protected more from the public view than in the United States. German governmental bodies tended to be viewed as acting as a guardian of the public's interest, and, therefore, provision of detailed information to the public was not perceived as necessary.[16] There is currently no equivalent in Germany to SARA Title III, requiring annual reports of emissions and releases to environmental media to be prepared and made directly available to the public. Despite some indications of change, the idea of sharing detailed data with the public on a facility's environmental performance is still foreign to the typical German company.

Even if willing to provide general data on emissions and waste generation, German companies are reluctant to publish what are regarded as internal management issues, which they regard as easily misinterpreted by the public, especially when conducted in the framework of official requirements. The issue appears to be less the lack of familiarity with the concept of publishing company data and more the fundamental objection to the linkage between communication of the results of corporate environmental performance activities and the perceived dictation of internal management processes, with accompanying disclosures.[17] As a practical concern, doubts have been raised as to whether an audit can be fully effective if the necessary atmosphere of openness is jeopardized by perceptions that the results could be publicized.[18]

Participation in EMAS involves much more than implementation of an auditing program, and the regulation was retitled from the original Eco-Audit designation to Eco-Management and Audit in order to reflect this reality. Companies wishing to register will be obliged to commit to the establishment of an extensive environmental management system and to a policy of continuous improvement of environmental performance. Full achievement of the requirements for an environmental policy, objectives, program, and management system as conceptualized by the commission presents an ambitious task for the majority of companies. This realization may well have privately contributed to the hesitant attitude of German and other European firms. More serious was the concern that the attempt to set down management structures and procedures, along with ensuing interpretations, overstepped the bounds of governmental intrusion into a corporation's abilities to organize and manage its own affairs.[19]

The auditing aspects of the regulation, however, have met with the brunt of the opposition. German companies often pointed to the high level of supervision and attention paid to environmental issues in their facilities, as described earlier, and declared that the scheme represented an unnecessary and burdensome additional level of control. The idea might be beneficial, it was argued, in countries where a comparable system of checks and balances did not exist.

The issue of suitability of the scheme to different member states is related to one of the commonest complaints of German industry: that the E.U. regulation would create market distortions. German companies would be evaluated against the high German standards, it was feared, whereas facilities located in countries with less stringent standards would find it much easier to earn the right to use the "statement of participation." The same high German standards, it was also argued, made the scheme unnecessary in Germany even if the concept could be beneficial elsewhere.

The fear of market distortions has become a sensitive issue as international shifts in production have been, since the beginning of 1993, simpler in the Single European Market system. Germans fear being on the losing side of production-location decisions as the business world takes an increasingly unified view of Europe.

A great number of issues were raised on the detail of the proposed wording. Many of these concerns reflected the fundamental objection to the idea of the nature of an auditing program, or indeed of an environmental management system, being dictated by law. The European Commission's response in making participation voluntary was not regarded as a complete solution in that market pressures could effectively force companies to participate. Many also believed that the provision for review of the scope of the scheme after a period of up to five years provided the mechanism by which the commission planned to make participation compulsory.

Auditing with and without EMAS

In summary, certain sectors of German industry raised doubts about the usefulness of auditing itself, whether or not the audits are implemented on an entirely voluntary basis or within a framework specified by law.[20] Others indicated a positive attitude toward the concept of auditing, while expressing serious reservations on the appropriateness of an E.U. auditing regulation.[21,22] A growing number of those who regard auditing as a useful tool, both in government and in the private sector, see the EMAS regulation as a mechanism to encourage all parties to improve their environmental performance.[23] Even among those who viewed the regulation favorably were many who questioned how successfully and appropriately the scheme could actually be implemented. Environmental activists were also critical of the regulation, although generally the charge was that the wording was insufficiently comprehensive and precise.[24]

The EMAS audit regulation was put to a vote at a meeting of the Council of Environment Ministers in December, 1992. Although the other E.U. countries approved the wording, it was perhaps not surprising that, given the above concerns and against the background of mixed feelings within the country, Germany filed a general reservation on the proposed regulation. Germany's intention was to improve some aspects of the wording, not to suppress the regulation; upon ratification of the Maastricht Treaty modifying the terms under which the E.U. exists, the regulation could have been adopted by majority vote. As it was, the agreement of all 12 member states was finally reached, after some further changes were made to the wording, in March 1993, and the regulation was formally adopted in June 1993.

The Need for Auditing

A review of the above considerations could lead to the conclusion that, owing to manifold differences between the United States and Germany in corporate culture, the historical development of environmental and safety practices, awareness, regulation, and national characteristics among other factors, auditing, at least in the form envisaged in the United States, does not have its place in Germany. There could perhaps be a temptation to lean toward this view; German industrial facilities often give a spotless, orderly appearance, with modern, well-maintained equipment and efficient, properly trained personnel.

Nevertheless, companies that have introduced auditing programs in Germany have reported the process to be beneficial and that auditing plays a useful role in identifying areas for potential improvement.[25–27] The auditing experience of the author's company, in common with other consulting firms in Germany, is that positive action items for improvement almost invariably stem from the audit process.[28] Even the cleanest and most orderly of companies generally reveal some deficiencies, albeit minor. Some items may in fact relate to German regulatory requirements that do not exist in the United States. As may be expected, larger companies with their greater resources tend to have in place a higher level of development of the necessary systems and procedures to achieve their environmental, health, and safety performance goals.

German companies have a reputation for the use of strong engineering skills in achieving solutions, and this tendency can be observed in the technology used in, for example, air pollution control, wastewater engineering, and hazardous-waste management. However the management systems in place to ensure ongoing achievement of company performance goals do not always match up to these high engineering standards.

Typical weak points found in environmental, health, and safety management systems include:

- Insufficiently detailed delineation of responsibility for supervision of environmentally related activities

- Incomplete training of those with responsibility for environmentally related activities, particularly with regard to regulatory requirements

- Lack of written procedures including plans for contingencies such as releases of hazardous material

Some examples of the types of deficiency through which the weak points in the management system might be identified are:

- Inadequate or incorrect documentation of waste disposal

- Improper storage of hazardous materials and improper on-site accumulation or staging of hazardous waste[29]

- Delayed compliance with technical standards for wastewater discharge and delayed submission of permit applications[30]

Some companies have reported finding[31] that personnel responsible for activities involving a deficiency usually have some notion of its existence and that the audit served as a catalyst to take corrective action. In other cases it has been evident that personnel, possibly owing to insufficient training, were unaware of the existence of the situation but took their responsibilities seriously and were quite ready to implement the necessary measures.

The overwhelmingly positive auditing experiences reported to date in Germany suggest that auditing is indeed an appropriate and useful management tool for German companies, even if many differences exist between the United States and Germany in the underlying regulatory, technical, political, and cultural conditions. Indeed, a great number of those in the private and public sector now view auditing favorably even when expressing certain apprehensions about the nature of EMAS and the uncertainties surrounding its implementation.

Changes in Motivating Factors

A great deal of change has been under way in Germany, particularly in the early 1990s, that has profoundly affected the fate of auditing and the nature of auditing itself. Some of the most significant participating factors in this change are reviewed below.

Expanding Liabilities

Potential liability is increasing for the results of activities resulting in harm to public health or the environment. German enforcement agencies are vigorous in the pursuit of environmentally related wrongdoing. The number of environmental offenses pursued by the authorities in 1990 totaled 21,412 compared with 2321 in 1973.[32] The 1990 so-called Lederspray decision by a federal court[33] was an example of the intent to place responsibility for corporate actions squarely on the shoulders of individuals in management positions. The personal liability of corporate management was further focused in 1990 by the requirement introduced into the Federal Air Quality Act to nominate the person at the management level with responsibility for environmental affairs.[34]

The 1991 Environmental Liability Act increased potential corporate liabilities for operators of a broad range of facility types. The act gave new powers to potential plaintiffs and established a system of "presumption of cause" where a facility is considered capable of causing the environmental impairment. Significantly, the presumption of cause does not apply if the facility can demonstrate that it was operating normally in compliance with permit conditions and other relevant requirements—providing an incentive for operators to ensure that regulatory requirements are being met. Furthermore, efforts are being undertaken at the E.U. level to strengthen the legal framework for liability arising from environmental impairment.

The demand for insurance protection, mandated for many companies, to cover liabilities arising on account of the Environmental Liability Act will inevitably result in careful investigation by insurance companies of the risk potential presented by facility operations and the nature of the systems in place to minimize such risks.

The Catalyst Effect of EMAS Regulation Development

The huge amount of exposure and attention afforded in Germany to the debate surrounding the drafting of the EMAS regulation has resulted in a rapid increase in the level of awareness of German companies in regard to auditing concepts. This has been evidenced by the explosion in recent years in the number of seminars, conferences, and workshops devoted to auditing and the E.U. requirements. Companies are conscious of the activities of their national competitors, of the example of U.S. and other multinational subsidiaries in Germany, and the developments in other E.U. member states. Of particular interest in the latter context have been the activities of the United Kingdom in leading some of the European developments in auditing and management systems.

Increasing Volume of Regulatory Requirements

German industry, particularly the smaller and medium-sized companies, is finding it steadily more difficult to keep pace with the growing body of environmental legislation. In recent years the federal and state governments have introduced many important statutes and ordinances and have frequently revised existing legislation; nearly all of the major federal statutes have been revised in the last few years. The primary requirements are contained in federal and state statutes and ordinances, or in administrative orders that are directed at regulatory agencies. County and local requirements also exist. In addition, a large body of interpreting guidelines—themselves subject to periodic revision—are developed by officially sanctioned committees of experts. These interpreting guidelines add greatly to the complexity of environmental requirements.[35] Despite possible assistance from the local chamber of commerce or a trade association, a typical small or even medium-sized company with no team of regulatory specialists can frequently be overwhelmed by the details of the requirements.

Increasing Access to Corporate Operating Data

Changes are evident in public expectations of industrial openness. A gradual erosion of public trust in authorities is coupled with a trend to more public participation in decision making that was previously left to governmental bodies and experts.[36] The concept of "need-to-know" is gradually being replaced by "right-to-know." To implement the E.U. directive 90/313 on the "freedom of access to information on the environment," Germany was required to enact a law easing the way for information on the environment, which includes agency data on industrial environmental performance, to be made available to the public. Although the federal government failed to meet the December 31, 1992, deadline set by the E.U. for enactment of appropriate national legislation, the individual German Bundesländer have issued guidelines to agencies on responding to requests for information.

Other signs of a "right-to-know" trend are appearing at the state level: Beginning in 1993, any facility in Northrhein-Westphalia that exceeds an annual waste-generation threshold is required to prepare a waste-generation, recycling, and disposal summary that is to be made available to the public.

German companies are recognizing these changes, and some have redesigned their public relations strategy to provide for considerably enhanced openness on environmental activities and performance. Some German companies have begun to prepare annual environmental reports or have expanded the environmental section of their annual reports. These changes can be expected to make the preparation of "environmental statements" for the public, as foreseen by EMAS, gradually more acceptable for German companies.

Trend to Proactive Corporate Environmental Management

Owing to expanding potential liabilities, growing public expectations, increasing regulatory requirements, and other factors, many German corporations have opted to take a highly proactive stance in the management of environmental affairs. Evidence has been gathered of fundamental changes in management attitudes of the importance of environmental protection for the corporation: a 1990 study of 200 companies found that over 20 percent of the companies could be classified as "ecologically oriented innovators."[37] A more recent study of 592 companies, sponsored by the federal government, found that management considered environmental protection to be of high importance for the long-term security of the company.[38]

An early landmark pointing the way for these developments was the 1988 "Tutzinger Declaration on an Environmentally Compatible Corporate Policy," signed by thousands of interested parties from industry, government, academia, and the public.[39] The subject of environmental management has since then been intensively addressed by institutes and universities. Industry has supported the establishment of organizations, such as the Bundesdeutsche Arbeitskreis für umweltbewußtes Management and Förderkreis Umwelt future e.V., dedicated to the furtherance of corporate environmental management.

The proactive approach taken by many leading companies addresses all aspects of corporate activities such as the minimization of environmental impacts arising from products and suppliers, impacts of investment decisions, personnel hiring, energy efficiency, and many other components.[40] This concern for environmental affairs has inevitably led to consideration of all measures, including auditing, that may assist in the improvement of environmental performance.

Changes are also being experienced in some of the remaining factors mentioned earlier that have inhibited the development of auditing. External supervision of industry in the former West Germany has diminished somewhat in the early 1990s, giving government an additional incentive to encourage self-regulation via auditing and giving industry less reason to claim that external inspections render auditing superfluous. The reason for the decline is that cutbacks in inspections by regulatory agencies have been forced by the transfer of personnel and funds to the new states of former East Germany. Legal limits are imposed on regulatory agencies in the transfer of supervisory or inspection function to nongovernmental parties.[41]

Other influences have also contributed to the development of auditing. The good experience gained from audits of "good management practices," for example, in the pharmaceutical industry, has led to the extension of programs to cover environmental, health, and safety auditing. Companies supplying major industrial corporations are experiencing an increasing level of scrutiny not only of product quality but also now of the extent to which production processes are environmentally compatible. These examples point toward an increasingly international character of the motivating factors for auditing as companies must plan their strategies in the global marketplace.

Occupational Health and Safety Audits

The emphasis on much of the previous discussion has been on environmental auditing, although many of the considerations raised apply, to a greater or less extent, to health and safety audits. Occupational health and safety auditing has not been at the forefront of the auditing activities or developments in Germany. This is not to say that the health and safety of German workers is neglected; on the contrary health and safety has traditionally been awarded priority attention in Germany, as is the case in a number of other European countries. The emphasis on safe working conditions has its roots in a number of factors, one of the most important being the strength of worker representation in Germany. As an example, the "workers' council," which has a substantial influence on the daily operation at a facility, and the inclusion of representatives of the workforce on the management supervisory board are standard features of German industry that are virtually unknown in the United States.

A voluminous body of occupational health and safety requirements has been developed in Germany over the years, and facility inspections take place regularly. Both the Berufsgenossenschaften, the worker safety associations, and the Gewerbeaufsicht, the state commercial supervisory agencies, inspect for compliance with these requirements. The frequency of inspections by the Berufsgenossenschaften (in former West Germany) has not been greatly affected by reunification, as the organizations are funded directly by industry and have suffered relatively little from shifting of resources to the new German states.

In the areas of process safety, risk assessment, and emergency response planning, Germany has also taken a leadership position. In 1982, the E.U. issued a directive addressing these subjects, generally referred to as the "Seveso Directive" in reference to the release of 2,3,7,8-tetrachlorodibenzo-p-dioxin in Seveso, Italy, in 1976. The E.U.'s initiative was at the time regarded as a pioneering piece of legislation. The German ordinance (Störfallverordnung) and associated legislation that respond to the E.U. directive contain some of the strictest and most detailed provisions of their kind. The massive reaction by the media, the public, and the regulatory authorities to a series of incidents at a chemical plant in early 1993 provided testimony of the strength of concern for the safety of industrial plants in Germany.

Other reasons account for the focus on environmental auditing. One is the influence of U.S. multinational companies that concentrated initially on the introduction of environmental auditing programs. A second reason is the restriction of the E.U. EMAS regulation to environmental auditing. This focus on environmental auditing is also influencing other parties such as standards organizations to address environmental management and auditing separately from occupational health and safety.

Administratively, German companies frequently combine responsibilities for health, safety, and environmental affairs within a single department, or in the case of smaller companies, single individuals. Auditing of health and safety management is therefore often seen as combined with the task of environmental auditing. In fact, one main benefit of environmental auditing has sometimes been (too narrowly) viewed as the identification of risks of, for example, a release of hazardous material to the environment. As in the United States, the issues themselves, for instance, related to storage and accidental release of hazardous materials, sometimes overlap or boundaries between health and safety and environmental protection become blurred.

The New German States

On October 3, 1990, the five new Bundesländer[42]—comprising the area of the former German Democratic Republic (GDR)—became part of the Federal Republic of Germany. As a result of reunification, existing German law became applicable in the new states.

In the area of environmental protection, temporary exemptions and special conditions were established allowing for the gradual upgrading of existing facilities to meet the strict standards prevailing in the "old" states. Apart from these transition provisions, federal law became immediately enforceable, but the necessary state laws were still required to supplement federal law within the areas of responsibility of the states. Owing to the time required to establish the necessary legislative machinery, and the pressing needs in many directions, the body of enacted state environmental legislation in the new states is only now approaching the status of state legislation in former West Germany. The laws of the former GDR only remain to the extent that they do not conflict with federal law and until the respective new state laws have been adopted.

The disparity between the new and old states in the condition of industrial facilities, however, is far greater than any differences in legislation. Reunification revealed the full extent of the decay and neglect in the industrial facilities of the former GDR. Auditing of the management systems, or even pure compliance auditing, of these facilities was an almost pointless exercise in the initial years after reunification. More relevant were overall facility evaluations to determine which installations and systems could be retained, what level of investment would have to be made, and what staffing decisions were needed. Of particular importance for investors were environmental assessments to determine the potential for contamination so that, if necessary, an application for exemption from remediation costs could be submitted to the authorities.

A federal agency, the Treuhandanstalt, was created to administer and gradually to privatize those facilities that had been controlled by the former GDR government. "Due diligence" evaluations of these facilities were performed by potential investors in preparation for negotiations with the Treuhandanstalt on, for example, possible release from—or sharing of—remediation costs.

Therefore, the vast majority of environmental surveys or inspections have to date been of a "due diligence" nature to determine the overall status of the facility and the potential for subsurface contamination, rather than detailed management systems or compliance audits. Nevertheless, facility managers are faced with the responsibility of operating in compliance with a potentially bewildering array of statutes, ordinances, and standards. The German government recognized the need to assist fledgling management teams in upgrading their facilities to meet these requirements. A number of funding programs were set up to assist facilities in making the needed investments.

Support has also been provided in the performance of initial facility audits to determine the appropriate improvement measures to be undertaken. In 1990, the German government established a foundation, the Deutsche Bundesstiftung Umwelt, to promote environmental protection in business activities. One of the programs developed by the foundation is the funding of 85 percent of the cost of a facility environmental audit for small and medium-sized enterprises in the new states, up to a maximum of DM 3400.[43] Equivalent funding may also be provided for up to two follow-up audits focusing on specific areas of detail. The program is administered through the local chamber of commerce offices and the Association of German Crafts (Zentralverband des Deutschen Handwerks). A similar program exists for local community institutions. The foundation has furthermore sponsored seminars, training courses, and similar programs to assist facility personnel in becoming acquainted with the regulatory requirements.

Current Developments

Germany is now taking auditing under serious consideration. Many companies are weighing the merits of initiating audit programs, and—often independently—of registering under EMAS. National certification systems for verifiers of facility audits are now being established under EMAS.

Auditing is being regarded by many companies only as part of their deliberations on an overall corporate approach toward environmental, health, and safety management. A considerable body of literature has been established exploring environmental management concepts and techniques; in this respect—if not in the broad-scale implementation of such concepts and techniques—Germany is considered to be at least as active as the United States.[44]

Companies are also attentive to the external developments shaping the debate, such as the issuance in 1992 of the Specification for Environmental Management Systems by the British Standards Institute (BS 7750) or the requirements for management systems contained in Annex I of the EMAS regulation. These documents present no more than a framework, however, and the design and implementation of actual systems is a complex process involving numerous considerations such as the nature of facility operations, existing systems and procedures, and the corporate history and culture. Management concepts being discussed in Germany include modern techniques of "eco-controlling," and integrated environmental protection as a component of strategic planning. On completion of a management system design, in the view of many practitioners auditing then takes its place as appropriate to complement the series of checks and balances built into the system.

Some companies are exploring whether and how their existing quality management systems could be adapted to include environmental, health, and safety management, and proposals have been made for certification of such systems in a similar manner to the existing certification procedures for quality management systems.[45]

In contrast to the United States, where the concepts of auditing have emerged over many years, auditing has no comparable historical roots in Germany. In defining auditing concepts, therefore, Germans feel uninhibited in starting with a blank sheet of paper. While recognizing the importance of monitoring compliance, given the complexity of German regulations, the discussions have largely disregarded the historical origins of auditing as merely a measure of compliance with regulatory requirements.

A wide range of ideas, questions, and proposals have been debated, such as:

- Auditing of products as well as production
- Auditing of intracorporate activities unrelated to individual facilities
- Auditing as an internal management tool versus an external communications instrument
- Auditing versus certification of environmental management systems
- Auditing versus automatic feedback systems
- The effectiveness of auditing without national or international standards defining environmental management systems
- Environment, health, and safety management systems separate from a quality management system or an integral part of an overall management system based on the "9000 series" of quality management and quality assurance standards issued by the International Standards Organization (ISO)
- Auditing as a regular update of the Environmental Impact Statement prepared for the facility

Some of these items suggest auditing concepts that go well beyond those commonly accepted in the United States. For example, some German firms are considering the arguments for combining the traditional scope of an audit with elements of life-cycle analyses. Certainly, differences in approaches related to the listed items will not be resolved in the near future, as companies experiment with their own programs and approaches. Two factors will exert a major influence on the evolution of auditing in Germany:

- The implementation of EMAS at the national level
- The development of international standards by ISO and/or the European Standards Bureau, C.E.N.

A close eye is being kept in Germany on the progress of standardization efforts, in particular, those being undertaken by ISO's Technical Committee 207 for the planned 14000 series of standards. A keen awareness exists that, for better or worse, the international standards that are likely to evolve in the forthcoming years are destined to have a major impact on environmental management and auditing, just as the ISO 9000 series has profoundly influenced the discipline of quality management and quality assurance. For this reason, Germany—along with other countries—has shown a strong interest in participating in the standards development process.

The mechanics of the audit process are likely to be less controversial than the purpose and scope of the audit. An international auditing standard could be expected to exhibit similarities to the ISO standard 10011, which defines quality management audits, and to which reference is made in the E.U. regulation. The main interest lies in the discussions on a standard for environment management systems. While on the one hand attempting to influence the standards development process, German companies are aware of the advantages of working nationally in a direction that can be reconciled with eventual international standards.

Conclusions

In comparison with the United States, Germany has been slow in the adoption of environmental, health, and safety auditing techniques. The numerous differences between the two nations, which accounted for the delayed introduction of auditing in Germany, are beginning to fade. Numerous factors now are motivating German companies to consider auditing programs. The auditing debate has been stimulated, and heavily influenced, by the process of development of the E.U. Eco-Management and Audit regulation.

In a very environmentally conscious country, German companies have long recognized the importance of environmental protection in facility operations as well as other aspects of corporate activities. Environmental, health, and safety management systems have been and continue to be introduced in line with modern concepts and approaches. As part of these systems, auditing is increasingly finding a role whereby the character of the audits will depend on the manner of implementation of EMAS regulation in Germany and on the development of international standards for environmental management and auditing.

References and Notes

1. Federal Ministry for Environment, Nature Protection, and Reactor Safety, press release, March 26, 1993.

2. Martinson, Brian A., Corporate Environmental Audit Programmes: Emerging Trends, Standards and Public Policies, Energy and Resources Law '92, Graham and Trotman Ltd. and International Bar Association, 1992.

3. Conversely, one U.S. factor inhibiting auditing has not been of great significance in Germany: Owing to the quite different nature of the German legal system, German

companies have not shown a reluctance to initiate audit programs simply because of the potential for creation of material that might be subject to discovery in legal proceedings.

4. The European Commission, located in Brussels, is that part of the European Union legal system with primary responsibility for drafting of European Union legislation.

5. Process safety in this context concerns prevention and response related to major accident hazards. A separate type of requirement exists for health and safety officers.

6. The English title is only approximate; the Bundes-Immissionsschutzgesetz primarily addresses ambient air quality, but it also encompasses other concerns such as noise.

7. § 52a of the Bundes-Immissionsschutzgesetz of May 14, 1990 (BGBL.I S.880).

8. Feldhaus, Gerhard, Umweltschutzsichernde Betriebsorganisation, Neue Zeitschrift für Verwaltungsrecht, Volume 10, 1991.

9. Holoubek, Karl, and Geywitz, Jörg, Die Integration des Umweltschutzschutzes in das operative Betriebsgenchehen, in Steger, Ulrich (editor), Umwelt-Auditing—Ein neues Instrument der Risikovorsorge, Frankfurter Allgemeine Zeitung, Verlagsbereich Wirtschaftsbücher, Frankfurt am Main, 1991.

10. Henkel, Hans-Olaf, Umweltauditing bei der IBM, in Steger, Ulrich (editor) Umwelt-Auditing—Ein neues Instrument der Risikovorsorge, Frankfurter Allgemeine Zeitung, Verlagsbereich Wirtschaftsbücher, Frankfurt am Main, 1991.

11. The full title is: Council Regulation allowing voluntary participation by companies in the industrial sector in a Community Eco-Management and Audit scheme.

12. E.U. "regulations" are unusual in the environmental field. Normally, the E.U. issues "directives" that must then be transposed into national legislation by the individual member states; regulations, by contrast, are directly binding within the member states.

13. Union of Industrial and Employers' Confederations of Europe, letter of 3 April, 1992, to Mr. Laurens Jan Brinkhorst, Director General—DG XI, Commission of the European Communities.

14. Meurin, Gerhard, Umweltschutz-Audits contra Öko-Audit-System der EG, Zeitschrift für angewandte Umweltforschung, Issue 5, 1992.

15. Dittmann, Umweltschutz-Auditing—Ein Instrument zur umweltorientierten Unternehmensführung, Seminar Umweltschutz-Audits, Die Effizienz freiwilligen Ökomanagements, German Group of the International Chamber of Commerce, Cologne, October, 1991.

16. Renn, Ortwin, Risk Communication at the Community Level: European Lessons from the Seveso Directive, Journal of the Air and Waste Management Association, October, 1989.

17. Karl, Helmut, Öko-Audits—Ein sinnvolles Informationskonzept für Umweltbelastungen? Wirtschaftsdienst, Issue VII, 1992.

18. Meurin, Gerhard, Umweltschutz-Audits contra Öko-Audit-System der EG, Zeitschrift für angewandte Umweltforschung, Issue 5, 1992.

19. Meurin, Gerhard, Umweltschutz-Audits contra Öko-Audit-System der EG, Zeitschrift für angewandte Umweltforschung, Issue 5, 1992.

20. Holoubek, Karl, and Geywitz, Jörg, Die Integration des Umweltschutzschutzes in das operative Betriebsgeschehen, in Steger, Ulrich (editor), Umwelt-Auditing—Ein

neues Instrument der Risikovorsorge, Frankfurter Allgemeine Zeitung, Verlagsbereich Wirtschaftsbücher, Frankfurt am Main, 1991.

21. Meurin, Gerhard, Umweltschutz-Audits contra Öko-Audit-System der EG, Zeitschrift für angewandte Umweltforschung, Issue 5, 1992.

22. Dittmann, Umweltschutz-Auditing—Ein Instrument zur umweltorientierten Unternehmensführung, Seminar Umweltschutz-Audits, Die Effizienz freiwilligen Ökomanagements, German Group of the International Chamber of Commerce, Cologne, October, 1991.

23. Zink, Alexander, Rechtliche und politische Rahmenbedingungen der Umweltaudits, Seminar Öko-Auditing in der betrieblichen Praxis, Institute for International Research, Munich, January, 1993.

24. Kurz, Rudi, and Spiller, Achim, Umwelt-Auditing: Internes Risikocontrolling oder marktorientierte Umweltverträglichkeitsprüfung? Zeitschrift für angewandte Umweltforschung, Issue 5, 1992.

25. Henkel, Hans-Olaf, Umweltauditing bei der IBM, in Steger, Ulrich (editor), Umwelt-Auditing—Ein neues Instrument der Risikovorsorge, Frankfurter Allgemeine Zeitung, Verlagsbereich Wirtschaftsbücher, Frankfurt am Main, 1991.

26. Zeschmann, Ernst, Der Nutzen von Umweltschutzaudits für Unternehmen-Erfahrungen und Erwartungen aus den Umweltschutzaudits der Shell, Seminar Öko-Auditing in der betrieblichen Praxis, Institute for International Research, Munich, January, 1993.

27. Gozon, Istvan, Umweltschutz- und Sicherheits-Audits in Unternehmensverband Boehringer Ingelheim, Seminar Öko-Auditing in der betrieblichen Praxis, Institute for International Research, Munich, January, 1993.

28. Niemeyer, Adelbert, and Sartorius Bodo, Umwelt-Auditing, in Steger, Ulrich (editor), Handbuch des Umweltmanagements, Beck, 1992.

29. Stringent requirements have existed for many years in Germany specifying measures to prevent releases to surface or groundwater when storing hazardous materials.

30. These compliance deficiencies are mainly on account of legislation that was introduced in the early 1990s at the state (Bundesland) level for dischargers of wastewater to publicly owned treatment works.

31. Gozon, Istvan, Umweltschutz- und Sicherheits-Audits in Unternehmensverband Boehringer Ingelheim, Seminar Öko-Auditing in der betrieblichen Praxis, Institute for International Research, Munich, January, 1993.

32. Eidam, Gerd, Führungskräfte—Mit einem Bein im Gefängnis? in Die Organisation des betrieblichen Umweltschutzes, edited by Heinz Adams and Gerd Eidam, Frankfurter Allgemeine Zeitung, Verlagsbereich Wirtschaftsbücher, Frankfurt am Main, 1991.

33. Bundesgerichtshof Mainz, Decision of July 6, 1990.

34. § 52a of the Bundes-Immissionsschutzgesetz of May 14, 1990 (BGBL.I S.880).

35. Krieger, Stephan, Normkonkretisierung im Recht der wassergefährdenden Stoffe, Erich Schmidt Verlag, 1992.

36. Renn, Ortwin, Risk Communication at the Community Level: European Lessons from the Seveso Directive, Journal of the Air and Waste Management Association, October, 1989.

37. Steger, Ulrich, Normstrategien im Umweltmanagement, in Steger, Ulrich (editor), Handbuch des Umweltmanagements, Beck, 1992.

38. Antes, Ralf, et al., Umweltorientiertes Unternehmensverhalten—Ergebnisse aus einem Forschungsprojekt, in Steger, Ulrich (editor), Handbuch des Umweltmanagements, Beck, 1992.

39. Tutzinger Erklärung zur Umweltorientierten Unternehmenspolitik—Dokumentation der Aktion, Tutzinger Materialie 59/1989, Evangelische Akademie Tutzing, Tutzing, 1989.

40. Winter, Georg, Das umweltbewußte Unternehmen, Beck Munich, 1987.

41. Reinhardt, Michael, Wieweit muß die Kontrolle des Vollzuges staatlicher Auflagen durch staatliche Stellen erfolgen? Seminar Neue Ansätze im integrierten Umweltschutz—eine Herausforderung für Staat, Wirtschaft und Bürgher, Institut für Siedlungswasserwirtschaft at the Rheinisch-Westfalische technische Hochschule, Aachen, 1993.

42. Brandenburg, Mecklenburg-Vorpommern, Sachsen, Sachen-Anhalt and Thuringen. (East Berlin became part of the Bundesland of Berlin.)

43. Deutsche Bundesstiftung Umwelt, Richtlinien über die Förderung von Orientierungsberatungen im Bereich des Umweltschutzes für kleine und mittelere Unternehmen in den neuen Bundesländern und Berlin (Ost), 1 November 1991, modified on 1 January, 1993.

44. Antes, Ralf, et al., Umweltorientiertes Unternehmensverhalten—Ergebnisse aus einem Forschungsprojekt, in Steger, Ulrich (editor), Handbuch des Umweltmanagements, Beck, 1992.

45. Krieshammer, Gerd, and Rademacher, Helmut, Zertifizierung von Umweltschutzsystemen, in Adams, Heinz, and Eidam, Gerd (editors), Die Organisation des betrieblichen Umweltschutzes, Frankfurter Allgemeine Zeitung, Verlagsbereich Wirtschaftsbücher, Frankfurt am Main, 1991.

24

Environmental, Health, and Safety Auditing at Norsk Hydro

Stein-Ivar Aarsaether

Public Affairs Manager, Norsk Hydro

The Norsk Hydro group is one of the leading industrial enterprises in Scandinavia, with extensive operations in Europe and overseas. The company has about 34,000 employees, half of whom are based in Norway and half at production sites and sales offices in several countries.

The Norwegian state owns 51 percent of the shares in Norsk Hydro; the rest is spread over a large number of private shareholders in Norway, the United States, France, Germany, the United Kingdom, Switzerland, and other countries.

The company is principally involved in the production of oil and gas, refined oil products, petrochemicals, fertilizers, and light metals. Norsk Hydro is the largest producer of fertilizers and the second-largest producer of magnesium in the world. It is also the second-largest producer of aluminum in Europe. The company is also involved in numerous other activities, such as pharmaceuticals and fish farming, where it is also the second largest in Europe.

With regard to the overall organization of corporate management and the main business areas, it should be noted that each business area is autonomous and that the corporation mainly acts as a holding company, with a small central staff, including the central health, environment, and safety (HES) function.

Many of Norsk Hydro's business areas are environmentally sensitive, and some of them involve major potential process safety hazards. In many respects, the company was a forerunner in reducing its discharges to the environment through the 1970s and 1980s; however, public expectations have grown faster than the improvements. Also, a number of new environmental issues have been placed on the agenda, such as regional pollution problems, depletion of the ozone layer, and global warming.

Furthermore, the focus has partly shifted from stacks and discharge pipelines to the products themselves. Despite the improvements, Norsk Hydro was, at the end of the 1980s, still a main target for environmental groups, and the company's public image was not as good as desired.

Whereas the public focus was mainly on the external environment, it was realized by corporate management that there were also major challenges with respect to the internal environment, the safety of personnel, and the safe operation of the plants. Furthermore, other international companies had demonstrated that by focusing on improvement in these areas, they had also achieved good results in the overall management of their plants and business in general.

One of these companies was DuPont, which had achieved remarkably good safety records. A collaboration with DuPont was established, and a major process was started to change attitudes, to create motivation, and to set objectives for continuous improvement.

A more systematic and comprehensive approach was needed. There were three key features of the new policies that were adopted in the late 1980s:

- HES issues should be met by being proactive rather than reactive; i.e., to assess the problems and start to address them before being forced to do so by public pressure and authority demands.

- The HES issues are line management responsibilities and need their involvement.

- There should be openness in the communication internally and with the public.

Even if these policies had been in place as statements for some time—and had been practiced to a certain extent—they were now given full backing from top management. In order to achieve a proactive attitude throughout the corporation, some key corporate directives and standards were issued, and in order to monitor that the message was received, understood, and was being implemented, a scheme of internal auditing was established.

As a result of the openness policy Norsk Hydro published a detailed report of the environmental status of all its Norwegian operations in 1989 and a more general report for the whole corporation in 1990. A report concerning its U.K. operations, which was published in 1990, caught the attention of both the public and industry. What they found particularly interesting was the fact that the report had been validated by an external auditor.

Some major improvements have been achieved during the last 5 to 7 years, as exemplified by the frequency of lost-time incidents and by the reduction of fluoride emissions from the primary aluminum smelters. Obviously, auditing is but one of many factors that have contributed to this, but its role has been significant.

Framework of the Audits

What Is an Audit?

From the ISO definition of quality audits (ISO 10011), the following definition of HES audits can be derived:

> A systematic and independent examination to determine whether HES activities and related results comply with regulatory and company requirements and objectives, and whether the systems to achieve these objectives are implemented effectively and are suitable for the purpose.

The first part of this definition relates to *compliance* with requirements. The last part relates to the adequacy and efficiency of *management systems* that should be in place to ensure compliance, i.e., how good is the *performance* (beyond the strict requirements). The focus of the audit may be on either or both of these elements.

A compliance audit necessitates thorough checks of results versus requirements, and the outcome is usually a relatively clear-cut "yes" or "no." Management system audits or performance audits are open for a more qualitative judgment.

Reference Requirements

Among the four main statements that constitute the basic business philosophy of Norsk Hydro, there is one which relates to environmental aspects:

- In all its activities Hydro will emphasize quality, efficient use of resources, and care for the environment.

This is further elaborated in the key strategic principles of the company:

- We will demonstrate a strong sense of responsibility for people and the environment.

- Hydro will be in the forefront in environmental care and industrial safety.

From such basic strategies, the company has issued policies, directives, standards, procedures, and instructions. As with other steering documents, these are interlinked in a hierarchy.

A basic requirement is that regulatory requirements should be met. In addition, the corporate directives for HES management systems comprise the following main elements:

- Each site should have HES policy statements, annual HES goals, and action plans.

- The HES responsibility lies in the line management organization.

- Each site should make an assessment of its emissions and wastes, their process hazards, and the occupational risks. Improvement potentials should be identified and evaluated for inclusion in the action plans.

- The management system should include systems for providing information to employees about HES issues and should encourage participation by the employees in the improvement process.

- All activities that may involve risks to personnel, equipment, or the environment should be regulated in procedures.

- The safety and environmental consequences of modifications, expansions, acquisitions, and termination of operations should be assessed before they are approved for implementation.

- All employees should be given relevant training and education for their tasks.

- All accidents and incidents should be investigated and reported.

- There should be emergency plans and regular drills to test preparedness.

- Each division should monitor the performance of its production sites, using audits as one of the tools to check the performance. Audits should be included in the annual action plans.

In addition to these requirements, there are corporate standards relating to such things as plant modifications, accident reporting, systematic maintenance, work permit systems, control room integrity, and storage of hazardous liquids.

Reference Guidelines for Audits

The Norsk Hydro HES audits are mainly based on the ISO 10011 standard—Guidelines for Auditing Quality Systems. These stipulate the main requirements to the planning, execution, and reporting of audits, requirements to the auditors, and requirements for managing an audit program.

Another main reference is the development of the EU's Eco-Audit scheme, which is applicable to the environmental auditing. This involves a comprehensive scheme of:

- Establishing internal environmental protection programs
- Issuing environmental status reports
- Performing internal audits
- Having these reviewed by an external auditor

On a national level, there is supposed to be an accreditation institute that will certify auditors.

So far it has not been requested that the auditors for the internal HES audits at Norsk Hydro be formally certified. However, there are strong qualification requirements for the audit leader and the team as a whole regarding both auditing as such and the topic of the audit. Formal certification may be a further step in the development of the audit program.

The scope of the Eco-Audit scheme is wider than the present auditing program within Norsk Hydro. At the moment of writing, there is considerable discussion among chemical industries in Europe as to whether to adopt the Eco-Audit scheme, and Norsk Hydro has not decided on it yet.

Regarding safety, some audit systems such as ISRS (International Safety Rating System) are based on comprehensive checklists, which aim at a quantitative rating. Some units within Norsk Hydro have adopted this system, but they are not used on a corporate level.

The Present Audit Program

Whereas the earliest corporate audits were mainly directed toward production sites—and to a certain extent had the character of being "missionaries" who were sent to communicate the message of the new policies and requirements—the approach is now more incorporated into the line management hierarchy. This means that the corporate audits are mainly directed toward the divisions, or groups within the divisions, whereas the divisions perform internal auditing on their production sites.

The division audits are usually system audits covering all HES aspects. The focus is on management attention and systems for communication and follow-up of HES requirements and objectives. The site audits are often more specific, covering only one of the main HES aspects. However, the emphasis is still on management systems. There are only spot checks to gauge compliance against regulatory requirements.

The present audit program comprises about 12 corporate audits per year. At the site level, it is Norsk Hydro's objective to conduct an audit every 3 years, at least for the larger sites. The subject of the audit may be different from time to time, but generally the follow-up of the last audit is checked.

Objectives and Main Features of the Auditing Program

Objectives.   As mentioned, auditing is a management tool to check that (1) the necessary systems are in place to meet the company requirements and regulatory requirements and (2) they are effective in reaching these objectives.

An important objective of auditing is to enhance openness ("glasnost") in internal communications about HES issues. Levels of management hierarchy can act as filters, which may distort vertical communication. Through an audit, it may appear that something important has been lost in this process.

The ISO guidelines stipulate that the auditors are to be independent of the auditee. In Norsk Hydro's practice, the auditors are selected from other parts of the company. The risk of biased results is small, and it enhances communication between units and increases the internal competence in the company. This cross communication is considered to be very valuable.

An important aspect of making up the team from internal resources, and specifically including people dealing with the same kind of problems, is that the resulting recommendations are usually realistic and will be respected by the auditee.

Main Features of the Audit Program.   The main features of the audit process at site level are:

- Selection of audit subject, timing, and scope
- Selection of audit team and audit leader
- Setting up audit program
- Previsit information package
- Site visit with interviews and inspection
- Presentation and discussion of results
- Reporting
- Follow-up

The details are discussed below.

Audit Preparation

Selection of Audit Subject, Timing, and Scope

On a corporate level, the audit subjects are selected each year in collaboration between division management and the corporate HES staff. In addition, the division may select a number of units to be audited by the division.

The unit to be audited is informed and a tentative scope and date is agreed, usually no less than 3 months in advance. (There is no point in making the audits a surprise exercise; on the contrary, our experience is that the awareness of the coming audit is an inspiration to tie up some loose ends.)

Usually, the scope of an audit at site level is in one area—environment, process safety, or occupational safety and health—whereas some audits may cover all these fields. Additionally, it has to be defined whether the whole site or only a part of it is to be audited.

Selection of Audit Team and Audit Leader

The audit leader is usually either a person with senior experience in general quality auditing or an expert in the environmental or safety fields who has also been trained in auditing. The audit leader should be acknowledged by both the client (HES staff or division management) and the auditee, and his or her independence and integrity should be respected.

In addition to the audit leader, the audit team is supplemented with a senior person from another unit with production similar to that of the auditee. He or she should be either a production manager or HES coordinator of this unit. This is important, since he or she will then have relevant experience about the particular problems within the specific type of production. This may help in rapidly focusing the audit on the main issues.

Depending on the experience of the audit leader and the person from another unit, the team may be supplemented with a professional in the field relevant to the audit, i.e., either environment, process safety, or occupational hygiene, etc.

As appropriate, one of the above may act as audit secretary to take notes and draft the report, or a fourth person, who may be a trainee in auditing, can take this job.

The audits may reveal questions of interpretation of acts and regulations, which may require legal expertise to clarify. It has not been found worthwhile to include legal experts in the audit teams. Such clarifications may best take place as follow-up actions of the audit.

Setting Up the Audit Program

The audit leader contacts the site manager and gets a general overview of the organization, the production, and the main environmental and safety issues. Based on this, he or she drafts an audit program, which is then discussed and agreed with the site management. If necessary, the audit leader may pay a visit to the site some weeks in advance of the audit to finalize the details of the program and to get acquainted with the site.

It should be clarified in advance with the client or auditee whether there are particular issues that should be given special attention, so that the program may be tailored accordingly.

Previsit Information Package

The auditee should prepare a previsit information package, which is distributed to the audit team members. For all kinds of audits this package should contain:

- General information about the site, the production, and the organization
- Written policy statements, HES programs and action plans, goals, and objectives
- Previous audit reports

For an environmental audit, the following additional information is required:

- Permits relating to emissions and wastes
- Description of pollution control systems and waste management systems
- Programs for emissions monitoring and environmental monitoring
- Records of emissions and wastes
- Noise survey data
- Spill emergency plans and spill reports

- Environmental impact assessment report or environmental risk analysis, if available
- Strategies for development of the site with a view to future regulatory requirements

For a process safety audit, the following additional information is required:

- A summary of process accidents and incidents during the last 3 years
- Description of major hazards at the site (if a formal risk analysis has been performed, this should be forwarded)
- A summary of the current process safety plans

For an occupational safety and health audit, the following additional information is required:

- A summary for the last 3 years' statistics for:
  a. Absence due to illness (percent of total)
  b. Accident record, including lost time injuries frequency and absence frequency (due to accidents at work)
  c. Occupational illness (number of cases, types of illness
  d. Serious accidents (number of accidents, brief description)
  e. Survey reports regarding the occupational health and safety conditions, if available
- Reports regarding major improvement measures during the last 2 years

Execution of an Audit

General

The main elements of an audit program are:

- Opening meeting—auditors and auditees get acquainted and discuss the scope and the purpose of the audit. As appropriate, adjust the program.
- Brief interview and walk around the plant with the plant manager.
- Interview with the HES coordinator.
- Interviews with the production manager, different section managers, maintenance manager, laboratory manager, QA manager.
- Visits to specific areas of the site, meet operators.
- Meeting with the plant manager.
- Internal audit team summing-up meeting.
- Presentation of the results to the auditees, discussion.

The audit may also include a meeting with the responsible division to clarify matters relating to the communication between the unit and the division.

By studying the preaudit information, the audit team should get a good impression of the strong and weak points of the auditee in terms of complying with the company requirements and the regulatory requirements. However, a good appearance in the written documentation may be deceptive, and the auditors should check that the practice conforms with the writing. In the interviews with personnel from the different levels of the organization, auditors should check how management directives are interpreted and implemented to verify consistency and to check that the communication is effective.

One technique that is frequently used to check communication is to ask the personnel to be audited to rank on a scale from 0 to 5 some performance indicators, such as management involvement in HES issues, instruction and training, protective measures, and housekeeping before the audit interviews start. Low scores or large variation in scores may give clues that closer study is warranted. At the end of the audit, the scores are presented, together with the audit team's evaluation.

General checklists are useful for both interviews and inspections. However, these should be used with care. Being too bound to the checklists may prevent good dialogue. Also, irrelevant points should be deleted in advance to avoid wasting time and frustrating the auditees. The questions should be related to the responsibilities of those interviewed. The general checklists may also be supplemented with specific questions derived from the preaudit information.

It is important that conflicting statements and indications of nonconformance to requirements are clarified before the audit report is concluded. It may therefore be necessary to come back to some of the auditees for further clarification.

The interviews may preferably take place in the offices of each auditee. This will facilitate retrieval of documents that may be needed, and it may also make the auditee more comfortable.

As an example, the auditing procedure for environmental audits is presented.

Opening Meeting

In the opening meeting, the plant manager should give a general view of the environmental situation of the plant, the main requirements, main problems, and objectives and goals of the operation. There should be no detailed discussion of these items at this meeting.

Walk around the Plant

By making such a presentation, plant managers will give the auditors a good impression of how well informed they are about the HES issues and their attitude toward them. This should not be a lengthy inspection; specific items may be visited later.

Environmental Coordinator

During the interview with the coordinator for environmental affairs the following aspects should be discussed:

- How does the environmental management system work?
- Who is responsible for what?
- How are objectives and goals established?
- How is the organization informed and motivated to achieve these?
- What are the reporting routines?
- How are environmental aspects incorporated in plans and budgets?
- Are there problems with conforming to the regulatory requirements?
- How are noconformities reported and followed up? (It should be checked whether the nonconformities are more or less continuous, frequent, or nonfrequent and to what extent the permit conditions may allow for excursions above the specified level.)
- Are there self-imposed restrictions in the operation of the plant or in the waste management beyond the authority requirements? If so, how are violations of these treated?
- How is external communication (relations with media, neighbors, and authorities) handled?
- If there are elements missing in the preaudit information, or if there are questions about such information, they should be discussed here.
- General overview of air pollution control system, i.e., sources, treatment, monitoring, and specific problem areas.
- General overview of liquid effluents, i.e., sources, treatment, monitoring, and specific problem areas.
- Systems for waste management:
  a. Waste minimization
  b. Reuse
  c. Waste classification
  d. Labeling
  e. Transport and storage
  f. Waste disposal
  g. Treatment or destruction
  h. Own landfills
  i. External contractors
  j. Liabilities for old landfills
- Plans for phase-out or replacement of CFC and halons.
- Status with respect to contaminated soil, groundwater, and sediments.

- Noise—main sources, emission levels in populated areas.
- Environmental risks related to accidental discharges or spills, prevention, emergency plans.
- Is environmental information given or available to customers?
- Are there provisions for returning packing material and/or production waste from customers?

Norsk Hydro is operating in many countries, and the environmental legislation varies. In some countries, most of the requirements are specified in individual permits, whereas in other countries they are specified in general regulations. The coordinator should be asked to clarify how such general regulations apply to the plant and how general limitations are translated into specific, internal requirements to each part of the plant.

Production

The interviews with the production manager and the different section managers should focus on:

- How do they understand their responsibility with respect to environmental questions?
- How are interfaces with other departments such as waste management handled?
- Which are their main environmental requirements, and how is compliance controlled? How are noncompliances handled?
- Do they actively participate in formulating action plans and goals for environment? Ask them to present this year's action plan and to comment on its status.
- To what extent can the operators influence the emission levels, and how are they motivated to keep the performance high?
- How are trade-offs handled between different compartments of the environment (e.g., generation of liquid effluents or wastes by air emission treatment) or between environment and energy or economy?
- If lack of maintenance leads to deteriorating performance of treatment systems, who is responsible?
- How are emissions and wastes minimized during shutdowns and start-ups and other transient conditions?

Utilities

Depending on the nature of the production and the size of the organization, utilities may be operated by a separate department or as an integrated part of the production management. If the former is the case, wastewater treatment, waste incinerators, and waste handling are often the responsibility of the utilities department. Most of the questions above are relevant to this department; however, interfaces with the other production departments should be checked. For instance, is an overloaded wastewater treatment plant a utility problem or a production problem?

Maintenance

As with the utilities, the interview with the maintenance manager depends on how responsibilities are defined. The performance of emission treatment systems depends on the maintenance, but the responsibility to decide on whether maintenance is required usually lies with the production management. The interface here should be discussed.

Often, the maintenance department has a general responsibility for housekeeping in nonproduction areas, and in some cases takes care of the waste handling. This should be clarified. Depending on the responsibility, some of the questions above may also be relevant.

Laboratory

The laboratory is usually responsible for sampling and analyses of emissions, effluents, and wastes, and in some cases also takes samples and analyses of the surrounding environment. The following items should be discussed with the laboratory manager:

- How is the program for measurement, sampling, and analyses established and, as appropriate, revised?
- How much is due to regulatory requirements, and how much is voluntary?
- Is the frequency of sampling dependent on statistical analysis of the results?
- Is the regulatory conformance established through own measurements or by external control?
- Is advantage taken of the latest techniques in continuous monitoring (where this is feasible)?
- Are the results of continuous monitoring directly available to the operators concerned?
- How are the results of sampling and analyses communicated to plant management, production management, shift forepersons, and the environmental coordinator?
- Is the frequency of sampling regular or random?
- Do the operators know when sampling is done?
- Are emissions checked also during weekends?
- How is the quality of the sampling and analyses assured?

- Has the laboratory taken steps to obtain official accreditation?
- How are laboratory wastes handled?

Inspection Tours

In addition to the general guided site visit, inspection tours to more specific plant areas may be performed. This could be done in context with the interviews with the persons responsible for these areas.

What to Look for:

- General housekeeping and tidiness
- Visible smoke, dust, etc., from chimneys, and excessive flaring
- Losses and emissions from transport conveyors and open storage
- Fumes and odors
- State of storage tanks for volatile liquids (floating roofs, seals, etc.)
- Spot checks on sampling points: location, representativeness, and access
- Segregation of surface water (polluted, nonpolluted)
- Effluent treatment plant(s): load vs. capacity, instrumental monitoring, flow measurement, and sampling points
- Impression of final effluent: odor, color, suspended solids, foam, and visible oil
- Visible pollution of surface water: how is surface water and groundwater quality checked?
- Discharge pipeline, aesthetics
- Waste containers, labeling, actual contents, and location
- Hazardous-waste storage: design, protection from leakage, and fencing
- Old landfills: fencing, marking, and seepage control
- Berms or collection pits around storage tanks for oil products and other chemicals: how are these drained of rainwater?
- Transformers and other equipment containing PCB: labeling, collection pits
- CFC- and halon-containing equipment: labeling, maintenance status
- Main noise sources: noise level, pure tones, impact noise
- Surroundings: how is the plant situated with respect to dwelling areas, nature protection zones; vulnerability of receiving waters, etc.?

During the site tour, some operators should be asked if there are environmental constraints related to their work, how they ensure that these are met, and how deviations are reported. This could be done in the control room. For instance, ask the operator to mention the main parameters to be checked on a filter, a scrubber, or other equipment.

Engineering and Project Planning

If the plant has a dedicated engineering organization, it should be asked how they evaluate environmental aspects in project planning.

Plant Manager and Environmental Coordinator

This is partly a clarification meeting to tie up loose ends and to give initial feedback on the impressions of the audit. If the audit has revealed serious findings, these should be discussed.

Communication with the division regarding environmental policies and financing of environmental projects may also be discussed. Are directives and objectives given by the division also followed up with necessary budgets?

Internal Summing-up Meeting

The audit team meets and summarizes the findings and observations.

Findings.   These include noncompliance with authority or company requirements and may be classified as major or minor. If a lack of environmentally conscious attitudes has been found, this may also be called a finding, even if this has not resulted in violations of written limits. Any finding should be documented and have reference to evidence.

Observations.   These have the character of suggestions of areas where improvement can be made, while the current practice is not in conflict with requirements.

Summing-Up Meeting

In the summing-up meeting, all the auditees are gathered, and the audit leader presents the results. In addition to the findings and observations, he or she should also comment on the organization strong points and areas of excellence. The auditees should be encouraged to challenge the findings and observations if they feel they are inappropriate.

Process Safety Audits and Occupational Safety and Health Audits

These audits follow the same general lines as the environmental audits. However, the focus is on other issues.

Reporting

General

The report is an important part of the audit. For an audit as described above, the report should contain the following elements:

1. *Introduction.*   Which unit was audited, who requested it, who performed the audit, when was it performed, what was the scope?
2. *Reference requirements.*   A list of the main requirements against which the performance is to be measured
3. *General background information.*   A minimum of information regarding the plant and its operation, organization, main issues related to the audit scope, etc., to make the report readable for someone who is not acquainted with the plant
4. *Findings and observations.*   A list of findings and observations with reference to the evidence that they are based on
5. *General evaluations, recommendations.*   A general impression of the HES management system and recommendations for improvement
6. *Follow-up.*   A statement regarding who is responsible for responding to the report and following up its recommendations

Appendixes

- Scope of the audit
- Audit program
- Photographic impressions

Follow-Up

It is a line responsibility to follow up the audit findings and recommendations. This means that the auditee has a responsibility to take the necessary corrective actions. Usually, the client (i.e., the HES staff or the division) will want a report on how the recommendations are to be handled. The division will also be involved in the implementation if major investments are needed.

Experience and Challenges

Important Issues

Norsk Hydro was one of the first companies to begin HES auditing. Presently there is an almost overwhelming interest in the subject on the part of legislators, inspectorates, consultants, and also from the public. It may therefore be necessary to emphasize that auditing is only one link in the chain that constitutes the management system. Without genuine management attention to the issues—and a direct follow-up by line management—audits will not bring about significant results.

Audits do not go through every aspect of the operation at a site; they focus on management systems and take spot checks. An audit report cannot be regarded as a "clean bill of health" or a certificate ensuring that if certain things are corrected, everything will be OK. The use of audits (whether or not to use them, how to use them, and how often to use them, etc.) therefore should be considered in context with the objectives; there may be other ways and means to achieve the same results.

Some important issues to decide are whether the auditors should be internal or external, and whether the results should be made public. Norsk Hydro has tried both internal and external auditors and has also published the results of some audits, e.g., the U.K. environmental report. Our experience is mixed. While some external audits have been useful, others have been of low value. A general opinion is that if we have the competence internally, this should be used.

It is important that there be openness in communicating with the public, and it is important that what is presented has a high degree of credibility and confidence. In this context, it may be useful to use external auditors to validate what is presented, as was done in the United Kingdom.

Experiences

Some of our experiences with auditing to date:

- The audit teams have generally been met by openness and a genuine interest in the process. This has contributed to making the auditing an efficient tool in identifying areas of potential improvement.

- In a few cases, the audits have been more difficult; e.g., it has been difficult to engage management in the process. This is especially true where the organization has been in a turbulent state owing to restructuring, etc. In such cases, even if the HES issues have to be addressed properly, it may not be the best time for an audit. It should be borne in mind that audits are resource-demanding exercises.

- The cases where audits have been most successful are where the management has had the time to try out a management system and is interested in an independent review.

- Since the audit process has only been going on for a few years, few sites have been audited twice. It is therefore difficult to assess the degree of follow-up and improvement. There are "sunshine examples" of major improvements at some sites; however, the auditing at division level has indicated that the division managements do not always follow up the audits, as the performance profiles indicate.

- When auditing started, the sites that already had good records were audited first. This was done as part of the learning process for the auditors, i.e., what can be expected at a good site, and what are the key factors involved in achieving good results? It would also give those sites with a poor record some time to adjust to the new system.

Challenges

One of the major challenges ahead will be whether Norsk Hydro will decide to adopt the Eco-Audit scheme. Should the company do so, we must determine how to adopt it. In this context, it may also be a challenge to avoid duplicate efforts in auditing, e.g., by internal auditors, by external auditors, and possibly by authority auditors. There is also a danger that auditing may be overemphasized compared with other parts of the management systems. According to the present guidelines for the Eco-Audit scheme, audits will be required annually at major sites.

Another challenge will be to educate and maintain a good staff of auditors. As with other professions, there is a need for new blood, and those who are presently involved may want to change their field. Since the auditing is a resource-demanding process, it is important that the auditors are well qualified.

Finally, it is an objective to integrate the different HES audits into common audits. This will, however, mean that the emphasis will be greater on management systems and total quality, and that more specific issues may need to be reviewed in other contexts.

25

Auditing as an Integral Element of Environmental Policy at Volvo

Axel Wenblad

*Corporate Manager, Environmental
Auditing, AB Volvo*

Volvo's environmental policy was formulated originally at the time of the United Nations' conference on the environment, which was held in Stockholm in 1972. At that time, the Swedish Environmental Protection Act (which had come into force a few years earlier in 1969) was indicative of a new attitude with regard to the environmental impact of industry in that controls on all emissions and nuisances were embodied in a single piece of legislation. This approach is now a key area of discussion with the European Union under the heading of Integrated Pollution Prevention and Control. The Swedish legislation also required industry to adopt what was described as the "best available technology" under specified economic conditions.

As a result of the new legislation, Volvo's environmental protection program was well under way by 1972. At that time—and throughout most of the 1970s—attention was focused largely on the discharge of industrial effluents and certain types of environmentally hazardous waste. Purification plants were built and effluent monitoring systems introduced as part of officially approved programs, and regular reports (usually monthly) were submitted to the authorities.

Treatment facilities of basically the same standard as those in Sweden were also installed in the company's plants abroad, whether or not required by local legislation. As an example, the Volvo plants at Ghent in Belgium were equipped to treat all sanitary wastewaters, a facility still lacked by the city of Brussels.

During the 1980s, increasing attention was focused on airborne emissions, which in Volvo's case consisted mainly of solvent releases from painting operations. As environmental awareness grew, the related problems became increasingly politicized. According to opinion polls carried out prior to the 1988 Swedish general election, the environment was the main political issue.

Under these circumstances, it was natural that the Volvo Group should take the initiative of reviewing its policy in the area. In autumn 1988, Group President Pehr G. Gyllenhammar appointed a steering group, whose members included the heads of Volvo's product companies, to formulate proposals for a new policy and to suggest mechanisms for reviewing and monitoring its implementation on an ongoing basis.

At this time, a government commission was also reviewing environmental protection structures at the national level to achieve more efficient monitoring, for example, of industrial activities. Under the commission's terms of reference, none of the measures proposed was to involve additional state expenditure. Drawing an analogy from the accounting sector, it suggested the concept of compulsory, external environmental auditing of major industries by accredited auditors, the audit to be included in the environmental report submitted annually by each company to the appropriate authorities.

Not surprisingly, the report was met with stiff resistance from industry, which saw the proposed environmental auditor as a third party playing an undefined role in the relationship between industry and authorities. At that time (1988), environmental auditing, as practiced in the United States, was practically unheard of in Sweden.

Although the proposal never came to fruition, the Federation of Swedish Industries recognized that its responsibilities in the matter extended beyond mere rejection of the idea and that industry was required to take initiatives of its own to improve its control and monitoring activities. To this end, the federation initiated a fact-finding project to gather information on environmental auditing programs in the United States, including discussions with pioneers such as Allied-Signal and Arthur D. Little.[1]

In the spring of 1989, as a result of these activities, the federation organized an introductory training course in environmental auditing, with the assistance of training personnel from Arthur D. Little. Swedish industry responded positively, and companies were advised to introduce U.S.-style environmental auditing, tailored to suit Swedish conditions. Since this coincided with the finalization of Volvo's new environmental policy, it was natural that such auditing should be adopted as a means of monitoring policy implementation within Volvo.

Integration of Auditing with Environmental Policy

Volvo's environmental steering group submitted its report in spring 1989, and the company's environmental policy was finalized the following August. The report consisted of the following main headings:

- Environmental policy
- Three- to-five-year action program
- Minimum standards
- Training
- Organization
- Environmental auditing

The policy contains six statements expressing the company's commitment to environmental protection. Taken together, these represent Volvo's acceptance of responsibility for the environmental "behavior" of a product throughout its life cycle (Fig. 25-1).

The second commitment, which undertakes "to opt for manufacturing processes which have the least possible impact on the environment," is aimed directly at the production sector. Although the statement does not define the standards to be achieved in this area, the steering group report spells out the decision-making process in detail. In summary, the report states that every action and investment necessary to achieve the highest possible technical standards must be undertaken when making major production changes involving emissions. If this is not feasible for any reason, the decision must be referred to company management.

The last commitment states that Volvo "shall strive to attain a uniform, worldwide environmental standard for processes and products"—an objective that the company can achieve either by working for internationally harmonized legislation or by establishing its own worldwide environmental standards.

The policy document concludes by directing that the head of each individual company shall be responsible for implementing the appropriate measures. In a decentralized organization such as Volvo, this means that procedures for formulating specific goals and action plans must be put in place at different levels. Thus each company within the group prepares a three- to five-year action program and also sets "minimum" standards or quantitative environmental targets for its own operations. The ongoing planning process means that implementation of the minimum standards for the preceding year is evaluated at year's end, while those for the following year are defined at the same time.

In the larger companies, this requires a structure in which activities take place on several different levels. This is best described by the example dealing with Volvo Car Corporation (VCC) shown in Fig. 25-2, in which the group's overall environmental policy is shown at the top of the triangle, with VCC's environmental strategy and the three- to five-year action programs underneath. The latter are then broken down by division and, in appropriate cases, by production unit. Similar structures are employed by the other product companies.

Training is an essential element of the environmental policy. Training activities effectively began when Pehr G. Gyllenhammar invited Volvo's top 400 managers in Sweden to a one-day environmental seminar in January 1989, as part of the policy development program. This work was subsequently continued by the

VOLVO

THE ENVIRONMENTAL POLICY OF THE VOLVO GROUP

Industrial operations and transportation require resources in the form of capital, raw materials, labour -- and the environment.

Industrial operations have an impact on the surroundings. The appearance of the landscape is altered. Air, soil and water are affected. Problems connected with noise and waste can arise. Transportation by land, sea and air affects the environment mainly through emissions to the atmosphere and noise. Volvo's operations in the transportation sector are such that both our production and our products make demands on environmental resources.

A favourable earnings trend enables rational, ecologically oriented investments in production and products. Effective environmental efforts are, in turn, the basis for long-range profitability and favourable economic growth. Long- as well as short-term environmental investments must have ambitious goals and be adequate in time and size to contribute to Volvo's long-term competitiveness.

Volvo intends to minimize the environmental impact of its operations by

- developing and marketing products with superior environmental properties and which meet the highest efficiency requirements

- opting for manufacturing processes that have the least possible impact on the environment

- participating actively in, and conducting our own research and development in the environmental field

- selecting environmentally compatible and recyclable material in connection with the development and manufacture of our products and when we purchase components from our suppliers

- applying a total view regarding the adverse impact of our products on the environment

- striving to attain a uniform, world-wide environmental standard for processes and products.

The president of each company is responsible for implementing this policy.

Authorized as of August 25, 1989

Pehr G Gyllenhammar

Figure 25-1.

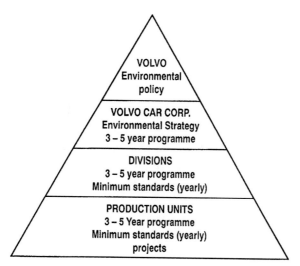

Figure 25-2. Environmental policy structure at Volvo.

individual product companies, all of which undertook comprehensive training programs for various categories of personnel.

Although certain guidelines were laid down in the steering group report, the organizational structures may vary depending on the nature of the activity. In the first instance, responsibility for the environment is part of operational responsibility, while legal responsibility can be delegated only in accordance with specific principles. Almost every production unit has an environmental coordinator with a monitoring and coordinating function, while each product company also has an environmental committee responsible for drafting action programs and minimum standards.

The ultimate objectives of environmental auditing are to monitor and evaluate the implementation of the environmental policy and to provide support for a dedicated environmental program. Thus the environmental auditor's remit is extremely wide, extending to production and products alike.

Environmental Auditing Program

Following the adoption of the company's environmental policy, environmental auditing commenced in the autumn of 1989 in the form of a pilot program designed to develop an organization and working procedures. A number of production plants, representative of activities in both Sweden and abroad and of different types of production, were selected. In general, the audits were carried out using the methodology of the Arthur D. Little training courses, adapted as appropriate to Swedish conditions.[2,3]

The pilot program was evaluated at the end of the trial period (1989–1990). The findings were extremely positive, as regards both the audits per se and the manner in which they were received, and it was decided to audit all production units as a first step. The results also supported the finding of the steering group report that auditing should focus on the company's environmental policy, with statutory compliance as one of several aspects.

In its program for 1991–1993, Volvo declared environmental auditing to be an integral part of environmental policy implementation. Noting that it must be developed in parallel with environmental activities as a whole, the authors go on to observe that auditing should be evaluated and refined by degrees as experience is gained in the field.

In more specific terms, the purposes of environmental auditing are:

- To verify compliance with existing legislation and to evaluate the conditions for complying with future legislation

- To evaluate the conditions for implementing action programs and attaining minimum standards

- To provide a basis for the assessment and improvement of management functions and procedures in the environmental area

- To identify areas and activities which may prove environmentally hazardous in the future

The verification of compliance with existing legislation—as expressed in the first of these objectives—is an obvious auditing prerequisite, which calls for expertise and thoroughness on the part of the auditors.

In Sweden, furthermore, most of the applicable environmental conditions are contained in the operating license, which governs the environmental impact of an activity as a whole. The application for an operating license forms an integral part of the license itself; thus, in addition to complying with the terms of the license, the activity must be conducted generally as described in the application. In the case of a large company (whose application may run to several volumes of documentation), this represents a substantial body of conditions. In addition, compliance with legislation other than the Environmental Protection Act is also required. In contrast with the United States, where the legislation is much more specific, these additional conditions are often expressed in relatively general terms.

Overall, this means that the demands imposed on the expertise of the auditing team are high. In the absence of general checklists, the members must familiarize themselves, in detail, with the conditions applicable to the specific plant and must also be capable of making assessments on the basis of statutory conditions that are sometimes fairly generalized.

Although existing legislation forms the primary basis of environmental auditing, it is equally important, in most cases, to be cognizant of future developments. Strictly speaking, it is open to question whether an audit based on present conditions should be projected forward in this manner. However, the purpose of this is to evaluate the current situation in the light of future statutory requirements—a consideration that is particularly important in a world of ever-stricter environmental controls. This is especially true of the European Union, several members of which are presently reviewing their entire national legislation in the field.

Although legal compliance is important, it is only one of the means used to monitor environmental policy implementation by auditing; action programs and minimum standards are the main instruments used for this purpose. Consequently, a review of the programs and standards that have been laid down is an important part of auditing. Minimum standards, which are defined annually and are often peculiar to one individual plant, must be specific and quantifiable. A typical formulation may be that the use of solvents in underbody treatment shall be reduced by 20 percent by a certain date or that all supervisors must undergo basic training in environmental matters and in the company's environmental policy.

The purpose of environmental auditing is to ascertain whether the action programs and minimum standards in question are reasonable and are in accord with Volvo's environmental policy. In addition it must be verified whether programs and standards already have been or should have been implemented. If auditing is performed during the current operating year, the implementation must be evaluated. This may be a matter of confirming that plans have been prepared and that sufficient resources have been allocated to the project. Ultimately, the auditor must make a judgment as to whether the established minimum standards are achievable.

The third objective—that of evaluating the related management functions—hardly merits detailed explanation. In a decentralized organization, it is important that environmental responsibilities be clearly defined and delineated in each individual case, especially since many environmental problems are delegated to plant managements, and that responsibility in this area is part of operational line responsibility.

The fourth, and final, objective implies an effort to be ahead of the requirements of future legislation. In this context, the professional experience of the auditing team can be used to identify processes or materials that, on the basis of research findings, may be expected to pose environmental problems in the future. This form of early-warning system often yields a recommendation either to pursue the particular development or to initiate a search for alternative materials or processes.

As indicated earlier, the scope of the 1991–1993 program extends to all production units. Other departments, such as laboratories, within the same geographical area are included in the audit, as are independent activities (large storage facilities, etc.) with a potential environmental impact.

As in most companies, environmental audits are carried out as part of acquisition and divestiture procedures. Audits of this nature are normally undertaken by external consultants who are independent of both vendor and purchaser.

Although provision is made in the program for environmental auditing of suppliers, this is carried out only occasionally. As an example, audits have been carried out in some instances in which an environmentally hazardous process, such as painting, is contracted out rather than being undertaken by Volvo itself. In such cases, the aims have been to ensure that this does not increase the environmental impact of the process and to verify that the supplier complies with the relevant legislation, ensuring that the operation is not interrupted by official intervention.

Volvo's environmental auditing organization consists of one head-office-based person with responsibility for the auditing program, backed by the resources and expertise available throughout the group at large. In practice, this means that about 20 experts from the various product companies and from the Technological Development Department are available to participate in auditing assignments. To meet the requirements of independence and objectivity, these individuals never take part in audits of their own companies.

This arrangement has a number of advantages. First, it creates a permanent auditing structure that is both small and flexible. Second, expert personnel with practical experience are available to it at all times, enabling the composition of the team to be adapted to the type of audit to be carried out in each particular case. Third, it promotes the spread of good practice since most of the participants are employed as environmental coordinators or in similar functions in their own organizations.

One drawback is that the team members often lack the opportunity of developing an auditing routine. This, in turn, imposes greater demands on the individual who, as team leader, is responsible for the program. To overcome this, the participants undergo an external, four-day, basic training course, supplemented by one-day courses held in-house each year.

A two-part company manual for environmental auditors has also been published. The first part contains practical instructions dealing with the various stages of an audit, while the second contains examples of good practice in different subareas as a form of "benchmarking."

Naturally, consultants are also employed in auditing assignments other than those associated with plant acquisition or disposal. In this case, they will have specialist roles but will also form part of the team in the normal manner. Typical instances include audits in which specialized know-how (for example, in the area of effluent purification) or a detailed knowledge of national and local environmental legislation may be required.

The practical performance of environmental auditing will be discussed in greater detail in the following section. In general, audits are carried out in accordance with the International Chamber of Commerce's Guide to Effective Environmental Auditing and the Swedish-language manual published by the Federation of Swedish Industries.[4,5]

Finally, it is the responsibility of plant management to follow up the audit, which means that an action plan must be drawn up in consultation with the appropriate parties within the product company. Although the plan is not communicated automatically to the individual responsible for the audit, he or she has the option of monitoring progress within six months to a year following its completion. This involves a review (held at the plant) of the audit report and the action plan, which should be in existence by that time. This has proved to be an effective intermediate stage between major audits, and, although it usually requires the presence of the team leader at the plant for only one day, it affords an excellent overview of the situation and permits fine tuning of the action plan in consultation with plant personnel.

Practical Performance

Since the practical performance of an environmental audit is described in comprehensive detail elsewhere in this handbook, the discussion in this section will be confined to certain practical aspects that Volvo's experience has shown to be of particular significance. Plants are selected for auditing by the environmental committee of the Volvo Group. Since only about 40 plants are involved, it has not been considered necessary to introduce special, formal selection procedures. The plants are chosen on the basis of a number of criteria, impending modifications and the plant's own requirements being important factors in this context.

Once the selection has been made, the plant environmental coordinator and product company environmental manager are contacted to draw up a rough timetable, while the plant manager and product company president are also advised formally in writing. Following this, the team leader appoints the auditing team, attaching particular weight to factors such as production orientation and the main environmental problems when choosing the specialists.

The next stage is to assemble plant documentation, a task that is normally carried out by the plant environmental coordinator. In this context, it is important to strike a judicious balance between the need for documentation to prepare the audit carefully and the work involved in producing and copying large quantities of documents and drawings.

When feasible, it may sometimes be advantageous for the team leader to pay a preliminary visit to the plant as part of the planning procedure. This will enable him or her to acquire an overall view of the operation and associated environmental problems in order to plan the audit program correctly in terms of time and personnel. A visit of this nature also permits the documentation to be reviewed at site. Preliminary visits of this nature usually take place 1 to 2 months before the audit proper.

Once the documentation begins to arrive, it is time to prepare an audit schedule, normally under the following headings:

- Purpose of audit
- Scope and definition in terms of time and areas

- Auditing team and allocation of responsibilities
- Notes from preliminary visit (if applicable)
- Detailed working schedule, including interview program
- Instructions for preparing audit documents and reports
- Reporting schedule

A satisfactory plan is indispensable to a successful audit. The plan is important not only to the performance of the audit at the plant but also to the ongoing preliminary planning activities. It familiarizes the team members with their particular areas of responsibility, enabling them to concentrate on their own sections of the documentation.

Communication with the inspection authorities forms part of the preparatory phase (at least in Sweden) and is always undertaken in consultation with the plant concerned. The main purpose of this is to ascertain whether the authority possesses any documentation in the nature of a "skeleton in the cupboard." However, it also enables Volvo to supply the authorities details of its environmental policy and auditing program. This type of direct liaison also provides an indication of the official perception of the plant's environmental activities.

Similar contacts may also be established with local and regional authorities abroad. In Volvo's experience, this often improves communication between plant and official agencies and has proved to be an excellent source of information on forthcoming legislation and its implications in specific cases.

The documentation review is an important part of the preparatory work. The auditors must become so familiar with the documentation that they can go straight to the heart of each problem and identify the potential problem areas immediately. This means that the team must meet once or twice in advance to discuss the documentation, checklists, interview schedules, and so on.

As already noted, it was considered neither feasible nor especially desirable to produce detailed, standardized checklists. Instead, it is left to the individual team members to combine their own expertise with suitable checklists. Figure 25-3 indicates the type of checklists that have nevertheless been developed.

All preparatory activities are carried out in close cooperation with plant personnel, in particular with the environmental coordinator, who plays a liaison role. For example, he or she is responsible for arranging interviews with various staff members, disseminating information to employee organizations, and making all practical arrangements necessary to ensure that the audit is successful.

Experience has shown that the selection of individuals for interview is of crucial importance. Since only a limited number of people can be interviewed in the course of an audit, the list must be drawn with care. Under normal circumstances, the most important categories for inclusion (in suitable proportions) are the following:

- Individuals with a specific environmental responsibility (e.g., plant manager, production manager, and environmental coordinator)

```
1     Responsibilities
2     Organization
3     Routines
4     Training
5     Action programmes and minimum standards
6     Planning legislation
7     Operating licence etc.
8     Self-monitoring
9     Chemicals procedures
10    Storage of oil products and chemicals
11    Risk management
12    Waste management
13    Effluents
a)    effluent systems
b)    treatment plants
14    Solvents
a)    general
b)    management
c)    paint shops etc.
15    Other airborne emissions
```

Figure 25-3. Volvo's environmental audit checklists.

- Specialists in areas such as production and effluent and waste treatment processes
- Personnel in charge of or responsible for carrying out practical operations (e.g., waste handling or treatment plant operation)

Every audit opens with a meeting with plant management to discuss the objectives, the problems that will be raised, and the auditing and reporting procedures. Following a tour of the plant (if considered necessary), interviews and meetings are held in accordance with the prepared schedule (subject to any running modifications that may be required).

The team normally meets once per day and in the evening to review progress to date and to revise the program as necessary. The evening sessions are designed to provide the group with an overview of the findings that have emerged, enabling the members to avail themselves of mutual assistance and monitoring of problems.

Interviews are normally conducted with one individual at a time, preferably in the interviewee's office or workplace. The interviews may be combined, as necessary, with visits to the plant to inspect a particular process or equipment installation. Experience has shown that $1^1/2$ hours should be allocated for each interview, including the time required for preparing and summarizing the session, since it is important that the interviewer's overall impressions be carried forward from one interview to the next.

Although interview psychology will not be dealt with specifically in this section, securing the confidence of the interviewee is of crucial importance to the achievement of beneficial results. Apart from the auditor's personal qualities, important factors include:

- The auditor must be seen as being competent in the particular area of specialization

- The audit must not be seen as an external inspection, rather as a form of professional assistance

- The audit must be carried out internally by people from the same organization

Each evening, the team assembles to evaluate and summarize the day's sessions and interviews. A picture of the plant's strong and weak points emerges in the course of these meetings, and an overall evaluation is made on the last evening. This is used as the basis for the verbal report to plant management that concludes each audit. The main points are recorded briefly in writing so that the team can submit an initial outline of its report before leaving the plant. The final meeting is also used to report positive observations and to present the team's overall impressions of plant operations.

An audit normally takes 2 to 5 days, while the auditing team consists of two to five members. Experience to date indicates that the period should be limited to a week, or five working days, $1^{1}/_{2}$ to 2 days being the minimum. The team should not number less than two for satisfactory discussion of the findings or more than five to avoid problems of coordination.

The aim is to submit a preliminary audit report to the plant manager and environmental manager of the product company not later than 1 month after completion of the audit. An effort is also made to keep the report as short as possible; ideally, the document should consist of 10 to 15 pages, with an absolute limit of 25. The report should also include an executive summary addressed to management.

In terms of structure, the report is divided into specialized areas, each of which is evaluated in relation to both statutory and in-house standards. A typical list of contents is shown in Fig. 25-4. The observations, evaluation, and conclusions and recommendations of the team are reported briefly under each main heading. The conclusions are highlighted in the text to make them easily recognizable. Figure 25-5 gives an example of an observation in an audit report.

The final report is issued when the plant manager and environmental officer of the product company have had the opportunity to comment on the draft. A copy of the final report is also forwarded to the president of the company concerned, with a summary to the chairman of the board. A summarized report is submitted one or twice annually to Volvo Group management.

The results of audits are not normally published outside Volvo. However, a brief summary of audits carried out during the year is included in the annual environmental report that each of the Swedish plants is legally required to produce, while a short commentary is also published in the annual report of the Volvo Group.

ENVIRONMENTAL AUDITING

Volvo XYZ plant
Gothenburg, Sweden

22-26 February 1993

CONTENTS

SUMMARY

    1.   BACKGROUND
    1.1 Scope
    1.2 Objectives

    2.   AUDIT

    3.   AUDIT REPORT
    3.1 Organization and procedures
        3.1.1 Organization
        3.1.2 Procedures

    3.2 Information and training

    3.3 Water-borne emissions
        3.3.1 Sewage system
        3.3.2 Industrial treatment plant
        3.3.3 Sanitary treatment plant

    3.4 Airborne emissions
        3.4.1 Painting operations
        3.4.2 Power installations
        3.4.3 Other airborne emissions

    3.5 Chemicals

    3.6 Waste
        3.6.1 Waste recycling
        3.6.2 Hazardous waste

Figure 25-4.

Volvo's Experience

The wholehearted support of management has been a major factor in the success of environmental auditing at Volvo. This backing, which has come not only from Pehr G. Gyllenhammar, who initially authorized the project to develop proposals for an environmental policy auditing program, but also from group management as a whole, has given the auditing program a flying start and gained it respect within the organization.

3.3.2. Industrial treatment plant

The plant seems to be working well due to well developed routines for maintenance, operation and control. The routines and instructions are also well documented. The monitoring system with daily process control at the treatment plant in combination with monthly monitoring by an external acredited laboratory is well implemented. The sampling equipment is, however, not cleaned according to instructions.

On five occasions during 1992 the target values for nickel and chromium have, according to monitoring data, been exceeded. This may be due to a slightly too low pH-value. Another reason may be contamination of the sampling equipment.

The limits for COD and BOD have been exceeded on two occassions in 1992. It also seems that the COD/BOD ratio is high (>3). This indicates that the water contains substances which are not readily biodegradable.

We recommend that an intensive monitoring of nickel and chromium is carried out following a thorough maintenance of the sampling equipment. The results from the monitoring should be used to optimize the flocculation process.

Furthermore, we recommend that an inventory of maintenance material in all factory units is carried out. The biodegradability of the products should be evaluated in order to replace those which are not easily biodegradable according to OECD recommendations.

Figure 25-5. Example of an audit observation.

During the project development phase, particular care was taken to emphasize that audit results were intended mainly to assist plant managers in improving environmental protection standards. In other words, environmental auditing is not an inspection procedure imposed by group management but rather a sophisticated assessment tool. Most plant managers have accepted it as such, welcoming a review of the operation for which they bear a major personal responsibility but which they are unable to monitor in every detail. Auditing provides an impartial assessment of the plant's environmental protection activities, highlighting the fact that environmental problems are inherent in the operation for which each manager is responsible.

The environmental coordinators have also given the program a uniformly positive reception without (as might be expected) regarding it as an unwanted form of control. This clearly reflects the fact that auditing enhances the status of in-plant environmental activities and focuses attention, for a time, on the problems involved. In addition, it affords the coordinator an opportunity of putting forward requirements and demands that may previously have gone unheard.

Management immediately below plant-manager level may be somewhat more guarded in its attitude. Managers who have not considered the environmental aspects of their operational responsibilities may see the environmental audit as an uncomfortable reminder of this fact. An example might be a production manager who has failed to incorporate environmental protection measures in the operation of a paint shop or a purchasing manager whose chemical-purchasing procedures do not include evaluation of the environmental implications. However, a properly performed audit may provide the incentive needed to correct this type of situation.

The decentralized nature of Volvo's organizational structure makes the review of environmental organization and procedures a particularly important aspect of auditing. In this context, numerous parallels can be drawn with quality-assurance auditing programs. Analogies such as this enable many individuals to adjust their orientation and grasp what is meant by integrating environmental considerations into their activities, to the extent that it may be sufficient to suggest that they substitute the term product quality with "environmental" quality.

Experience has also demonstrated the importance of a contribution by the auditing team itself, either in the form of information exchange or by encouraging innovative thinking among plant personnel. Obviously, this must not interfere with the primary task of auditing or compromise the independence of the auditor; however, Volvo has found that this is not a problem, provided that auditors are professionally qualified and experienced in their fields.

Sweden's long tradition in the field of environmental protection ensured that expertise of an extremely high order was available when the auditing program was initiated. Since the individuals considered for inclusion in the auditing team possessed both professional qualifications and wide practical experience in the solution of industrial pollution or production problems, the main requirement was to provide them with the requisite training in the use of auditing techniques as a working tool. In addition, the generalized nature of the Swedish legislation made it essential to build on the practical experience and expertise of the auditors.

Environmental Auditing in the Future

Environmental auditing has undoubtedly come to stay at Volvo; the experience of the last few years has shown that it is an extremely valuable tool for developing as well as monitoring environmental activities. The first round of audits will be completed early in 1994, by which time most of the follow-up programs will also have been concluded. Volvo has deliberately refrained from taking a stance on the next phase before the findings of the initial program have been evaluated.

The changing world in which we live has a particular influence on environmental policy; thus the major issues of a few years ago have now been superseded by other problems. The focus is shifting from production plant emissions to an emphasis on the environmental impact of the product itself, from the emission of individual pollutants to an overall view of environmental impact.

In common with many other companies, Volvo is a signatory of the ICC Business Charter for Sustainable Development, a 16-point document that takes an overall view of the environmental impact of industrial activity. This includes the use of environmental auditing as a means of verifying that corrective measures have genuinely been implemented.

This overall approach to environmental impact is the first of two trends that will influence Volvo's environmental auditing programs in the future. This will find expression inter alia in the development of life-cycle analysis (LCA) tools, and it is not impossible that tools of this type—in modified form—may prove useful when environmental auditing is extended to the product-development area. Areas not covered as yet include the after-sale and disposal and recycling aspects (the latter, in particular, will assume increased importance in the course of time).

The second trend will be the progressive integration of environmental issues in Total Quality Management (TQM). There are compelling reasons for regarding environmental protection as an element of overall quality activities. Among other factors, this eliminates the need for several implementation and monitoring systems operating in parallel, and also enables environmental auditing (integrated, in part, with quality auditing) to be built in at different levels. One interesting practice that has already been adopted is the review of suppliers' environmental programs as part of quality management system evaluation.

Volvo will keep abreast of and participate actively in the development of these two trends, the combination of which will almost certainly determine the future shape of environmental auditing.

References

1. Almgren, R., Bjallas, U., and Wenblad, A. Miljorevision Internationellt (International Environmental Auditing). Swedish Environment Protection Board, Report No. 3626, 1989.

2. Greeno, J. L., Hedstrom, G. S., and DiBerto, M. Environmental Auditing—Fundamentals and Techniques. Center for Environmental Assurance, Arthur D. Little, Inc., Cambridge, Mass., 1987.

3. Greeno, J. L., Hedstrom, G. S., and DiBerto, M. The Environmental Health and Safety Auditor's Handbook. Center for Environmental Assurance, Arthur D. Little, Inc., Cambridge, Mass., 1988.

4. International Chamber of Commerce. ICC Guide to Effective Environmental Auditing. ICC Publishing SA, 1991.

5. Almgren, R. Miljorevision (Environmental Auditing). Federation of Swedish Industries, Stockholm, 1990.

Index

About the Editor

Lee Harrison has been involved with the environment professionally since graduating from Northeastern University with a degree in civil engineering (environmental emphasis) in 1969. He has written extensively on environmental issues as environment editor of *Chemical Week* magazine, as publisher of the *Environmental Audit Letter*, and as a contributor to numerous publications, including *The New York Times*. Mr. Harrison is currently a consultant to corporate managers on crisis avoidance, management, and communications.

Supplement Available

This book will be kept up to date with periodic supplements. To add your name to the automatic shipment list, please return the coupon below. All supplements are sent on a 15-day free examination basis, and you will be notified before any are sent to you.

❏ YES, I want to automatically receive supplements to Harrison's *Environmental, Health, and Safety Auditing Handbook* (ISBN 0-07-026904-1) when they become available.

Name —————————————————————————————

Address ———————————————————————————

Company ——————————————————————————

City/State/Zip ————————————————————————

Work Telephone ————————————————————————

Mail to: McGraw-Hill Order Services, P.O. Box 545, Blacklick, Ohio 43004-0545 or fax to (614) 755-5645.